Fundamental Statistical Concepts and Techniques in the Biological and Environmental Sciences

Fundamental Statistical Concepts and Techniques in the Biological and Environmental Sciences: With jamovi is an introductory textbook for learning statistics. It starts with the very basics and prioritises helping the reader to develop a conceptual understanding of statistics, and apply the most fundamental statistical tools. New concepts are introduced with examples designed to be familiar to the reader, serving as a useful starting point for exploring more abstract concepts.

Key Features:

- Designed to be accessible for students learning statistics in biological and environmental sciences.
- Utilizes the statistical software jamovi to explore new concepts.
- Prioritizes good statistical judgement over adherence to protocols.

This book will be useful to students beginning their study of statistical concepts in biological and environmental sciences, whilst also acting as an insightful resource for teachers using jamovi in the classroom. *Fundamental Statistical Concepts and Techniques in the Biological and Environmental Sciences: With jamovi* is a valuable resource for anyone who wishes to understand and apply statistical techniques commonly used in the biological and environmental sciences.

A. Bradley Duthie is a Lecturer in Environmental Modelling at the University of Stirling, Scotland, UK. He completed undergraduate degrees in Biology and Philosophy at Southern Illinois University Edwardsville. He earned his PhD in Ecology and Evolutionary Biology from Iowa State University with a graduate minor in Statistics. His research focuses primarily on theory and modelling in ecology and evolutionary biology with particular interests in evolutionary ecology and community ecology. He contributes to several research projects as a biostatistician and wrote and maintains two R packages for agent-based modelling.

Fundamental Statistical Concepts and Techniques in the Biological and Environmental Sciences

With jamovi

A. Bradley Duthie

CRC Press is an imprint of the
Taylor & Francis Group, an **informa** business
A CHAPMAN & HALL BOOK

Designed cover image: © A. Catherine Duthie

First edition published 2025
by CRC Press
2385 NW Executive Center Drive, Suite 320, Boca Raton FL 33431

and by CRC Press
4 Park Square, Milton Park, Abingdon, Oxon, OX14 4RN

CRC Press is an imprint of Taylor & Francis Group, LLC

ISBN: 978-1-032-69237-1 (hbk)
ISBN: 978-1-032-68718-6 (pbk)
ISBN: 978-1-032-69238-8 (ebk)

DOI: 10.1201/9781032692388

Typeset in Latin Modern font
by KnowledgeWorks Global Ltd.

Publisher's note: This book has been prepared from camera-ready copy provided by the authors.

Contents

Preface

Nearly all research in the biological and environmental sciences relies on data analysis of some kind. Statistical literacy is therefore important, not just for *doing* research, but also for *understanding* and *evaluating* the research of other scientists. This book is an introductory textbook for learning statistics. It starts with the very basics and prioritises helping the reader develop a conceptual understanding of statistics and applying the most fundamental statistical tools. Mathematical details are generally placed in footnotes, and developing good statistical judgement is embraced over the adherence to rigid protocols. Wherever possible, I have tried to introduce new concepts with concrete examples that will likely be familiar to the reader and therefore serve as a useful starting point for understanding the more abstract concepts. After reading this book, the reader should be able to understand and communicate fundamental concepts of introductory statistics and apply them in the biological and environmental sciences.

This book originated as a workbook for the second-year university statistics class that I teach for biological and environmental science students at the University of Stirling in Stirling, Scotland. I wanted to rebuild the learning content from scratch after switching from proprietary statistical software to the free and open-source jamovi (The jamovi project, 2024). There are a lot of good statistics textbooks (many of which are cited throughout this book), but I could not find anything for biological or environmental science students that used jamovi. As a teacher, I have found jamovi to be the ideal tool for teaching introductory statistics. Its user friendly interface has made it possible to focus on teaching statistical concepts rather than the details of navigating software. It works on Windows, Mac, Linux, or Chrome and can even be used in a browser (`https://www.jamovi.org/`), so I know that my students will have access to jamovi after they are done with my class. I hope that this book is helpful for teachers wanting to use jamovi in the classroom, and that it is useful for anyone who wants to understand and apply statistical techniques that are widely used across the biological and environmental sciences.

How this book is structured

This book has 35 chapters, with 26 chapters introducing statistical concepts and 9 chapters focused on the practical skills required to apply them in jamovi. These chapters are interspersed throughout the book such that chapters introducing inter-related statistical concepts are followed by a practical skills chapter (the exception is a final chapter on randomisation, which is not followed by a practical skills chapter). Conceptual chapters are generally quite short, and longer chapters are broken down into manageable subsections. Practical skills chapters include multiple exercises in jamovi (except for Chapter 3, which uses spreadsheets outwith jamovi). Answers to exercise questions can be found in Appendix A. Together, a set of conceptual chapters and the practical skills chapter that follows (e.g., Chapters 1–3) could be used as one week of material in an introductory statistics class. In several places throughout the book, there are footnotes with links to interactive applications that I have written in shiny (Chang et al., 2024; Schloerke & Chang, 2023; Wickham, 2021). These applications should make it easier to visualise some of the more challenging concepts and reinforce the text.

Datasets used in this book

Datasets for all exercises (and code for shiny apps) are freely available to download from the Open Science Framework (`https://osf.io/dxwyv`). They are also stored on GitHub (`https://github.com/bradduthie/stats`), and can be accessed there using direct links in footnotes.

Datasets are mostly inspired by, or directly sourced from, real research in the biological and environmental sciences. A lot of data were collected from, or based on, my own doctoral work at Iowa State University. Other datasets are inspired by projects led by my colleagues at the University of Stirling. **These data are for pedagogical purposes only.** With some exceptions from my own data collected from the Sonoran Desert Rock Fig (*Ficus petiolaris*) system (see Duthie et al., 2015; Duthie & Nason, 2016), all data in this book were constructed solely to illustrate statistical concepts and techniques as effectively as possible. Any data introduced in this book should therefore be treated as entirely hypothetical, and the reader will be reminded of this in subsequent chapters.

Acknowledgements

Most of this book was written in the summer and autumn of 2022, and during a busy spring semester of teaching statistics in 2023. I am very grateful for the feedback that I have received from undergraduate students in the 2023 and 2024 Statistical Techniques class at the University of Stirling. I love teaching this class, and conversations that I have had with students during weekly practical sessions have made the text clearer and corrected numerous typographical errors. The book has also benefited from suggestions and corrections from numerous PhD students who served as teaching assistants on the class, including Adam Fell, Adrian Bach, Arianna Chiti, Benjamin Marshall, Chloe Pow, Daniel Atton Beckman, Eleri Kent, Rebecca Metcalf, and Shubham Pawar. I am especially grateful to my co-teachers Martina Quaggiotto and Ian Jones. Martina and Ian supported my plan to overhaul the learning material for our statistics class and put in a lot of work to fix teaching content that was not actually broken. I am also grateful to several colleagues at the University of Stirling who have provided images and inspired me with interesting projects and datasets, including Alan Law, Becky Boulton, Carmen Carmona, David Copplestone, Elisa Fuentes-Montemayor, Izzy Jones, Jens-Arne Subke, Jessica Burrows, Kirsty Park, Lidia de Sousa Teixeira, Matthew Tinsley, Nigel Willby, and Nils Bunnefeld.

I want to thank CRC Press, and especially Lara Spieker, for making the publishing process such a positive experience. This book has benefited from feedback from three anonymous reviewers. It was written using Bookdown (Xie, 2015, 2016, 2023) within RStudio (RStudio Team, 2020) in the R programming language (R Core Team, 2022) running on the Xubuntu open source operating system (`https://xubuntu.org/`).

I want to thank my kid, Emrys. They are a joy in my life and have taught me so much.

Lastly, I want to thank my wife Catherine. We have supported one another through so many challenges in life, through the uncertainties of career, immigration, and COVID-19. Her support has been indispensable for writing this book, and I will always be grateful for it.

About the author

A. Bradley Duthie (http://bradduthie.github.io) is a Lecturer in Environmental Modelling at the University of Stirling, Scotland, UK. He completed undergraduate degrees in Biology and Philosophy at Southern Illinois University Edwardsville. He earned his PhD in Ecology and Evolutionary Biology from Iowa State University with a graduate minor in Statistics. His research focuses primarily on theory and modelling in ecology and evolutionary biology with particular interests in evolutionary ecology and community ecology. He also contributes to several research projects as a biostatistician and wrote and maintains two R packages for agent-based modelling. As an educator, Brad completed a Graduate Student Teaching Certificate at Iowa State University and became a Fellow of the Higher Education Academy while working as a Postdoctoral Research Fellow at the University of Aberdeen. He has taught undergraduate classes in statistics, ecology, and evolution at the University of Stirling and is the Programme Director for Biology and Animal Biology.

1

Background mathematics

There are at least two types of mathematical challenges that come with first learning statistics. The first challenge is simply knowing the background mathematics upon which many statistical tools rely. Fortunately, while the *theory* underlying statistical techniques does rely on some quite advanced mathematics (e.g., see Mclean et al., 1991; Miller & Miller, 2004; Rencher, 2000), the *application* of standard statistical tools usually does not. This book focuses on the application of statistical techniques, so all that is required is a background in some fundamental mathematical concepts such as mathematical operations (addition, subtraction, multiplication, division, and exponents), simple algebra, and probability. This chapter will review mathematical operations and the symbols used to communicate them.

The second mathematical challenge that students face when learning statistics for the first time is a bit more subtle. Students with no statistical background sometimes have an expectation that statistics will be similar to previously learnt mathematical topics such as algebra, geometry, or trigonometry. In some ways, this is the case, but in a lot of ways statistics is a much different way of thinking than any of these topics. A lot of mathematical subjects focus on questions that have very clear right or wrong answers (or, at least, this is how they are often taught). If, for example, we are given the lengths of two sides of a right triangle, then we might be asked to calculate the hypotenuse of the triangle using Pythagorean theorem ($a^2 + b^2 = c^2$, where c is the hypotenuse). If we know the length of the two sides, then the length of the hypotenuse has a clear correct answer (at least, on a Euclidean plane). In statistics, answers are not always so clear-cut. Statistics, by its very nature, deals with uncertainty. While all of the standard rules of mathematics still apply, statistical questions such as, 'Can I use this statistical test on my data?', 'Do I have a large enough sample size?', or even 'Is my hypothesis well-supported?' often do not have unequivocal 'correct' answers. Being a good statistician often means making well-informed, but ultimately at least somewhat subjective, judgements about how to make inferences from data.

For now, we will move on to looking at numbers and operations, logarithms, and order of operations. These topics will be relevant throughout the book, so it is important to understand them and be able to apply them when doing calculations.

1.1 Numbers and operations

Calculating statistics and reading statistical output requires some knowledge of numbers and basic mathematical operations. This section is a summary of the basic mathematical tools that will be used in introductory statistics. Much of this section is inspired by Courant et al. (1996) and chapter 2 of Pastor (2008). This section will focus on only the numbers and mathematical operations relevant to this book. The objective here is to present some very well-known ideas in an interesting way, and to intermix them with bits of information that might be new and interesting. For doing statistics, what you really *need* to know here are the operations and the notation, that is, how operations such as addition, multiplication, and exponents are calculated and represented mathematically.

We can start with the *natural* numbers, which are the kinds of numbers that can be counted using fingers, toothpicks, pebbles, or any discrete sets of objects.

$$1, 2, 3, 4, 5, 6, 7, 8, ...$$

There are an infinite number of natural numbers (we can represent the set of all of them using the symbol $\mathbb{N}$). For any given natural number, we can always find a higher natural number using the operation of addition. For example, a number higher than 5 can be obtained by simply adding 1 to it,

$$5 + 1 = 6.$$

This is probably not that much of a revelation, but it highlights why the natural numbers are countably infinite (for any number you can think of, N, there is always a higher number $N + 1$). It also leads to a reminder about two other important mathematical symbols for this book (in addition to $+$, which indicates addition), greater than ($>$) and less than ($<$). We know that the number 6 is greater than 5, and we express this mathematically as the **inequality**, $6 > 5$. Note that the large end of the inequality faces the higher number, while the pointy end (i.e., the smaller end) faces the lower number. Inequalities are used regularly in statistics, e.g., to indicate when a probability of something is less than a given value (e.g., $P < 0.05$, which can be read 'P is less than 0.05'). We might also use the symbols $\geq$ or $\leq$ to indicate when something is greater than or equal to ($\geq$) or less than or equal to ($\leq$) a particular value. For example, $x \geq 10$ indicates that some number x has a value of 10 or higher.

Whenever we add one natural number to another natural number, the result is another natural number, a sum (e.g., $5 + 1 = 6$). If we want to go back from the sum to one of the values being summed (i.e., get from 6 to 5), then we need to subtract,

$$6 - 1 = 5.$$

This operation is elementary mathematics, but a subtle point that is often missed is that the introduction of subtraction creates the need for a broader set of numbers than the natural numbers. We call this broader set of numbers the *integers* (we can represent these using the symbol $\mathbb{Z}$). If, for example, we want to subtract 5 from 1, we get a number that cannot be represented on our fingers,

$$1 - 5 = -4.$$

The value -4 is an integer (but *not* a natural number). Integers include 0 and all negative whole numbers,

$$..., -4, -3, -2, -1, 0, 1, 2, 3, 4, ...$$

Whenever we add or subtract integers, the result is always another integer.

Now, suppose we wanted to add the same value up multiple times. For example,

$$2 + 2 + 2 + 2 + 2 + 2 = 12.$$

The number 2 is being added 6 times in the equation above to get a value of 12. But we can represent this sum more easily using the operation of multiplication,

$$2 \times 6 = 12.$$

The 6 in the equation just represents the number of times that 2 is being added up. The equation can also be written as $2(6) = 12$, or sometimes, $2 \star 6 = 12$ (i.e., the asterisk is sometimes used to indicate multiplication). Parentheses (i.e., round brackets) indicate multiplication when no other symbol separates them from a number. This rule also applies to numbers that come immediately before variables. For example, $2x$ can be interpreted as *two times* x. When multiplying integers, we always get another integer. Multiplying two positive numbers always equals another positive number (e.g., $2 \times 6 = 12$). Multiplying a positive and a negative number equals a negative number (e.g., $-2 \times 6 = -12$). And multiplying two negative numbers equals a positive number (e.g., $-2 \times -6 = 12$). There are multiple ways of thinking about why this last one is true (see, e.g., Askey, 1999 for one explanation), but for now we can take it as a given.

As with addition and subtraction, we need an operation that can go back from multiplied values (the product) to the numbers being multiplied. In other words, if we multiply to get $2 \times 6 = 12$ (where 12 is the product), then we need something that goes back from 12 to 2. Division allows us to do this, such that $12 \div 6 = 2$. In statistics, the symbol $\div$ is rarely used, and we would more often express the calculation as either $12/6 = 2$ or,

$$\frac{12}{6} = 2.$$

As with subtraction, there is a subtle point that the introduction of division requires a new set of numbers. If instead of dividing 6 into 12, we divided 12 into 6,

$$\frac{6}{12} = \frac{1}{2} = 0.5.$$

We now have a number that is not an integer. We therefore need a new broader set of numbers, the *rational* numbers (we can represent these using the symbol $\mathbb{Q}$). The rationals include all numbers that can be expressed as a *ratio* of integers. That is, p/q, where both p and q are in the set $\mathbb{Z}$.

We have one more set of operations relevant for introductory statistics. Recall that we introduced 2×6 as a way to represent $2 + 2 + 2 + 2 + 2 + 2$. We can apply the same logic to multiplying a number multiple times. For example, we might want to multiply the number 2 by itself 4 times,

$$2 \times 2 \times 2 \times 2 = 16.$$

We can represent this more compactly using an **exponent**, which is written as a superscript,

$$2^4 = 16.$$

The 4 in the equation above indicates that the 2 should be multiplied 4 times to get 16. Sometimes this is also represented by a caret in writing or code, such that 2^4 = 16. Very occasionally, some authors will use two asterisks in a row, 2**4 = 16, probably because this is how exponents are represented in some statistical software and programming languages. One quick note that can be confusing at first is that a negative in the exponent indicates a reciprocal. For example,

$$2^{-4} = \frac{1}{16}.$$

This can sometimes be useful for representing the reciprocal of a number or unit in a more compact way than using a fraction (we will come back to this in Chapter 6).

As with addition and subtraction, and multiplication and division, we also need an operation to get back from the exponentiated value to the original number. That is, for $2^4 = 16$, there should be an operation that gets us back from 16 to 2. We can do this using the **root** of an equation,

$$\sqrt[4]{16} = 2.$$

The number under the radical symbol $\sqrt{}$ (in this case 16) is the one that we are taking the root of, and the index (in this case 4) is the root that we are calculating. When the index is absent, we assume that it is 2 (i.e., a square root),

$$\sqrt[2]{16} = \sqrt{16} = 4.$$

Note that $4^2 = 16$ (i.e., 4 squared equals 16).

Instead of using the radical symbol, we could also use a fraction in the exponent. That is, instead of writing $\sqrt[4]{16} = 2$, we could write $16^{1/4} = 2$ or $16^{1/2} = 4$. In statistics, however, the $\sqrt{}$ is more often used. Either way, this yet again creates the need for an even broader set of numbers. This is because expressions such as $\sqrt{2}$ do not equal any rational number. In other words, there are no integers p and q such that their *ratio*, $p/q = \sqrt{2}$ (the proof for why is very elegant!). Consequently, we can say that $\sqrt{2}$ is *irrational* (not in the colloquial sense of being illogical or unreasonable, but in the technical sense that it cannot be represented as a ratio of two integers). Irrational numbers cannot be represented as a ratio of integers, or with a finite or repeating decimal. Remarkably, the set of irrational numbers is larger than the set of rational numbers (i.e., rational numbers are countably infinite, while irrational numbers are uncountably infinite, and there are more irrationals; you do not need to know this or even believe it, but it is true!).

Perhaps the most famous irrational number is π, which appears throughout science and mathematics and is most commonly introduced as the ratio of a circle's circumference to its diameter. Its value is $\pi \approx 3.14159$, where the symbol $\approx$ means 'approximately.' Actually, the decimal expansion of π is infinite and non-repeating; the decimals go on forever and never repeat themselves in a predictable pattern. As of 2019, over 31 trillion (i.e., 31000000000000) decimals of π have been calculated (Yee, 2019).

The rational and irrational numbers together comprise a set of numbers called *real* numbers (we can represent these with the symbol $\mathbb{R}$), and this is where we will stop. This story of numbers and operations continues with imaginary and complex numbers (Courant et al., 1996; Pastor, 2008), but these are not necessary for introductory statistics.

1.2 Logarithms

There is one more important mathematical operation to mention that is relevant to introductory statistics. Logarithms are important functions, which will appear in multiple places (e.g., statistical transformations of variables). A logarithm tells us the exponent to which a number needs to be raised to get another number. For example,

$$10^3 = 1000.$$

Verbally, 10 raised to the power of 3 equals 1000. In other words, we need to raise 10 to the power of 3 to get a value of 1000. We can express this using a logarithm,

$$\log_{10}(1000) = 3.$$

Again, the same relationship is expressed in $10^3 = 1000$ and $\log_{10}(1000) = 3$. For the latter, we might say that the base 10 logarithm of 1000 is 3. This is actually extremely useful in mathematics and statistics. Mathematically, logarithms have the very useful property,

$$\log_{10}(ab) = \log_{10}(a) + \log_{10}(b).$$

Historically, this has been used to make calculations easier by converting multiplication to addition (Stewart, 2008). In statistics, and across the biological and environmental sciences, we often use logarithms when we want to represent something that changes exponentially on a more convenient scale. For example, suppose that we wanted to illustrate the change in global CO_2 emissions over time (Friedlingstein et al., 2022). We could show year on the x-axis and emissions in billions of tonnes of CO_2 on the y-axis (Figure 1.1).

We can see from Figure 1.1 that global CO_2 emissions go up exponentially over time, but this exponential relationship means that the y-axis has to cover a large range of values. This makes it difficult to see what is actually happening in the first 100 years. Are CO_2 emissions increasing from 1750 to 1850, or do they stay about the same? If instead of plotting billions of tonnes of CO_2 on the y-axis, we plotted the logarithm of these values, then the pattern in the first 100 years becomes a bit clearer (Figure 1.2).

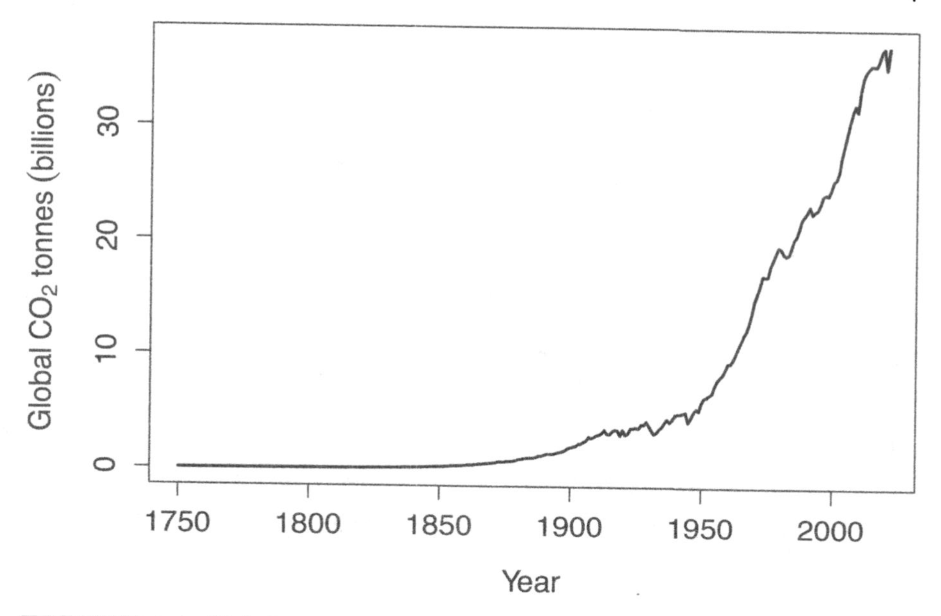

FIGURE 1.1 Global carbon dioxide emissions 1750–2021.

It appears from the logged data in Figure 1.2 that global CO_2 emissions were indeed increasing from 1750 to 1850. Note that Figure 1.2 presents the *natural logarithm* of CO_2 emissions on the y-axis. The natural logarithm uses Euler's number, $e \approx 2.718282$, as a base. Euler's number e is an irrational number (like π), which corresponds to the intrinsic rate of increase of a population's size in ecology (Gotelli, 2001), or, in banking, interest compounded continuously (like π, e actually shows up in a lot of different places throughout science and mathematics). We probably could have just as easily used 10 as a base, but e is usually the default base to use in science (bases 10 or 2 are also often used). Note that we can convert back to the non-logged scale by raising numbers to the power of e. For example, $e^{-4} \approx 0.018$, $e^{-2} \approx 0.135$, $e^0 = 1$, and $e^2 = 7.390$.

1.3 Order of operations

Every once in a while, a maths problem like the one below seems to go viral online,

$$x = 8 \div 2\,(2 + 2)\,.$$

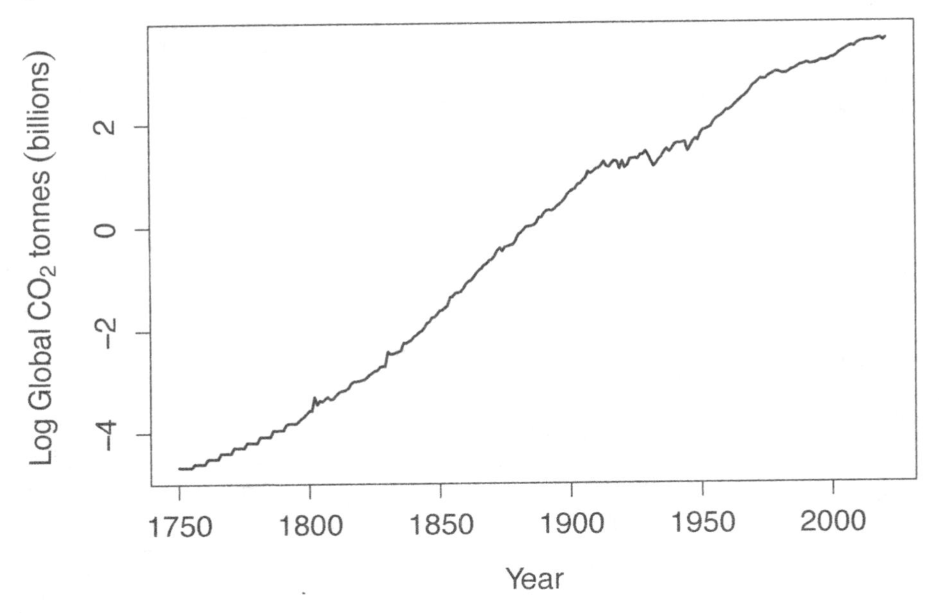

FIGURE 1.2 Natural logarithm of global carbon dioxide emissions 1750–2021.

Depending on the order in which calculations are made, some people will conclude that $x = 16$, while others conclude that $x = 1$ (Chernoff & Zazkis, 2022). The confusion is not caused by the above calculation being difficult, but by peoples' differences in interpreting the rules for the order by which calculations should be carried out. If we first divide $8/2$ to get 4, then multiply by $(2 + 2)$, we get 16. If we first multiply 2 by $(2 + 2)$ to get 8, then divide, we get 1. The truth is that even if there is a 'right' answer here (Chernoff & Zazkis, 2022), the equation could be written more clearly. We might, for example, rewrite the above to more clearly express the intended order of operations,

$$x = \frac{8}{2}(2 + 2) = 16.$$

We could write it a different way to express a different intended order of operations,

$$x = \frac{8}{2(2 + 2)} = 1.$$

The key point is that the order in which operations are calculated matters, so it is important to write equations clearly, and to know the order of operations to calculate an answer correctly. By convention, there are some rules for the order in which calculations should proceed.

1. Anything within parentheses should always be calculated first.
2. Exponents and radicals should be applied second.
3. Multiplication and division should be applied third.
4. Addition and subtraction should be done last.

These conventions are not really rooted in anything fundamental about numbers or operations (i.e., we made these rules up), but there is a logic to them. First, parentheses are a useful tool for being unequivocal about the order of operations. We could, for example, always be completely clear about the order to calculate by writing something like $(8/2) \times (2+2)$ or $8/(2(2+2))$, although this can get a bit messy. Second, rules 2–4 are ordered by the magnitude of operation effects; for example, exponents have a bigger effect than multiplication, which has a bigger effect than addition. In general, however, these are just standard conventions that need to be known for reading and writing mathematical expressions. In this book, you will not see something ambiguous like $x = 8 \div 2\,(2+2)$, but you should be able to correctly calculate something like this,

$$x = 3^2 + 2\,(1+3)^2 - 6 \times 0.$$

First, remember that parentheses come first, so we can rewrite the above,

$$x = 3^2 + 2\,(4)^2 - 6 \times 0.$$

Exponents come next, so we can calculate those,

$$x = 9 + 2\,(16) - 6 \times 0.$$

Next comes multiplication and division,

$$x = 9 + 32 - 0.$$

Lastly, we calculate addition and subtraction,

$$x = 41.$$

In this book, you will very rarely need to calculate something with this many different steps. But you will often need to calculate equations like the one below,

$$x = 20 + 1.96 \times 2.1.$$

It is important to remember to multiply 1.96×2.1 *before* adding 20. Getting the order of operations wrong will usually result in the calculation being completely off.

One last note is that when operations are above or below a fraction, or below a radical, then parentheses are implied. For example, we might have something like the fraction below,

$$x = \frac{2^2 + 1}{3^2 + 2}.$$

Although rules 2–4 still apply, it is implied that there are parentheses around both the top (numerator) and bottom (denominator), so you can always read the above equation like this,

$$x = \frac{(2^2 + 1)}{(3^2 + 2)} = \frac{(4 + 1)}{(9 + 2)} = \frac{5}{11}.$$

Similarly, anything under the $\sqrt{}$ can be interpreted as being within parentheses. For example,

$$x = \sqrt{3 + 4^2} = \sqrt{(3 + 4^2)} \approx 4.47.$$

This can take some getting used to, but with practice, it will become second nature to read equations with the correct order of operations.

2

Data organisation

In the field or the lab, data collection can be messy. Often data need to be recorded with a pencil and paper, and in a format that is easiest for writing in adverse weather or a tightly controlled laboratory. Sometimes data from a particular sample, such as a bird nest (Figure 2.1), cannot all be collected in one place.

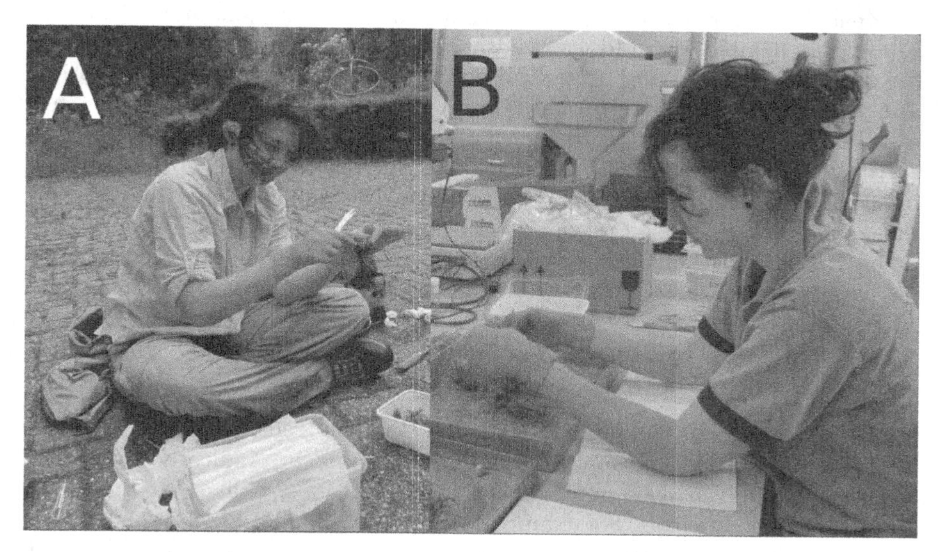

FIGURE 2.1 Dr Becky Boulton collects data from nest boxes in the field (A), then she processes nest material in the lab (B).

Data are sometimes missing due to circumstances outwith the researcher's control, and data are usually not collected in a format that is immediately ready for statistical analysis. Consequently, we often need to reorganise data from a lab or field book to a spreadsheet on the computer. Fortunately, there are some generally agreed upon guidelines for formatting data for statistical analysis. This chapter introduces the tidy format (Wickham, 2014), which is widely used for structuring data files. This chapter will provide an example of how to put data into a tidy format, and how to save a dataset into a file that can be read and used in jamovi (The jamovi project, 2024).

2.1 Tidy data

After data are collected, they need to be stored digitally (i.e., in a computer file, such as a spreadsheet). This should happen as soon as possible so that back-up copies of the data can be made. Retaining field and lab notes as a record of the originally collected data is also a good idea. Sometimes it is necessary to return to these notes, even years after data collection. Often we will want to double-check to make sure that we copied a value or observation correctly from handwritten notes to a spreadsheet. Note that sometimes data can be input directly into a spreadsheet or mobile application, bypassing handwritten notes altogether, but it is usually helpful to have a physical copy of collected data.

Most biological and environmental scientists store data digitally in the form of a spreadsheet. Spreadsheets enable data input, manipulation, and calculation in a highly flexible way. Most spreadsheet programs even have some capacity for data visualisation and statistical analysis. For the purposes of statistical analysis, spreadsheets are probably most often used for inputting data in a way that can be used by more powerful statistical software. Commonly used spreadsheet programs are MS Excel, Google Sheets, and LibreOffice Calc. The interface and functions of these programs are very similar, nearly identical for most purposes. They can all open and save the same file types (e.g., XLSX, ODS, CSV), and they all have the same overall look, feel, and functionality for data input, so the program used is mostly a matter of personal preference. In this book, we will use LibreOffice because it is completely free and open source, and easily available to download[1] (`http://libreoffice.org`). Excel and Google Sheets[2] are also completely fine to use.

Spreadsheets are separated into individual rectangular cells, which are identified by a specific column and row (Figure 2.2). Columns are indicated by letters, and rows are indicated by numbers. We can refer to a specific cell by its letter and number combination. For example, the active cell in Figure 2.2 is F3, which has a value of '3' indicating the value recorded in that specific measurement (in this case, foundress pollinators in a fig fruit). We will look more at how to interact with the spreadsheet in Chapter 3, but for now we will focus on how the data are organised.

There are a lot of potential ways that data could be organised in a spreadsheet. For good statistical analysis, there are a few principles that are helpful to follow. Whenever we collect data, we record observations about different units. For example, we might make one or more measurements on a tree, a patch of land, or in a sample of soil. In this case, trees, land patches, or soil samples are our **units of observation**. Each attribute of a unit that we are measuring is

[1] `https://www.libreoffice.org/download/download-libreoffice/`

[2] `https://docs.google.com/spreadsheets`

Site	Tree	Fruit	Tree_Lat	Tree_Lon	Foundress_Pollinators	Fruit_Volume_mm	Poll	SO1	SO2
S70	T70	F1	23.73629	-109.83987	4	1978.2	138	0	0
S70	T70	F2	23.73629	-109.83987	3	2535.55	97	3	0
S70	T70	F3	23.73629	-109.83987	1	3692.64	58	3	0
S70	T70	F4	23.73629	-109.83987	1	NA	39	0	0
S70	T70	F5	23.73629	-109.83987	1	1758.4	129	0	0
S70	T70	F6	23.73629	-109.83987	1	2009.6	77	0	0
S70	T70	F7	23.73629	-109.83987	1	1648.5	74	0	0

FIGURE 2.2 LibreOffice spreadsheet showing data from fig fruits collected in 2010. Each row is a unique sample (fruit), and columns record properties of the fruit.

a **variable**. These variables might include tree heights and leaf lengths, forest cover in a patch of land, or carbon and nitrogen content of a soil sample. Tidy datasets that can be used in statistical analysis programs are defined by three characteristics (Wickham, 2014):

1. Each variable gets its own column.
2. Each observation gets its own row.
3. Different units of observation require different data files.

If, for example, we were to measure the heights and leaf lengths for 4 trees, we might organise the data as in Table 2.1.

TABLE 2.1 Hypothetical tidy dataset in which each column of data is a variable, and each row of data is an observational unit (tree).

Tree	Species	Height (m)	Leaf Length (cm)
1	Oak	20.3	8.1
2	Oak	25.4	9.4
3	Maple	18.2	12.5
4	Maple	16.7	11.3

By convention (Wickham, 2014), variables tend to be in the left-most columns if they are known in advance or fixed in some way by the data collection or experiment (e.g., tree number or species in Table 2.1). In contrast, variables that are actually measured tend to be in the right-most columns (e.g., tree height or leaf length). This is more for readability of the data; jamovi will not care about the order of data columns.

2.2 Data files

Data can be saved using many different file types. File type is typically indicated by an extension following the name of a file and a full stop. For example, 'photo.png' would indicate a PNG image file named 'photo'. A peer-reviewed journal article might be saved as a PDF, e.g., 'Wickham2014.pdf'. A file's type affects what programs can be used to open it. One relevant distinction to make is between text files and binary files.

Text files are generally very simple. They only allow information to be stored as plain text; no colour, bold, italic, or anything else is encoded. All of the information is just made up of characters on one or more lines. This sounds so simple as to be almost obsolete; what is the point of not allowing anything besides plain text? The point is that text files are generally much more secure for long-term storage. The plain text format makes data easier to recover if a file is corrupted, is readable by a wider range of software, and is more amenable to version control (version control is a tool that essentially saves the whole history of a folder, and potentially different versions of it in parallel; it is not necessary for introductory statistics but is often critical for big collaborative projects). There are many types of text files with extensions such as TXT, CSV, HTML, R, CPP, or MD. For data storage, we will use comma-separated value (CSV) files. As the name implies, CSV files include plain text separated by commas. Each line of the CSV file is a new row, and commas separate information into columns. These CSV files can be opened in any text editor, but they are also recognised by nearly all spreadsheet programs and statistical software. The data shown in Figure 2.2 are from a CSV file called 'wasp_data.csv'. Figure 2.3 shows the same data when opened with a text editor.

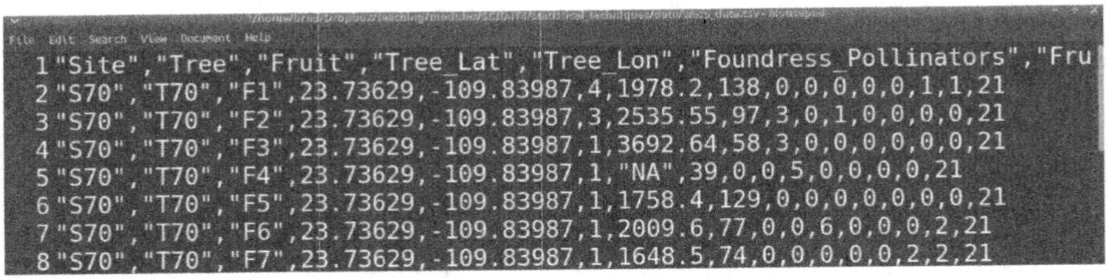

```
1 "Site","Tree","Fruit","Tree_Lat","Tree_Lon","Foundress_Pollinators","Fru
2 "S70","T70","F1",23.73629,-109.83987,4,1978.2,138,0,0,0,0,0,1,1,21
3 "S70","T70","F2",23.73629,-109.83987,3,2535.55,97,3,0,1,0,0,0,0,21
4 "S70","T70","F3",23.73629,-109.83987,1,3692.64,58,3,0,0,0,0,0,0,21
5 "S70","T70","F4",23.73629,-109.83987,1,"NA",39,0,0,5,0,0,0,0,21
6 "S70","T70","F5",23.73629,-109.83987,1,1758.4,129,0,0,0,0,0,0,0,21
7 "S70","T70","F6",23.73629,-109.83987,1,2009.6,77,0,0,6,0,0,0,2,21
8 "S70","T70","F7",23.73629,-109.83987,1,1648.5,74,0,0,0,0,0,2,2,21
```

FIGURE 2.3 Plain text comma-separated value (CSV) file showing data from fig fruits collected in 2010. Each line is a unique row and observation (fruit), and commas separate the data into columns in which the variables of fruit are recorded. The file has been opened in a program called 'Mousepad', but it could also be opened in any text editor such as gedit, Notepad, vim, or emacs. It could also be opened in spreadsheet programs such as LibreOffice Calc, MS Excel, or Google Sheets, or in any number of statistical programs.

The data shown in Figure 2.3 are not easy to read or work with, but the format is highly effective for storage because all of the information is in plain text. The information will therefore always look *exactly* the same, and it can be easily recovered by any text editor, even after years pass and old software inevitably becomes obsolete.

Binary files are different from text files and contain information besides just plain text. This information could include formatted text (e.g., bold, italic), images, sound, or video (basically, anything that can be stored in a file). The advantages of being able to store this kind of information are obvious, but the downside is that the information needs to be interpreted in a specific way, usually using a specific program. Examples of binary files include those with extensions such as DOC, XLS, PNG, GIF, MP3, or PPT. Some file types such as DOCX are not technically binary files, but a collection of zipped files (which, in the case of DOCX, include plain text files). Overall, the important point is that saving data in a text file format such as CSV is generally more secure.

2.3 Managing data files

Managing data files (or any files) effectively requires some understanding of how files are organised on a computer or cloud storage. In mobile phone applications, file organisation is often hidden, so it is not obvious where a file actually goes when it is saved on a device. Many people find files in these applications using a search function. The ability to search for files like this, or at least the tendency to do so regularly, is actually a relatively new phenomenon. And it is an approach to file organisation that does not work quite as well on non-mobile devices (i.e., anything that is not a phone or tablet), especially for big projects. On laptop and desktop computers, it is really important to know *where* files are being saved, and to ideally have an organisational system that makes it easy to find specific files without having to use a search tool.

On a computer, files are stored in a series of nested folders. You can think of the storage space on a computer, cloud, or network drive, as a big box. The big box can contain other smaller boxes (folders, in this analogy), or it can contain items that you need (files, in this analogy). Figure 2.4 shows the general idea. On this computer, there is a folder called 'brad', which has inside it five other folders (Figure 2.4A). Each of the five inner folders is used to store more folders and files for a specific class from 2006. Clicking on the 'Biostatistics' folder leads to the sub-folders inside it, and to files saved specifically for a biostatistics class (e.g., homework assignments, lecture notes, and an exam review document). Files on a computer therefore have a location that we can find using a particular **path**. We can write the path name using slashes to indicate nested folders. For example, the file 'HW9.scx' in Figure 2.4B would have the path name '/home/brad/Spring_2006/Biostatistics/HW9.scx'. Each folder is contained within slashes, and the file name itself is after the last slash.

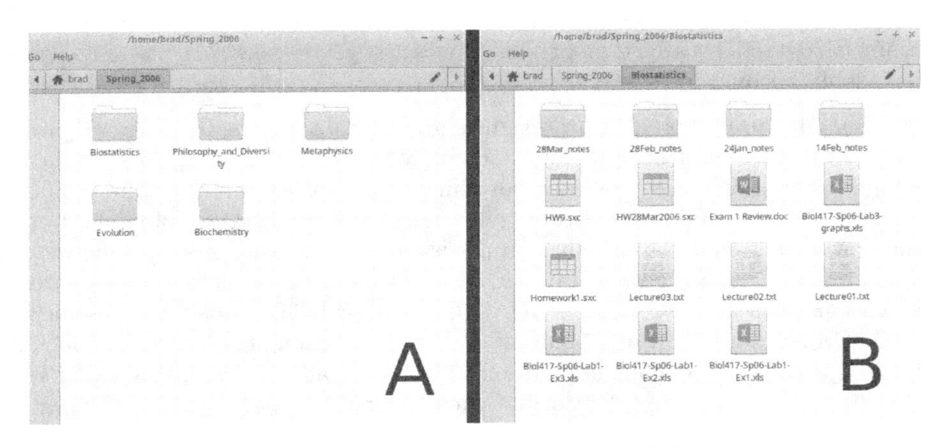

FIGURE 2.4 File directory of a computer showing (A) the file organisation of classes taken during spring 2006. Within one folder (B), there are multiple sub-folders and files associated with a biostatistics class.

These path names might look slightly different depending on the computer operating system that you are using. But the general idea of files nested within folders is the same. Figure 2.5 shows the same folder 'Spring_2006' saved in a different location, on OneDrive.

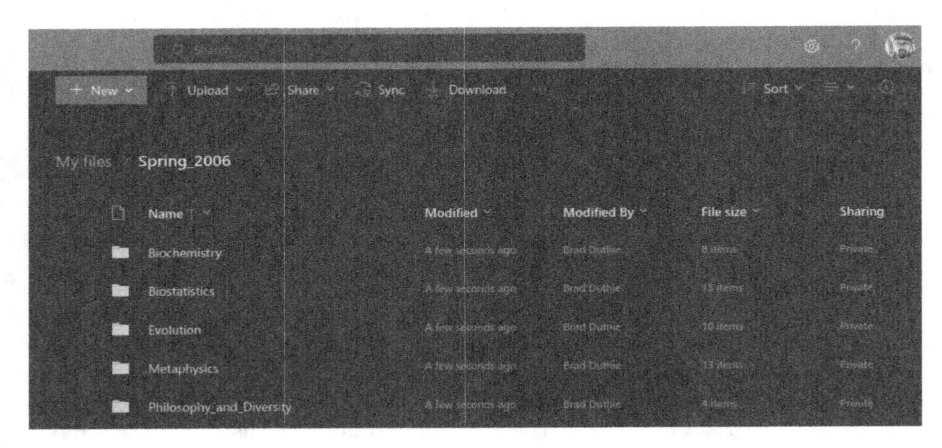

FIGURE 2.5 OneDrive file directory showing the file organisation of classes taken during spring 2006.

The path style '/home/user/folder/file.txt' is used for Linux and macOS. Windows has the same general file organisation (Figure 2.6). Path names for storing files on the hard drive of a Windows computer look something like

'C:\Users\MyName\Desktop\Spring_2006\Biostatistics\HW9.scx'. The 'C:\' is the root directory of the hard drive; it is called 'C' for historical reasons ('A:\' and 'B:\' used to be for floppy disks; the 'A:\' floppy disks had about 1.44 MB of storage, and 'B:\' had even less, so these are basically obsolete).

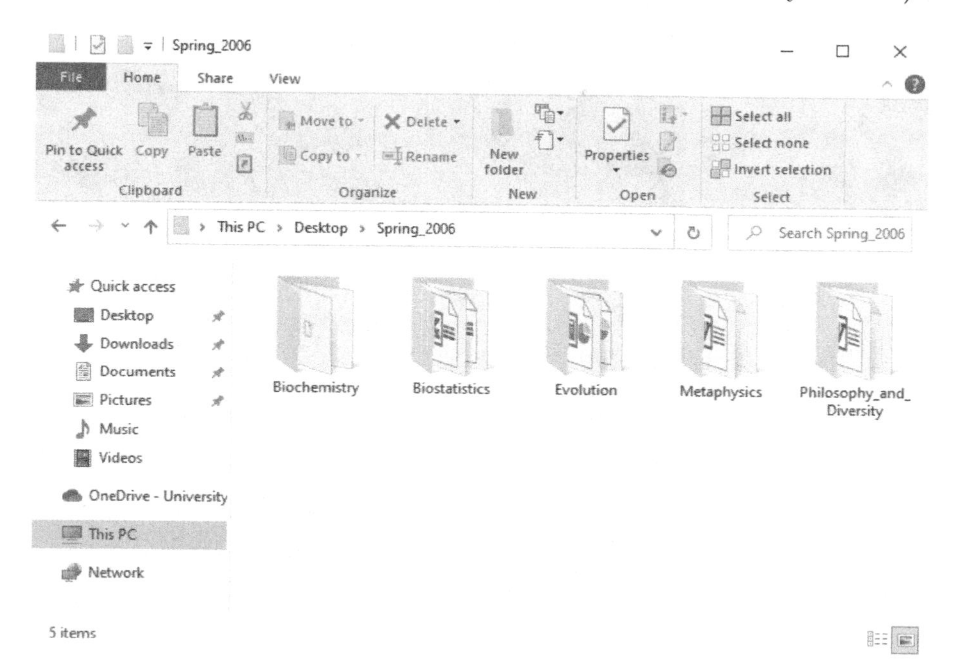

FIGURE 2.6 Windows file directory showing the file organisation of classes taken during spring 2006. In this case, the 'Spring_2006' folder is located on the desktop; the path to the folder is visible in the toolbar above the folders.

The details are not as important as the idea of organising files in a logical way that allows you to know roughly where to find important files on a computer or cloud drive. It is usually a good idea to give every unique project or subject (e.g., university class, student group, holiday plans, health records) its own folder. This makes it much easier to find related files such as datasets, lecture notes, or assignments when necessary. It is usually possible to right-click somewhere in a directory to create a new folder. In Figure 2.6, there is even a 'New folder' button in the toolbar with a yellow folder icon above it. It takes some time to organise files this way, and to get used to saving files in specific locations, but it is well worth it in the long-term.

3

Practical. *Preparing data*

In this chapter, we will use a spreadsheet to organise datasets following the tidy approach explained in Chapter 2, then save these datasets as CSV files to be opened in jamovi statistical software (The jamovi project, 2024). The data organisation in this chapter can be completed using LibreOffice Calc[1], MS Excel, or Google Sheets[2]. The screenshots below will be from LibreOffice Calc, but the instructions provided will work on any of the three aforementioned spreadsheet programs.

There are four data exercises in this chapter. All of these exercises will focus on organising data into a tidy format. Exercise 3.1 uses handwritten field data that need to be entered into a spreadsheet in a tidy format. These data include information shown in Figure 2.2, plus tallies of seed counts. The goal is to get all of this information into a tidy format and save it as a CSV file. Exercise 3.2 presents some data on the number of eggs produced by five different fig wasp species (more on these in Chapter 8). The data are in an untidy format, so the goal is to reorganise them and save them as a tidy CSV file. Exercise 3.3 presents counts of the same five fig wasp species as in Exercise 3.2, which need to be reorganised in a tidy format. Exercise 3.4 presents data that are even more messy. These are morphological measurements of the same five species of wasps, including lengths and widths of wasp heads, thoraxes, and abdomens. The goal in this exercise is to tidy the data, then estimate total wasp volume from the morphological measurements using mathematical formulas, keeping in mind the order of operations from Chapter 1.

3.1 Transferring data to a spreadsheet

Exercise 3.1 focuses on data collected from the fruits of fig trees collected from Baja, Mexico in 2010 (Duthie et al., 2015; Duthie & Nason, 2016). Due to the nature of the work, the data needed to be recorded in notebooks and collected in two different locations. The first location was the field, where data were

[1] https://www.libreoffice.org/discover/calc/
[2] https://docs.google.com/spreadsheets/

collected identifying tree locations and fruit dimensions. Baja is hot and sunny (Figure 3.1).

FIGURE 3.1 Fully grown Sonoran Desert Rock Fig in the desert of Baja, Mexico.

Fruit measurements were made with a ruler and recorded in a field notebook. These measurements are shown in Figure 3.2.

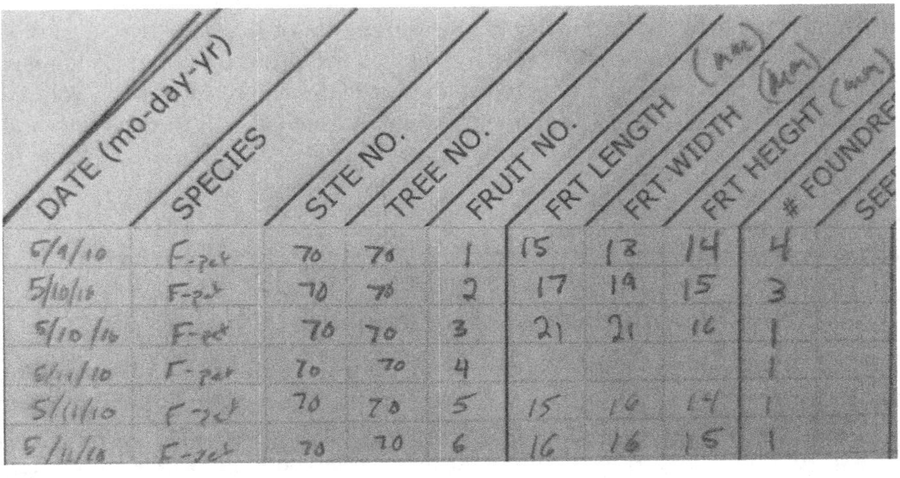

DATE (mo-day-yr)	SPECIES	SITE NO.	TREE NO.	FRUIT NO.	FRT LENGTH (mm)	FRT WIDTH (mm)	FRT HEIGHT (mm)	# FOUNDRE	SEE
5/9/10	F-pet	70	70	1	15	13	14	4	
5/10/10	F-pet	70	70	2	17	19	15	3	
5/10/10	F-pet	70	70	3	21	21	16	1	
5/11/10	F-pet	70	70	4				1	
5/11/10	F-pet	70	70	5	15	16	14	1	
5/11/10	F-pet	70	70	6	16	16	15	1	

FIGURE 3.2 Portion of a lab notebook used to record measurements of fig fruits from different trees in 2010.

The second location was in a lab in Iowa, USA. Fruits were dried and shipped to Iowa State University so that seeds could be counted under a microscope. Counts were originally recorded as tallies in a lab notebook. The goal of Exercise 3.1 is to get all of this information into a single tidy spreadsheet.

The best place to start is with an empty spreadsheet, so open a new one in LibreOffice Calc, MS Excel, or Google Sheets. Remember that each row will be a unique observation; in this case, a unique fig fruit from which measurements were recorded. Each column will be a variable of that observation. Fortunately, the data in Figure 3.2 are already looking quite tidy. The information here can be put into the spreadsheet mostly as written in the notebook. But there are a few points to keep in mind:

1. It is important to start in column A and row 1; do not leave any empty rows or columns because when we get to the statistical analysis in jamovi, jamovi will assume that these empty rows and columns signify missing data.
2. There is no need to include any formatting (e.g., bold, underline, colour) because it will not be saved in the CSV or recognised by jamovi.
3. Missing information, such as the empty boxes for the fruit dimensions in row 4 in the notebook (Figure 3.2), should be indicated with an 'NA' (capital letters, but without the quotes). This will let jamovi know that these data are missing.
4. The date is written in an American style of month-day-year, which might get confusing. It might be better to have separate columns for year, month, and day, and to write out the full year (2010).

The column names in Figure 3.2 are (1) Date, (2) Species, (3) Site number, (4) Tree number, (5) Fruit length in millimetres, (6) Fruit width in millimetres, and (7) Fruit height in millimetres. All of the species are *Ficus petiolaris*, which is abbreviated to 'F-pet' in the field notebook. How you choose to write some of this information down is up to you (e.g., date format, capitalisation of column names), but when finished, the spreadsheet should be organised like the one in Figure 3.3.

Year	Month	Day	Species	Site number	Tree number	Fruit number	Fruit length (mm)	Fruit width (mm)	Fruit height (mm)
2010	5	9	F_petiolaris	70	70	1	15	18	14
2010	5	10	F_petiolaris	70	70	2	17	19	15
2010	5	10	F_petiolaris	70	70	3	21	21	16
2010	5	11	F_petiolaris	70	70	4	NA	NA	NA
2010	5	11	F_petiolaris	70	70	5	15	16	14
2010	5	11	F_petiolaris	70	70	6	16	16	15

FIGURE 3.3 Spreadsheet with data organised in a tidy format and nearly ready for analysis.

This leaves us with the data that had to be collected later in the lab. Small seeds needed to be meticulously separated from other material in the fig fruit, then tallied under a microscope. Tallies from this notebook are recreated below.

Site 70, Tree 70, Fruit 1: 238 total

𝍸 𝍸 𝍸 𝍸 𝍸 𝍸 𝍸 𝍸 𝍸 𝍸 𝍸 𝍸 𝍸 𝍸 𝍸 𝍸 𝍸 𝍸 𝍸 𝍸

𝍸 𝍸 𝍸 𝍸 𝍸 𝍸 𝍸 𝍸 𝍸 𝍸 𝍸 𝍸 𝍸 𝍸 𝍸 𝍸 𝍸 𝍸 𝍸 𝍸

𝍸 𝍸 𝍸 𝍸 𝍸 𝍸 𝍸 |||

Site 70, Tree 70, Fruit 2: 198 total

𝍸 𝍸 𝍸 𝍸 𝍸 𝍸 𝍸 𝍸 𝍸 𝍸 𝍸 𝍸 𝍸 𝍸 𝍸 𝍸 𝍸 𝍸 𝍸 𝍸

𝍸 𝍸 𝍸 𝍸 𝍸 𝍸 𝍸 𝍸 𝍸 𝍸 𝍸 𝍸 𝍸 𝍸 𝍸 𝍸 𝍸 𝍸 𝍸 |||

Site 70, Tree 70, Fruit 3: 220 total

𝍸 𝍸 𝍸 𝍸 𝍸 𝍸 𝍸 𝍸 𝍸 𝍸 𝍸 𝍸 𝍸 𝍸 𝍸 𝍸 𝍸 𝍸 𝍸 𝍸

𝍸 𝍸 𝍸 𝍸 𝍸 𝍸 𝍸 𝍸 𝍸 𝍸 𝍸 𝍸 𝍸 𝍸 𝍸 𝍸 𝍸 𝍸 𝍸 𝍸

𝍸 𝍸 𝍸 𝍸

Site 70, Tree 70, Fruit 4: 169 total

𝍸 𝍸 𝍸 𝍸 𝍸 𝍸 𝍸 𝍸 𝍸 𝍸 𝍸 𝍸 𝍸 𝍸 𝍸 𝍸 𝍸 𝍸 𝍸 𝍸

𝍸 𝍸 𝍸 𝍸 𝍸 𝍸 𝍸 𝍸 𝍸 𝍸 𝍸 𝍸 𝍸 ||||

Site 70, Tree 70, Fruit 5: 188 total

𝍸 𝍸 𝍸 𝍸 𝍸 𝍸 𝍸 𝍸 𝍸 𝍸 𝍸 𝍸 𝍸 𝍸 𝍸 𝍸 𝍸 𝍸 𝍸 𝍸

𝍸 𝍸 𝍸 𝍸 𝍸 𝍸 𝍸 𝍸 𝍸 𝍸 𝍸 𝍸 𝍸 𝍸 𝍸 𝍸 𝍸 |||

Site 70, Tree 70, Fruit 6: 139 total

𝍸 𝍸 𝍸 𝍸 𝍸 𝍸 𝍸 𝍸 𝍸 𝍸 𝍸 𝍸 𝍸 𝍸 𝍸 𝍸 𝍸 𝍸 𝍸 𝍸

𝍸 𝍸 𝍸 𝍸 𝍸 𝍸 𝍸 ||||

Fortunately, the summed tallies have been written next to the site, tree, and fruit, which makes inputting them into a spreadsheet easier. But it is important to also recognise this step as a potential source of human error in data collection. It is possible that the tallies were counted inaccurately, meaning that the tallies do not sum to the numbers reported above. It is always good to be able to go back and check. There are at least two other potential sources of human error in counting seeds and inputting them into the spreadsheet, one before, and one after counting the tallies. Fill in 1 and 3 below with potential causes of error.

1.
2. Tallies are not counted correctly in the lab notebook
3.

Next, create a new column in the spreadsheet and call it 'Seeds' (use column K). Fill in the seed counts for each of the six rows. The end result will be a tidy dataset that is ready to be saved as a CSV.

What you do next depends on the spreadsheet program that you are using and how you are using it. If you are using LibreOffice Calc or MS Excel on your computer, then you should be able to simply save your file as something like 'Fig_fruits.csv', and the program will recognise that you intend to save as a CSV file (in MS Excel, you might need to find the pull-down box for 'Save as type:' under the 'File name:' box and choose 'CSV'). If you are using Google Sheets, you can navigate in the toolbar to 'File → Download → Comma-separated values (.csv)', which will start a download of your spreadsheet in CSV format. If you are using MS Excel in a browser online, then it is a bit more tedious. At the time of writing, the online version of MS Excel does not allow users to save or export to a CSV. It will therefore be necessary to save as an XLSX, then convert to CSV later in another spreadsheet program (local version of MS Excel, LibreOffice Calc, or Google Sheets).

Save your file in a location where you know that you can find it again. It might be a good idea to create a new folder on your computer or your cloud storage online for files in this book. This will ensure that you always know where your data files are located and can access them easily.

3.2 Making spreadsheet data tidy

Exercise 3.2 is more self-guided than Exercise 3.1. After reading Chapter 2 and completing Exercise 3.1, you should have a bit more confidence in organising data in a tidy format. Here we will work with a dataset that includes counts of the number of eggs collected from fig wasps, which are small species of insects that lay their eggs into the ovules of fig flowers (Weiblen, 2002). You can download this dataset online[3] or recreate it from Table 3.1.

Using what you have learnt in Chapter 2 and Exercise 3.1, create a tidy version of the wasp egg loads dataset. For a helpful hint, it might be most efficient to open a new spreadsheet and copy and paste information from the old to the new.

[3] `https://bradduthie.github.io/stats/data/wasp_egg_loads_untidy.xlsx`

TABLE 3.1 Untidy dataset of egg loads from fig wasps of five different species, including two unnamed species of the genus *Heterandrium* (Het1 and Het2) and three unnamed species of the genus *Idarnes* (LO1, SO1, and SO2).

Het1	Het2	LO1	SO1	SO2
35	51	72	50	44
32	55	76	47	44
34	52	77	48	46
38	54	78	54	36
34	55	76	54	51
34	54	72	46	50
34	56	79	50	36
34	53	76	50	56
32	54	77	52	58
30	54	75	51	45
				49
				39
				54
				52

How many columns did you need to create the new dataset? ____________

Are there any missing data in this dataset? ____________

Save the tidy dataset to a CSV file.

3.3 Making data tidy again

Exercise 3.3, like Exercise 3.2, is self-guided. The data are presented in a fairly common, but untidy, format, and the challenge is to reorganise them into a tidy dataset that is ready for statistical analysis. Table 3.2 shows the number of different species of wasps counted in five different fig fruits. Rows list all of the species and columns list the fruits, with the counts in the middle. This is an efficient way to present the data so that they are all easy to see, but this will not work for running statistical analysis.

This exercise might be a bit more challenging than Exercise 3.2. The goal is to use the information in Table 3.2 to create a tidy dataset. Remember that each observation (wasp counts, in this case) should get its own row, and each variable should get its own column. Try creating a tidy dataset from the information in Table 3.2, then save the dataset to a CSV file.

TABLE 3.2 Efficient but untidy way to present count data. Counts of different species of fig wasps (rows) are from five different fig fruits (columns). Data were originally collected from Baja, Mexico in 2010.

Species	Fruit_1	Fruit_2	Fruit_3	Fruit_4	Fruit_5
Het1	0	0	0	1	0
Het2	0	2	3	0	0
LO1	4	37	0	0	3
SO1	0	1	0	3	2
SO2	1	12	2	0	0

3.4 Tidy data and spreadsheet calculations

Exercise 3.4 requires some restructuring and calculations. The dataset that will be used in this exercise includes morphological measurements from five species of fig wasps, the same species used in Exercises 3.2 and 3.3. The dataset for this exercise can be downloaded online[4]. This dataset is about as untidy as it gets. First note that there are multiple sheets in the spreadsheet, which is not allowed in a CSV file. You can see these sheets by looking at the very bottom of the spreadsheet, which will have separating tabs called Het1, Het2, LO1, SO1, and SO2 (Figure 3.4).

FIGURE 3.4 Spreadsheets can include multiple sheets. This image shows that the spreadsheet containing information for fig wasp morphology includes five separate sheets, one for each species.

You can click on all of the different tabs to see the measurements of head length, head width, thorax length, thorax width, abdomen length, and abdomen width for wasps of each of the five species. All of the measurements are collected in millimetres. Note that the individual sheets contain text formatting (titles highlighted and in bold), and there is a picture of each wasp in its respective sheet. The formatting and pictures are a nice touch for providing some context, but they cannot be used in statistical analysis. The first task is to create a tidy version of this dataset. Probably the best way to do this is to create a new

[4] `https://bradduthie.github.io/stats/data/wasp_morphology_untidy.xlsx`

spreadsheet entirely and copy-paste information from the old. It is a good idea to think about how the tidy dataset will look before getting started. What columns should this new dataset include? Write your answer below.

How many rows are needed? ____________________

When you are ready, create the new dataset. Your dataset should have all of the relevant information about wasp head, thorax, and abdomen measurements.

Next comes a slightly more challenging part, which will make use of some of the background mathematics reviewed in Chapter 1. Suppose that we wanted our new dataset to include information about the volumes of each of the three wasp body segments, and wasp total volume. To do this, let us assume that the wasp head is a sphere (it is not, exactly, but this is probably the best estimate that we can get under the circumstances). Calculate the head volume of each wasp using the following formula,

$$V_{\text{head}} = \frac{4}{3}\pi \left(\frac{Head_{\text{L}} + Head_{\text{W}}}{4} \right)^3 .$$

In the equation above, $Head_{\text{L}}$ is head length (mm) and $Head_{\text{W}}$ is head width (note, $(Head_{\text{L}}+Head_{\text{W}})/4$ estimates the radius of the head). You can replace π with the approximation $\pi \approx 3.14$. To make this calculation in your spreadsheet, find the cell in which you want to put the head volume. By typing in the `=` sign, the spreadsheet will know to start a new calculation or function in that cell. Try this with an empty cell by typing '$= 5 + 4$' in it (without quotes). When you hit 'Enter', the spreadsheet will make the calculation for you, and the number in the new cell will be 9. To see the equation again, you just need to double-click on the cell.

To get an estimate of head volume into the dataset, we can create a new column of data. To calculate V_{head} for the first wasp in row 2 of the spreadsheet, we could select the spreadsheet cell H2 and type the code, `=(4/3)*(3.14)*((B2+C2)/4)^3`. Notice that the code recognises B2 and C2 as spreadsheet cells, and takes the values from these cells when doing these calculations. If the values of B2 or C2 were to change, then so would the calculated value in H2. Also notice that we are using parentheses to make sure that the order of operations is correct. We want to add head length and width before dividing by 4, so we type `((B2+C2)/4)` to ensure with the innermost parentheses that head length and width are added before dividing. Once all of this is completed, we raise everything in parentheses to the third power using the `^3`, so `((B2+C2)/4)^3`. Different mathematical operations can be carried out using the symbols in Table 3.3.

TABLE 3.3 List of mathematical operations available in a spreadsheet.

Symbol	Operation
`+`	Addition
`-`	Subtraction
`*`	Multiplication
`/`	Division
`^`	Exponent
`sqrt()`	Square-root

The last operation in Table 3.3 is a function that takes the square-root of anything within the parentheses. Other functions are also available that can make calculations across cells (e.g., `=SUM` or `=AVERAGE`).

Once head volume is calculated for the first wasp in cell H2, it is very easy to do the rest. One nice feature of a spreadsheet is that it can usually recognise when the cells need to change (B2 and C2, in this case). To get the rest of the head volumes, we just need to select the bottom right of the H2 cell. There will be a very small square in this bottom right (see Figure 3.5), and if we click and drag it down, the spreadsheet will do the same calculation for each row (e.g., in H3, it will use B3 and C3 in the formula rather than B2 and C2).

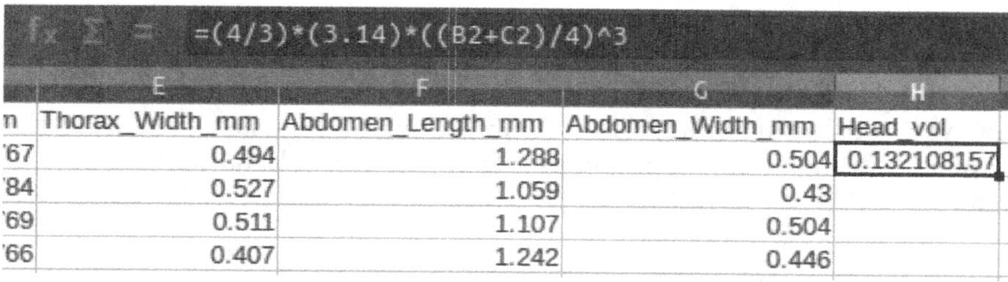

`=(4/3)*(3.14)*((B2+C2)/4)^3`

	E	F	G	H
n	Thorax_Width_mm	Abdomen_Length_mm	Abdomen_Width_mm	Head_vol
'67	0.494	1.288	0.504	0.132108157
'84	0.527	1.059	0.43	
'69	0.511	1.107	0.504	
'66	0.407	1.242	0.446	

FIGURE 3.5 Dataset of wasp morphological measurements from five species of fig wasps collected from Baja, Mexico in 2010. Head volume (column H) has been calculated for row 2, and to calculate it for the remaining rows, the small black square in the bottom right of the highlighted cell H2 can be clicked and dragged down to H27.

Another way to achieve the same result is to copy the contents of cell H2, highlight cells H3–H27, then paste. However you do it, you should now have a new column of calculated head volumes.

Next, suppose that we want to calculate thorax and abdomen volumes for all wasps. Unlike wasp heads, wasp thoraxes and abdomens are clearly not spheres. But it is perhaps not entirely unreasonable to model them as ellipsoids. To calculate wasp thorax and abdomen volumes assuming an ellipsoid shape, we can use the formula,

$$V_{\text{thorax}} = \frac{4}{3}\pi \left(\frac{Thorax_{\text{L}}}{2}\right)\left(\frac{Thorax_{\text{W}}}{2}\right)^2.$$

In the equation above, $Thorax_{\text{L}}$ is thorax length (mm) and $Thorax_{\text{W}}$ is thorax width. Substitute $Abdomen_{\text{L}}$ and $Abdomen_{\text{W}}$ to instead calculate abdomen volume (V_{abdomen}). What formula will you type into your empty spreadsheet cell to calculate V_{thorax}? Keep in mind the order of operations indicated in the equation above.

Now fill in the columns for thorax volume and abdomen volume. You should now have three new columns of data from calculations of the volumes of the head, thorax, and abdomen of each wasp. Lastly, add one final column of data for total volume, which is the sum of the three segments.

There are a lot of potential sources of error and uncertainty in these final volumes. What are some reasons that we might want to be cautious about our calculated wasp volumes? Explain in 2–3 sentences.

Save your wasp morphology file as a CSV. This was the last exercise of the chapter. You should now be comfortable formatting tidy datasets for use in jamovi.

3.5 Summary

Completing this practical should give you the skills that you need to prepare datasets for statistical analysis. There are many additional features of spreadsheets that were not introduced (mainly because we will do them in jamovi) but could be useful to learn. For example, if we wanted to calculate the sum of all head lengths, we could use the function `=sum(B2:B27)` in any spreadsheet cell (where B2 is the head length of the first wasp, and B27 is the head length of the last wasp). Other functions such as `=count()`, `=min()`, `=max()`, or `=average()` can be similarly used for calculations.

4

Populations and samples

When we collect data, we are recording some kind of observation or measurement. If we are working in a forest, for example, we might want to measure the heights of different trees, or measure the concentration of carbon in the soil. The idea might be to use these measurements to make some kind of inference about the forest. But as scientists, we are almost always limited in the amount of data that we can collect. We cannot measure everything, so we need to collect a *sample* of data and use it to make inferences about the *population* of interest. For example, while we probably cannot measure the height of every tree in a forest, nor can we measure the concentration of carbon at every possible location in the forest's soil, we can collect a smaller number of measurements and still make useful conclusions about overall forest tree height and carbon concentration.

Statistics thereby allows us to approximate properties of entire populations from a limited number of samples. This needs to be done with caution, but before getting into the details of how, it is important to fully understand the difference between a **population** and a **sample** to avoid confusing these two concepts. A **population** is the entire set of possible observations that could be collected (Sokal & Rohlf, 1995). Some examples will make it easier to understand:

- All of the birch trees in Scotland
- All genes in the house mouse (*Mus musculus*)
- All wrens in the United Kingdom

These populations might be important for a particular research question. For example, we might want to know something about the feeding behaviours of wrens in the UK. But there is no way that we can find and observe the behaviour of every single wren, so we need to take a subset of the population (sample) instead (Fowler et al., 1998). Examples of samples include the following:

- 25 birch trees from a Scottish forest
- 10 genes from the *Mus musculus* genome
- 60 caught wrens in UK nest boxes

It is important to recognise that the word 'population' means something different in statistics than it does in biology. A biological population, for example, could be defined as all of the individuals of the same species in the

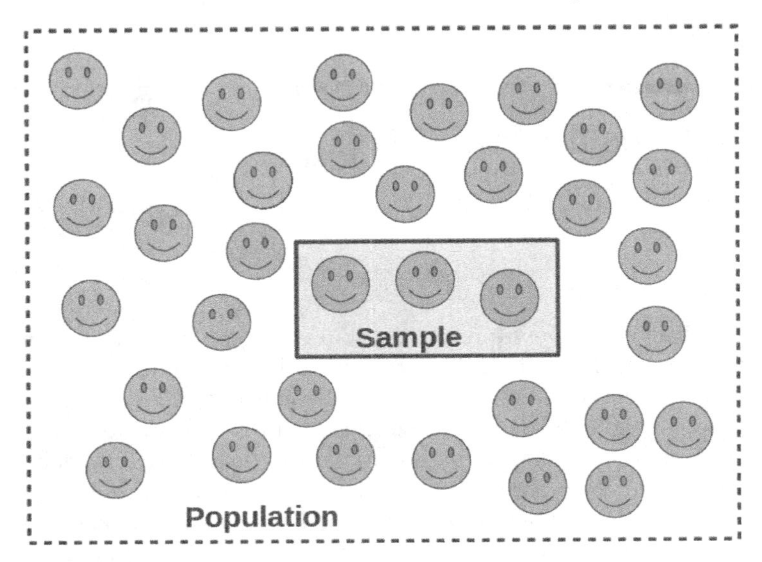

FIGURE 4.1 Conceptual figure illustrating how a statistical population relates to a statistical sample. The population is represented by 35 smiling faces enclosed within a dashed box. The sample is represented by a solid box within the dashed box, within which there are three smiling faces. Hence, we have a sample of three measurements from the total population.

same general location. A statistical population, in contrast, refers to a set of observations (i.e., things that we can measure). Sokal & Rohlf (1995) provide a more technical definition for 'population',

---

> In statistics, population always means the *totality of individual observations about which inferences are to be made, existing anywhere in the world or at least within a definitely specified sampling area limited in space and time* [p. 9, emphasis theirs].

They define a sample to be 'a collection of individual observations selected by a specified procedure' (Sokal & Rohlf, 1995). For our purposes, it is not necessary to be able to recite the technical definitions, but it is important to understand the relationship between a population and a sample. When we collect data, we are almost always taking a small sample of observations from a much larger number of possible observations in a population (Figure 4.1).

5

Types of variables

A variable is any property that is measured in an observation (Sokal & Rohlf, 1995), i.e., anything that varies among things that we can measure (Dytham, 2011). We can summarise how these measurements vary with summary statistics, or visually with figures. Often, we will want to predict one variable from a second variable. In this case, the variable that we want to predict is called the **response variable**, also known as the **dependent variable** or ***Y*** variable ('dependent' because it *depends* on other variables, and '*Y*' because this is the letter we often use to represent it). The variable that we use to predict our response variable is the **explanatory variable**, also known as the **independent variable** or ***X*** variable ('independent' because it does not depend on other variables, and '*X*' because this is the letter most often used to represent it). There are several different types of variables:

- **Categorical** variables take on a fixed number of discrete values (Spiegelhalter, 2019). In other words, the measurement that we record will assign our data to a specific category. Examples of categorical variables include species (e.g., 'Robin', 'Nightingale', 'Wren') or life history stage (e.g., 'egg', 'juvenile', 'adult'). Categorical variables can be either nominal or ordinal.
  - **Nominal** variables do not have any inherent order (e.g., classifying land as 'forest', 'grassland', or 'urban').
  - **Ordinal** variables do have an inherent order (e.g., 'low', 'medium', and 'high' elevation).
- **Quantitative** variables are represented by numbers that reflect a magnitude. That is, unlike categorical variables, we are collecting numbers that really mean something tangible (in contrast, while we might represent low, medium, and high elevations with the numbers 1, 2, and 3, respectively, this is just for convenience; a value of '2' does not always mean 'medium' in other contexts). Quantitative variables can be either discrete or continuous.
  - **Discrete** variables can take only certain values (Dytham, 2011). For example, if we want to record the number of species in a forest, then our variable can only take discrete counts (i.e., integer values). There could conceivably be any natural number of species (1, 2, 3, etc.), but there could not be 2.51 different species in a forest; that does not make sense.

- **Continuous** variables can take any real value within some range of values (i.e., any number that can be represented by a decimal). For example, we could measure height to as many decimals as our measuring device will allow, with a range of values from zero to the maximum possible height of whatever it is we are measuring. Similarly, we could measure temperature to any number of decimals, at least in theory, so temperature is a continuous variable.

The reason for organising variables into all of these different types is that different types of variables need to be handled in different ways. For example, it would not make sense to visualise a nominal variable in the same way as a continuous variable. Similarly, the choice of statistical test to apply to answer a statistical question will almost always depend on the types of variables involved. If presented with a new dataset, it is therefore very important to be able to interpret the different variables and apply the correct statistical techniques.

6

Accuracy, precision, and units

The science of measurements is called 'metrology', which, among other topics, focuses on measurement accuracy, precision, and units (Rabinovich, 2013). We will not consider these topics in depth, but they are important for statistical techniques because measurement, in the broadest sense of the word, is the foundation of data collection. When collecting data, we want measurements to be accurate, precise, and clearly defined.

6.1 Accuracy

When we collect data, we are trying to obtain information about the world. We might, for example, want to know the number of seedlings in an area of forest, the temperature of the soil at some location, or the mass of a particular animal in the field. To get this information, we need to make measurements. Some measurements can be collected by simple observation (e.g., counting seedlings), while others will require measuring devices such as a thermometer (for measuring temperature) or scale (for measuring mass). All of these measurements are subject to error. The *true* value of whatever it is that we are trying to measure (called the 'measurand') can differ from what we record when collecting data. This is true even for simple observations (e.g., we might miscount seedlings), so it is important to recognise that the data we collect come with some uncertainty. The **accuracy** of a measurement is defined by how close the measurement is to the *true* value of what we are trying to measure (Rabinovich, 2013).

6.2 Precision

The **precision** of a measurement is how consistent it will be if measurement is replicated multiple times. In other words, precision describes how similar measurements are expected to be (Wardlaw, 1985). If, for example, a scale measures an object to have the exact same mass every time it is weighed (regardless of whether the mass is accurate), then the measurement is highly precise. If, however, the scale measures a different mass each time the object is weighed (for this hypothetical, assume that the true mass of the object does not change), then the measurement is not as precise.

One way to visualise the difference between accuracy and precision is to imagine a set of targets, with the centre of the target representing the true value of what we are trying to measure (Figure 6.1)[1].

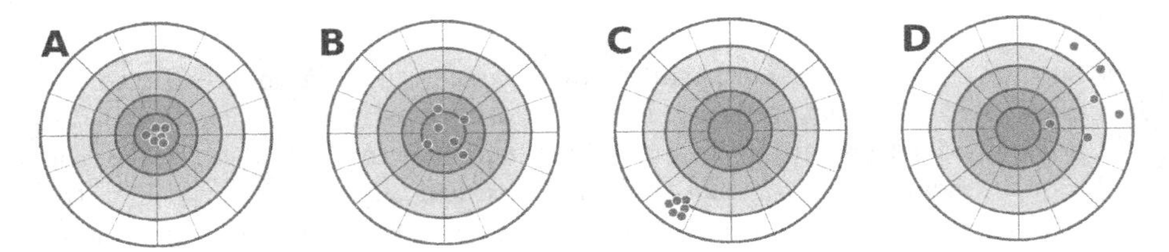

FIGURE 6.1 Conceptual figure illustrating the difference between accuracy and precision. Points in A are both accurate and precise, points in B are accurate but not precise, points in C are precise but not accurate, and points in D are neither accurate nor precise.

Note again that accuracy and precision are not necessarily the same (Dytham, 2011). Measurement can be accurate but not precise (Figure 6.1B) or precise but not accurate (Figure 6.1C).

6.3 Systems of units

Scientific units are standardised with the Système International D'Unités (SI). Having standardised units of measurement is highly important to ensure measurement accuracy (Quinn, 1995). Originally, these units were often defined in terms of physical artefacts. For example, the kilogram (kg) was once defined by a physical cylinder of metal housed in the Bureau International des Poids et Mesures (BIPM). In other words, the mass of a metal sitting at the BIPM *defined* what a kilogram was, with the mass of every other measurement being based on this physical object (Freedman et al., 2011; Quinn, 1995). This can potentially present a problem if the mass of that one object changes over time, thereby causing a change in how a kilogram is defined. Where possible, it is therefore preferable to define units in terms of fundamental constants of nature. In 2019, for example, the kilogram was redefined in terms of the Planck constant, a specific atomic transition frequency, and the speed of light (Stock

[1]This figure was released into the public domain by Egon Willighagen on 8 March 2014: `https://commons.wikimedia.org/wiki/File:Accuracy-vs-precision-nl.svg`

et al., 2019). This ensures that measurements of mass remain accurate over time because what a kilogram represents in terms of mass cannot change.

We can separate units into base units and derived units. Table 6.1 below lists some common base units for convenience (Quinn, 1995). You do not need to memorise these units, but it is good to be familiar with them. We will use these units throughout the book.

TABLE 6.1 Base units of SI measurements. For details see Quinn (1995).

Measured Quantity	Name of SI Unit	Symbol
Mass	kilogram	kg
Length	metre	m
Time	second	s
Electric current	ampere	A
Temperature	kelvin	K
Amount of a substance	mole	mol
Luminous intensity	candela	cd

We can also define derived SI units from the base units of Table 6.1 (Gupta, 2020). Examples of these derived SI units are provided in Table 6.2. Again, you do not need to memorise these units, but it is good to be aware of them.

TABLE 6.2 Examples of derived SI units.

Measured Quantity	Name of Unit	Symbol	Definition in SI Units
Area	square metre	A	m^2
Volume	cubic metre	V	m^3
Speed	metre per second	v	$\mathrm{m\ s^{-1}}$
Mass density	kilogram per metre cube	ρ	$\mathrm{kg\ m^{-3}}$
Amount concentration	mole per cubic metre	c	$\mathrm{mol\ m^{-3}}$
Force	newton	N	$\mathrm{m\ kg\ s^{-2}}$
Pressure	pascal	Pa	$\mathrm{m^{-1}\ kg\ s^{-2}}$
Energy	joule	J	$\mathrm{m^2\ kg\ s^{-2}}$

When numbers are associated with units, it is important to recognise that the units must be carried through and combined when calculating an equation. As a very simple example, if we want to know the speed at which an object is moving, and we find that it has moved 10 metres in 20 seconds, then we calculate the speed and report the correct units as below,

$$speed = \frac{10\ \mathrm{m}}{20\ \mathrm{s}} = 0.5\ \mathrm{m/s} = 0.5\ \mathrm{m\,s^{-1}}.$$

Notice that the final units are in metres per second, which can be written as m/s or $\mathrm{m\,s^{-1}}$ (remember that raising s to the -1 power is the same as 1/s; see Chapter 1 for a reminder about superscripts). Recognising that units are also part of calculations is important.

7

Uncertainty propagation

Nothing can be measured with perfect accuracy, meaning that every measurement has some associated error. The measurement error might be caused by random noise in the measuring environment, or by mistakes made by the person doing the measuring. The measurement error might also be caused by limitations or imperfections associated with a measuring device. The device might be limited in its measurement precision, or perhaps it is biased in its measurements due to improper calibration, manufacture, or damage from previous use. All of this generates uncertainty with respect to individual measurements.

Recall from Chapter 6 the difference between precision and accuracy. We can evaluate the precision and accuracy of measurements in different ways. Measurement precision can be estimated by replicating a measurement (i.e., taking the same measurement over and over again). The more replicate measurements made, the more precisely a value can be estimated. For example, if we wanted to evaluate the precision with which the mass of an object is measured, then we might repeat the measurement with the same scale multiple times and see how much mass changes across different measurements. To evaluate measurement accuracy, we might need to measure a value in multiple different ways (e.g., with different measuring devices). For example, we might repeat the measurement of an object's mass with a different scale (i.e., a different physical scale used for measuring the mass of objects).

Sometimes it is necessary to combine different measured values. For example, we might measure the mass of two different bird eggs in a nest separately, then calculate the total mass of both eggs combined. The measurement of each egg will have its own error, and these errors will propagate to determine the error of the total egg mass for the nest. How this error propagates differs depending on if errors are being added or subtracted, or if errors are being multiplied or divided.

7.1 Adding or subtracting errors

In the case of our egg masses, we can assign the mass of the first egg to the variable X and the mass of the second egg to the variable Y. We can assign the total mass to the variable Z, where $Z = X + Y$. The measurement errors associated with the variables X, Y, and Z can be indicated by E_X, E_Y, and E_Z, respectively. In general, if the variable Z is calculated by adding (or subtracting) 2 or more values together, then this is the formula for calculating E_Z (Allie et al., 2003),

$$E_Z = \sqrt{E_X^2 + E_Y^2}.$$

Hence, for the egg masses, the error of the combined masses (E_Z) equals the square root of the error associated with the mass of egg 1 squared (E_X^2) plus the error associated with the mass of egg 2 squared (E_Y^2). It often helps to provide a concrete example. If the error associated with the measurement of egg 1 is $E_X^2 = 2$, and the error associated with the measurement of egg 2 is $E_Y^2 = 3$, then we can calculate,

$$E_Z = \sqrt{2^2 + 3^2} \approx 3.61.$$

Note that the units of E_Z are the same as Z (e.g., grams).

7.2 Multiplying or dividing errors

Multiplying or dividing errors works a bit differently. As an example, suppose that we need to measure the total area of a rectangular field. If we measure the length (L) and width (W) of the field, then the total area is the product of these measurements, $A = L \times W$. Again, there is going to be error associated with the measurement of both length (E_L) and width (E_W). How the error of the total area (E_A) is propagated by E_L and E_W is determined by the formula,

$$E_A = A\sqrt{\left(\frac{E_L}{L}\right)^2 + \left(\frac{E_W}{W}\right)^2}.$$

Notice that just knowing the error of each measurement (E_L and E_W) is no longer sufficient to calculate the error associated with the measurement of the total area. We also need to know L, W, and A. If our field has a length of

$L = 20$ m and width of $W = 10$ m, then $A = 20 \times 10 = 200\,\text{m}^2$. If length and width measurements have associated errors of $E_L = 2$ m and $E_W = 1$ m, then,

$$E_A = 200\sqrt{\left(\frac{2}{20}\right)^2 + \left(\frac{1}{10}\right)^2} \approx 28.3\,\text{m}^2.$$

Of course, not every set of measurements with errors to be multiplied will be lengths and widths (note, however, that the units of E_A are the same as A, metres squared). To avoid confusion, the general formula for multiplying or dividing errors is below, with the variables L, W, and A replaced with X, Y, and Z, respectively, to match the case of addition and subtraction explained above,

$$E_Z = Z\sqrt{\left(\frac{E_X}{X}\right)^2 + \left(\frac{E_Y}{Y}\right)^2}.$$

Note that the structure of the equation is the exact same, just with different letters used as variables. It is necessary to be able to apply these equations correctly to estimate combined measurement error. It is not necessary to understand why the equations for propagating different types of errors are different, but a derivation is provided in Appendix B for the curious (see chapter 3 of Navidi, 2006, for more).

8

Practical. *Introduction to jamovi*

This chapter focuses on learning how to work with datasets in jamovi (The jamovi project, 2024). You can download jamovi (`https://www.jamovi.org/`) for free or run it directly from a browser using the jamovi cloud (`https://www.jamovi.org/cloud.html`). In this chapter, we will work with two datasets.

The first dataset includes some hypothetical measurements of soil organic carbon (grams of carbon per kilogram of soil) from topsoil and subsoil collected in a national park. Such data might be collected to understand how pyrogenic carbon (i.e., carbon produced by the charring of biomass during a fire) is stored in different landscape areas (Preston & Schmidt, 2006; Santín et al., 2016). These data can be downloaded online[1].

The second dataset includes measurements of figs from trees of the Sonoran Desert Rock Fig (*Ficus petiolaris*) in Baja, Mexico (Figure 8.1). I collected these data in an effort to understand coexistence in a fig wasp community (Duthie et al., 2015; Duthie & Nason, 2016). Measurements include fig lengths, widths, and heights in centimetres from four different fig trees, and the number of seeds in each fruit. This dataset can also be downloaded online[2].

This chapter will use the soil organic carbon dataset in Exercise 8.1 for summary statistics. The fig fruits dataset will be used for Exercise 8.2 on transforming variables and Exercise 8.3 on computing a variable. Some of these exercises will be similar to those of Chapter 3, but in jamovi rather than a separate spreadsheet.

8.1 Summary statistics

Once jamovi is open, you can import the soil organic carbon dataset by clicking on the three horizontal lines in the upper left corner of the tool bar, then selecting 'Open' (Figure 8.2).

[1] `https://bradduthie.github.io/stats/data/soil_organic_carbon.csv`
[2] `https://bradduthie.github.io/stats/data/fig_fruits.csv`

FIGURE 8.1 Three images showing the process of collecting data for the dimensions of figs from trees of the Sonoran Desert Rock Fig in Baja, Mexico. (A) Processing fig fruits, which included measuring the diameter of figs along three different axes of length, width, and height, (B) a fig still attached to a tree with a fig wasp on top of it, and (C) a sliced open fig with seeds along the inside of it.

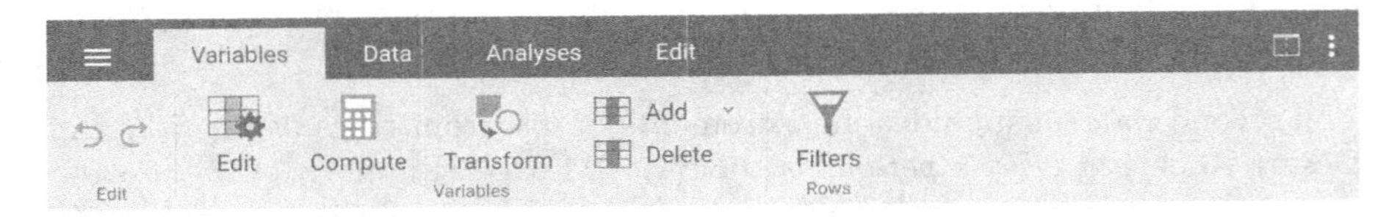

FIGURE 8.2 The jamovi toolbar including tabs for opening files, Variables, Data, Analyses, and Edit. To open a file, select the three horizontal lines in the upper left.

You might need to click 'Browse' in the upper right of jamovi to find the file. Once the data are imported, you should see two separate columns. The first column will show soil organic carbon values for topsoil samples, and the second column will show soil organic carbon values for subsoil samples. These data are not formatted in a tidy way. We need to fix this so that each row is a unique observation and each column is a variable (see Chapter 2). It might be easiest to reorganise the data in a spreadsheet such as LibreOffice Calc or Microsoft Excel. But you can also edit the data directly in jamovi by clicking on the 'Data' tab in the toolbar (Figures 8.3). The best way to reorganise the data in jamovi is to double-click on the third column of data next to 'subsoil' (see Figure 8.3).

After double-clicking on the location shown in Figure 8.3, there will be three buttons visible. You can click the 'New Data Variable' to insert a new variable named 'soil_type' in place of the default name 'C'. Keep the 'Measure type' as 'Nominal', but change the 'Data type' to 'text'. When you are done, click the > character to the right so that the variable is fixed (Figure 8.4).

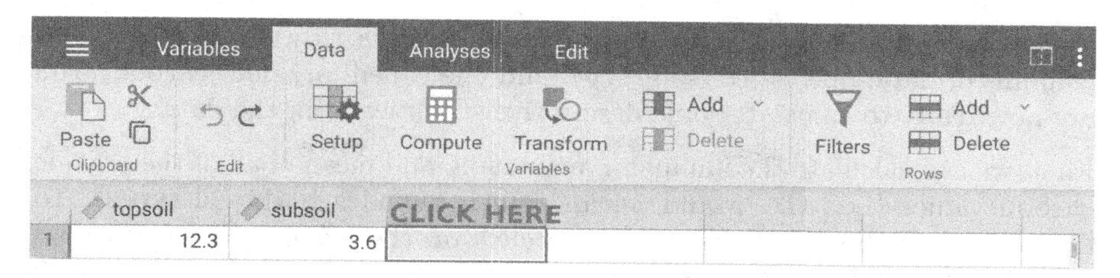

FIGURE 8.3 The jamovi toolbar is shown with the soil organic carbon dataset. In jamovi, double-clicking above column three where it says 'CLICK HERE' will allow you to input a new variable.

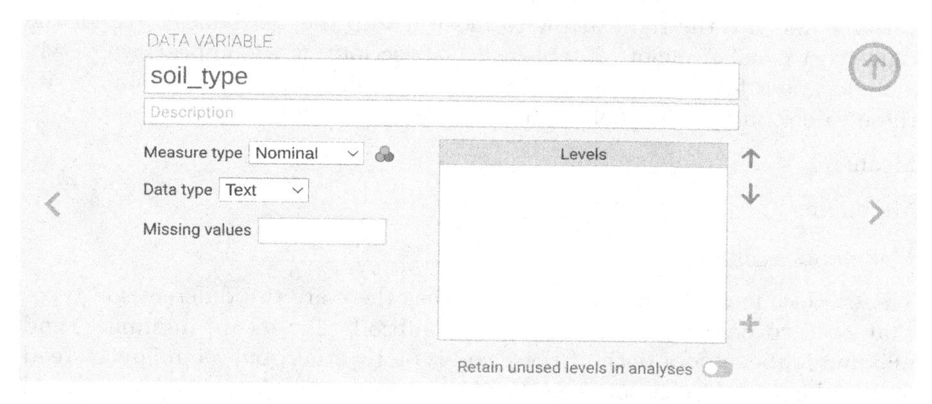

FIGURE 8.4 The jamovi toolbar is shown with the input for creating a new data variable. The new variable added is to indicate the soil type (topsoil or subsoil), so it needs to be a nominal variable with a data type of text.

After typing in the new variable 'soil_type', add another variable called 'organic_carbon'. The organic_carbon variable should have a measure type of 'Continuous' and a data type of 'Decimal'. After both soil_type and organic_carbon variables have been set, you can click the up arrow with the upper right circle (Figure 8.4) to get the new variable window out of the way.

With the two new variables created, we can now rearrange the data in a tidy format. The first 19 rows of soil_type should be 'topsoil', and the remaining 15 rows should be 'subsoil'. To do this quickly, you can write 'topsoil' in the first row of soil_type and copy-paste into the remaining rows. You can do the same to write 'subsoil' in the remaining rows 20–34. Next, copy all of the topsoil values in column 1 into the first 19 rows of column 4, and copy all of the subsoil values in column 2 into the next 15 rows. After doing all of this, your column 3 (soil_type) should have the word 'topsoil' in rows 1–19 and 'subsoil' in rows 20–34. The values from columns 1 and 2 should now fill rows 1–34 of column 4. You can now delete the first column of data by right clicking on the column name 'topsoil' and selecting 'Delete Variable'. Do the same for the second column 'subsoil'. Now you should have a tidy dataset with two columns of data, one called 'soil_type' and one called 'organic_carbon'. You are now ready to calculate some descriptive statistics from the data.

First, we can calculate the minimum, maximum, and mean of all of the organic carbon values (i.e., the 'grand' mean, which includes both soil types). To do this, select the 'Analyses' tab, then click on the left-most button called 'Exploration' in the toolbar.

After clicking on 'Exploration', a pull-down box will appear with an option for 'Descriptives'. Select this option, and you will see a new window with our two columns of data in the left-most box. Click once on the 'organic_carbon' variable and use the right arrow to move it into the 'Variables' box. In the right-most panel of jamovi, a table called 'Descriptives' will appear, which will include values for the organic carbon mean, minimum, and maximum. Write these values on the lines below, and remember to include units.

Mean: ____

Minimum: ____

Maximum: ____

These values might be useful, but recall that there are two different soil types that need to be considered: topsoil and subsoil. The mean, minimum, and maximum above pool both of these soil type together, but we might instead want to know the mean, minimum, and maximum values for topsoil and subsoil separately. Splitting organic carbon by soil type is straightforward in jamovi. To do it, go back to the Exploration → Descriptives option and again put 'organic_carbon' in the Variables box. This time, however, notice the 'Split by' box below the Variables box. Now, click on 'soil_type' in the descriptives and click on the lower right arrow to move soil type into the 'Split by' box. The table of descriptives in the right window will now break down all of the summary statistics by soil type. First, write the mean, minimum, and maximum topsoil values below.

Topsoil mean: ____

Topsoil minimum: ____

Topsoil maximum: ____

Next, do the same for the mean, minimum, and maximum subsoil values.

Subsoil mean: ____

Subsoil minimum: ____

Subsoil maximum: ____

From the values above, the mean of organic carbon sampled from the topsoil appears to be greater than the mean of organic carbon sampled from the subsoil. Assuming that jamovi has calculated the means correctly, we can be confident that the topsoil *sample* mean is higher. But what about the *population* means? Think back to concepts of populations versus samples from Chapter 4. Based on these samples in the dataset, can we really say for certain that the population mean of topsoil is higher than the population mean of subsoil? Think about this, then write a sentence below about how confident we can be about concluding that topsoil organic carbon is greater than subsoil organic carbon.

What would make you more (or less) confident that topsoil and subsoil population means are different? Think about this, then write another sentence below that answers the question.

Note that there is no right or wrong answer for the above two questions. The entire point of the questions is to help you reflect on your own learning and better link the concepts of populations and samples to the real dataset in this practical. Doing this will make the statistical hypothesis testing that comes later in this book more clear.

8.2 Transforming variables

In this next exercise, we will work with the fig fruits dataset. Open this dataset into jamovi. Note that there are five columns of data, and all of the data appear to be in a tidy format. Each row represents a separate fig fruit, while each column represents a measured variable associated with the fruit. The first several rows should look like Table 8.1.

TABLE 8.1 First six rows of the fig fruits dataset.

Tree	Length_cm	Width_cm	Height_cm	Seeds
A	1.5	1.8	1.4	238
A	1.7	1.9	1.5	198
A	2.1	2.1	1.6	220
A	1.5	1.6	1.4	188

Tree	Length_cm	Width_cm	Height_cm	Seeds
A	1.6	1.6	1.5	139
A	1.5	1.4	1.5	173

The dataset includes the tree from which the fig was sampled in column 1 (A, B, C, and D), then the length, width, and height of the fig in centimetres. Finally, the last column shows how many seeds were counted within the fig. Use the Descriptives option in jamovi to find the grand (i.e., not split by Tree) mean length, width, and height of figs in the dataset. Write these means down below (remember the units).

Grand mean length: ______________________________

Grand mean height: ______________________________

Grand mean width: ______________________________

Now look at the different rows in the Descriptives table of jamovi. Note that there is a row for 'Missing', and there appears to be one missing value for fig width and fig height. This is very common in real datasets. Sometimes practical limitations in the field prevent data from being collected, or something happens that causes data to be lost. We therefore need to be able to work with datasets that have missing data. For now, we will just note the missing data and find them in the actual dataset. Go back to the 'Data' tab in jamovi and find the figs with a missing width and height value. Report the rows of these missing values below.

Missing width row: ______________________________

Missing height row: ______________________________

Next, we will go back to working with the actual data. Note that the length, width, and height variables are all recorded in centimetres to a single decimal place. Suppose we want to transform these variables so that they are represented in millimetres instead of centimetres. We will start by creating a new column 'Length_mm' by transforming the existing 'Length_cm' column. To do this, click on the 'Data' tab at the top of the toolbar again, then click on the 'Length_cm' column name to highlight the entire column. Your screen should look like the image in Figure 8.5.

With the 'Length_cm' column highlighted, click on the 'Transform' button in the toolbar. Two things happen next. First, a new column appears in the dataset that looks identical to 'Length_cm'; ignore this for now. Second, a box appears below the toolbar allowing us to type in a new name for the transformed variable. We can call this variable 'Length_mm'. Below, note the first pull-down menu 'Source variable'. The source value should be 'Length_cm', so we can leave this alone. The second pull-down menu 'using transform' will

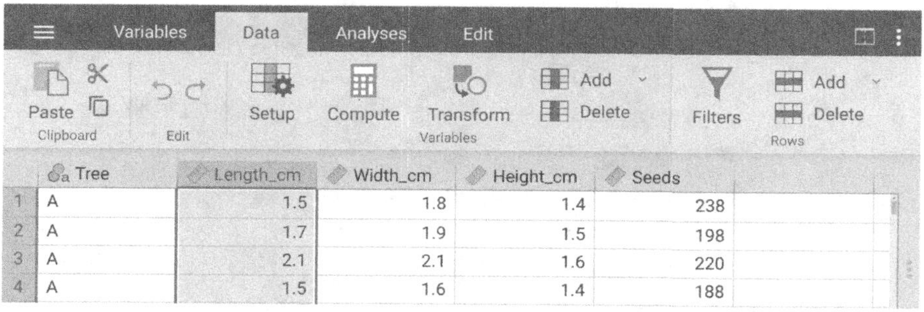

FIGURE 8.5 The jamovi toolbar where the tab 'Data' is selected. The length (cm) column is highlighted and will be transformed by clicking on the Transform button in the toolbar above.

need to change. To change the transform from 'None', click the arrow and select 'Create New Transform' from the pull-down. A new box will pop up allowing us to name the transformation. It does not matter what we call it (e.g., 'cm_to_mm' is fine). Note that there are 10 mm in 1 cm, so to convert from centimetres to millimetres, we need to multiply the values of 'Length_cm' by 10. We can do this by appending a '`* 10`' to the lower box of the transform window, so that it reads '`= $source * 10`' (Figure 8.6).

When we are finished, we can click the down arrow inside the circle in the upper right to get rid of the transform window, then the up arrow inside the circle in the upper right to get rid of the transformed variable window. Now we have a new column called 'Length_mm', in which values are 10 times greater than they are in the adjacent 'Length_cm' column, and therefore represent fig length in millimetres. If we want to, we can always change the transformation by double-clicking the 'Length_mm' column. For now, apply the same transformation to fig width and height, so we have three new columns of length, width, and height all measured in millimetres (note, if you want to, you can use the saved transformation 'cm_to_mm' that you used to transform length, saving some time). At the end of this, you should have eight columns of data, including three new columns that you just created by transforming the existing columns of Length_cm, Width_cm, and Height_cm into the new columns Length_mm, Width_mm, and Height_mm. Find the means of these three new columns and write them below.

Grand mean length (mm): ____________________

Grand mean height (mm): ____________________

Grand mean width (mm): ____________________

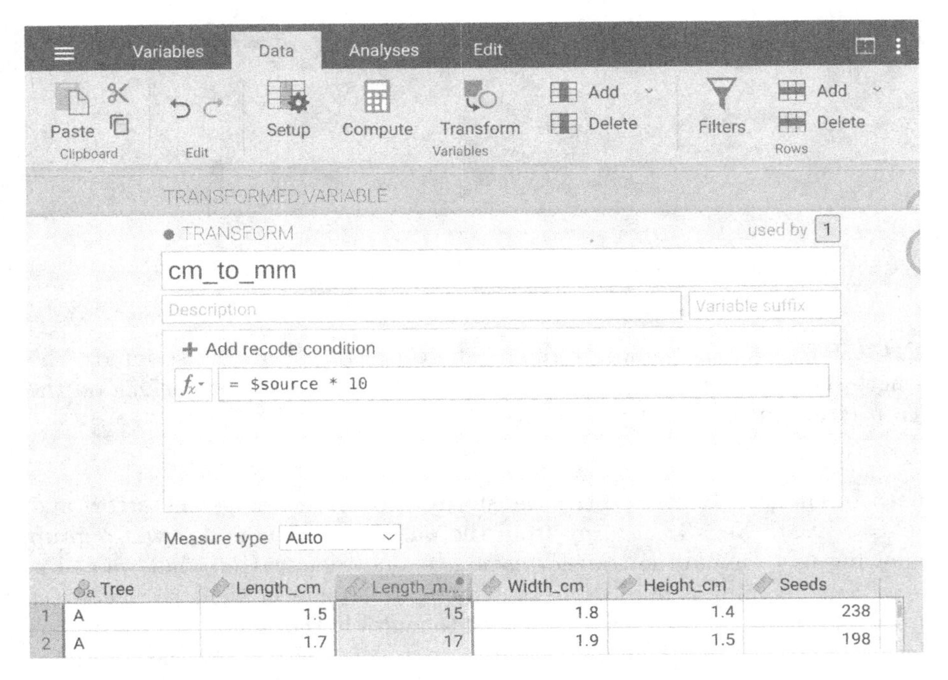

FIGURE 8.6 The jamovi toolbar where the tab 'Data' is selected. The box below shows the transform, which has been named 'cm_to_mm'. The transformation occurs by multiplying the source (Length_mm) by 10. The dataset underneath shows the first few rows with the transformed column highlighted (note that the new 'Length_mm' column is 10 times the length column).

Compare these means to the means calculated above in centimetres. Do the differences between means in centimetres and the means in millimetres make sense?

8.3 Computing variables

In this last exercise, we will compute a new variable 'fig_volume'. Because of the way that the dimensions of the fig were measured in the field, we need to make some simplifying assumptions when calculating volume. We will assume that fig fruits are perfect spheres, and that the radius of each fig is half of its measured width (i.e., 'Width_mm / 2'). This is obviously not ideal, but sometimes practical limitations in the field make it necessary to make these kinds of simplifying assumptions. In this case, how might assuming that figs are perfectly spherical affect the accuracy of our estimated fig volume? Write a sentence of reflection on this question below, drawing from what you learnt in Chapter 6 about accuracy and precision of measurements.

Now we are ready to make our calculation of fig volume. The formula for the volume of a sphere (V) given its radius r is,

$$V = \frac{4}{3}\pi r^3.$$

In words, sphere volume equals four-thirds times π, times r cubed (i.e., r to the third power). If this equation is confusing, remember that π is approximately 3.14, and taking r to the third power means that we are multiplying r by itself 3 times. We could therefore rewrite the equation above,

$$V = \frac{4}{3} \times 3.14 \times r \times r \times r.$$

This is the formula that we can use to create our new column of data for fig volume. To do this, double-click on the first empty column of the dataset, just to the right of the 'Seeds' column header. You will see a pull-down menu in jamovi with three options, one of which is 'NEW COMPUTED VARIABLE'. This is the option that we want. We need to name this new variable, so we can call it 'fig_volume'. Next, we need to type in the formula for calculating volume. First, in the small box next to the f_x, type in the (4/3) multiplied by 3.14 as below.

```
= (4/3) * 3.14
```

Next, we need to multiply by the variable 'Width_mm' divided by 2 (to get the radius) three times. We can do this by clicking on the f_x box to the left. Two new boxes will appear: the first is named 'Functions', and the second is named 'Variables'. Ignore the functions box for now, and find 'Width_mm' in the list of variables. Double-click on this to put it into the formula, then divide it by 2. You can repeat this two more times to complete the computed variable as shown in Figure 8.7.

Note that we can get the cube of 'Width_mm' more concisely by using the caret character (^). That is, we would get the same answer shown in Figure 8.7 if we instead typed the below in the function box.

```
= (4/3) * 3.14 * (Width_mm/2)^3
```

Note that the order of operations is important here, which is why there are parentheses around `Width_mm/2`. This calculation needs to be done before taking

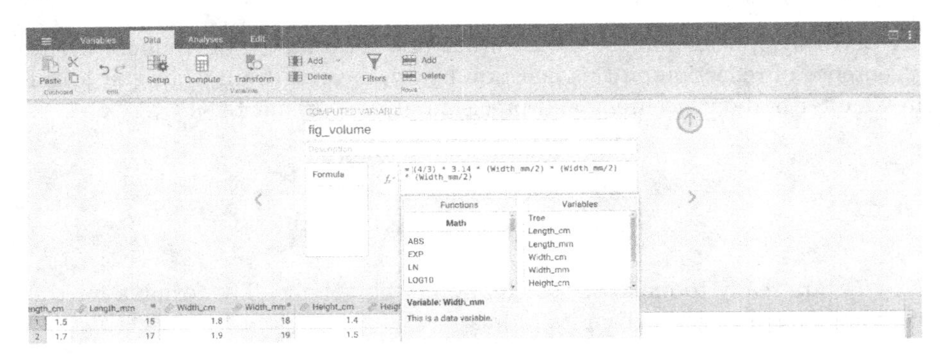

FIGURE 8.7 The jamovi toolbar where the tab 'Data' is selected. The box below shows the new computed variable 'fig_volume', which has been created by calculating the product of 4/3, 3.14, and Width_mm/2 three times.

the value to the power of 3. If we instead had written, `Width_mm/2^3`, then jamovi would first take the cube of 2 ($2 \times 2 \times 2 = 8$), then divided `Width_mm` by this value giving a different and incorrect answer. When in doubt, it is always useful to use parentheses to specify what calculations should be done first.

You now have the new column of data 'fig_volume'. Remember that the calculations apply to the units too. The width of the fig was calculated in millimetres, but we have taken width to the power of 3 when calculating the volume. In the spaces below, find the mean, minimum, and maximum volumes of all figs and report them in the correct units.

Mean: ______________________________

Minimum: ______________________________

Maximum: ______________________________

Finally, it would be good to plot these newly calculated fig volume data. These data are continuous, so we can use a histogram to visualise the fig volume distribution. To make a histogram, go to the Exploration → Descriptives window in jamovi (the same place where you found the mean, minimum, and maximum). Now, look on the lower left-hand side of the window and find the pull-down menu for 'Plots'. Click 'Plots', and you should see several different plotting options. Check the option for 'Histogram' and see the new histogram plotted in the window to the right. Draw a rough sketch of the histogram in the area below.

We should save the file that we have been working on. There are two ways to save a file in jamovi, and it is a good idea to save both ways. The first way is to use jamovi's own (binary) file type, which has the extension OMV. This will not only save the data (including the calculated variables created within jamovi), but also any analyses that we have done (e.g., calculation of minimums, maximums, and means) or graphs that we have made (e.g., the histogram). To do this, click on the three horizontal lines in the upper left of the jamovi toolbar, then select 'Save As'. Choose an appropriate name (e.g., 'chapter_8_exercises.omv'), then save the file in a location where you know that you will be able to find it again. Like all binary files, an OMV file cannot be opened as plain text. Hence, it might be a good idea to save the dataset as a CSV file (note, this will not save any of the analyses or graphs). To do this, click on the three horizontal lines in the upper left of the toolbar again, but this time click 'Export'. Give the file an appropriate name (e.g., 'chapter_8_dataset'), then choose 'CSV' from the pull-down menu below. Make sure to choose a save location that you know you will be able to find again (to navigate through file directories, click 'Browse' in the upper right). To save, click on 'Export' in the upper right.

8.4 Summary

You should now know some of the basic tools for working with data, calculating some simple descriptive statistics, plotting a histogram, and saving output and data in jamovi. These skills will be used throughout the book, so it is important to be comfortable with them as the analyses become more complex.

9

Decimal places, significant figures, and rounding

When making calculations, it is important that any numbers reported are communicated with the appropriate accuracy and precision. This means reporting numbers with the correct number of digits. This chapter focuses on correctly interpreting the decimal places and significant figures of a number, and correctly rounding.

9.1 Decimal places and significant figures

A higher number of digits communicates a greater level of accuracy. For example, the number 2.718 expresses a higher precision than 2.7 does. Reporting 2.718 implies that we know the value is somewhere between 2.7175 and 2.1785. But reporting 2.7 only implies that we know the value is somewhere between 2.65 and 2.75 (Sokal & Rohlf, 1995). These numbers therefore have a different number of *decimal places* and a different number of *significant figures.* Decimal places and significant figures are related, but not the same.

Decimal places are conceptually easier to understand. These are just the number of digits to the right of the decimal point. For example, 2.718 has 3 decimal places, and 2.7 has 1 decimal place.

Significant figures are a bit more challenging. These are the number of digits that you need to infer the accuracy of a value (Rahman, 1968). For example, the number 2.718 has 4 significant figures and 2.7 has 2 significant figures. This sounds straightforward, but it can get confusing when numbers start or end with zeros. For example, the number 0.045 has only 2 significant figures because the first two zeros only serve as placeholders (note that if this were a measurement of 0.045 m, then we could express the exact same value as 45 mm, so the zeros are not really necessary to indicate measurement accuracy). In contrast, the measurement 0.045000 has 5 significant figures because the last 3 zeros indicate a higher degree of accuracy than just 0.045 would (i.e., we know the value is somewhere between 0.044995 and 0.045005, not just 0.0445 and 0.0455). Lastly, the measurement 4500 has only 2 significant figures because the last 2 zeros are only serving as a placeholder to indicate magnitude, not accuracy (if we wanted to represent 4500 with 4 significant figures, we could use scientific notation and express it as 4.500×10^3).

Table 9.1 shows some examples of numbers, their decimal places, and their significant figures.

TABLE 9.1 Numbers are presented with columns indicating their decimal places and significant figures.

Number	Decimal Places	Significant Figures
3.14159	5	6
0.0333	4	3
1250	0	3
50000.0	1	6
0.12	2	2
1000000	0	1

It is a good idea to double-check that the values in these tables make sense. Make sure you are confident that you can report numbers to a given number of decimal places or significant figures.

9.2 Rounding

Often if you want to report a number to a specific number of decimals or significant figures, you will need to round the number. Rounding reduces the number of significant figures in a number, which might be necessary if a number that we calculate has more significant figures than we are justified in expressing. There are different rules for rounding numbers, but this book will follow Sokal & Rohlf (1995). When rounding to the nearest decimal, the last decimal written should not be changed if the number that immediately follows is 0, 1, 2, 3, or 4. If the number that immediately follows is 5, 6, 7, 8, or 9, then the last decimal written should be increased by 1.

For example, if we wanted to round the number 3.141593 to 2 significant figures, then we would write it as 3.1 because the digit that immediately follows (i.e., the third digit) is 4. If we wanted to round the number to 5 significant figures, then we would write it as 3.1416 because the digit that immediately follows is 9. And if we wanted to round 3.141593 to 4 significant figures, then we would write it as 3.142 because the digit that immediately follows is 5. Note that this does not just apply for decimals. If we wanted to round 1253 to 3 significant figures, then we would round by writing it as 1250.

Table 9.2 shows some examples of numbers rounded to a given significant figure.

TABLE 9.2 Numbers to be rounded are presented in rows of the first column. The significant figures to which rounding is desired is in the second column, and the third column shows the correctly rounded number.

Original Number	Significant Figures	Rounded Number
23.2439	4	23.24
10.235	4	10.24
102.39	2	100
5.3955	3	5.40
37.449	3	37.4
0.00345	2	0.0035

10

Graphs

Graphs are useful tools for visualising and communicating data. Graphs come in many different types, and different types of graphs are effective for different types of data. This chapter focuses on four types of graphs: (1) histograms, (2) pie charts, (3) barplots, and (4) box-whisker plots.

After collecting or obtaining a new dataset, it is almost always a good idea to plot the data in some way. Visualisation can often highlight important and obvious properties of a dataset more efficiently than inspecting raw data, calculating summary statistics, or running statistical tests. When making graphs to communicate data visually, it is important to ensure that the person reading the graph has a clear understanding of what is being presented. In practice, this means clearly labelling axes with meaningful descriptions and appropriate units, including a descriptive caption, and indicating what any graph symbols mean. In general, it is also best to make the simplest graph possible for visualising the data, which means avoiding unnecessary colour, three-dimensional display, or distractions from the information being conveyed (Dytham, 2011; Kelleher & Wagener, 2011). It is also important to ensure that graphs are as accessible as possible, e.g., by providing strong colour contrast and appropriate colour combinations (Elavsky et al., 2022), and alternative text for images where possible. As a guide, the histogram, pie chart, barplot, and box-whisker plot below illustrate good practice when making graphs.

10.1 Histograms

Histograms illustrate the distribution of continuous data. They are especially useful visualisation tools because it is often important to assess data at a glance and make a decision about how to proceed with a statistical analysis. The histogram shown in Figure 10.1 provides an example using the fig fruits dataset from the practical in Chapter 8 (an interactive application[1] shows a step-by-step demonstration of how a histogram is built).

The histogram in Figure 10.1 shows how many fruits there are for different

[1] https://bradduthie.github.io/stats/app/build_histogram/

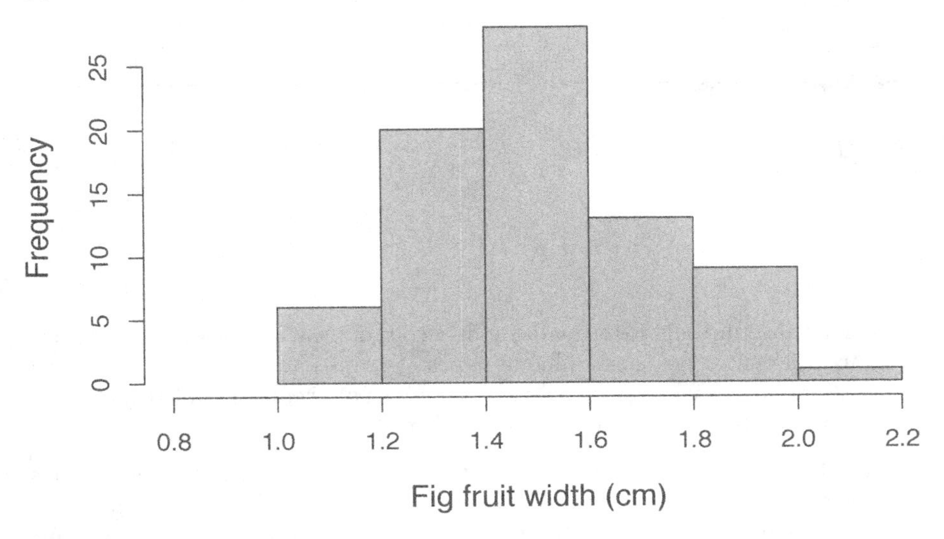

FIGURE 10.1 Example histogram fig fruit width (cm) using data from 78 fig fruits collected in 2010 from Baja, Mexico.

intervals of width, i.e., the frequency with which fruits within some width interval occur in the data. For example, there are 6 fruits with a width between 1.0 and 1.2, so for this interval on the x-axis, the bar is 6 units in height on the y-axis. In contrast, there is only 1 fig fruit that has a width greater than 2.0 cm (the biggest is 2.1 cm), so we see that the height of the bar for the interval between 2.0 and 2.2 is only 1 unit in frequency. The bars of the histogram touch each other, which reinforces the idea that the data are continuous (Dytham, 2011; Sokal & Rohlf, 1995).

It is especially important to be able to read and understand information from a histogram, because it is often necessary to determine if the data are consistent with the assumptions of a statistical test. For example, the *shape* of the distribution of fig fruit widths might be important for performing a particular test. For the purposes of this book, the *shape* of the distribution just means what the data look like when plotted like this in a histogram. In this case, there is a peak toward the centre of the distribution, with fewer low and high values (this kind of distribution is quite common). Different distribution shapes will be discussed more in Chapter 15.

10.2 Barplots and pie charts

While histograms are an effective way of visualising continuous data, barplots (also known as 'bar charts' or 'bar graphs') and pie charts can be used to visualise categorical data. For example, in the fig fruits dataset from Chapter 8, 78 fig fruits were collected from four different trees (A, B, C, and D). A barplot could be used to show how many samples were collected from each tree (see Figure 10.2).

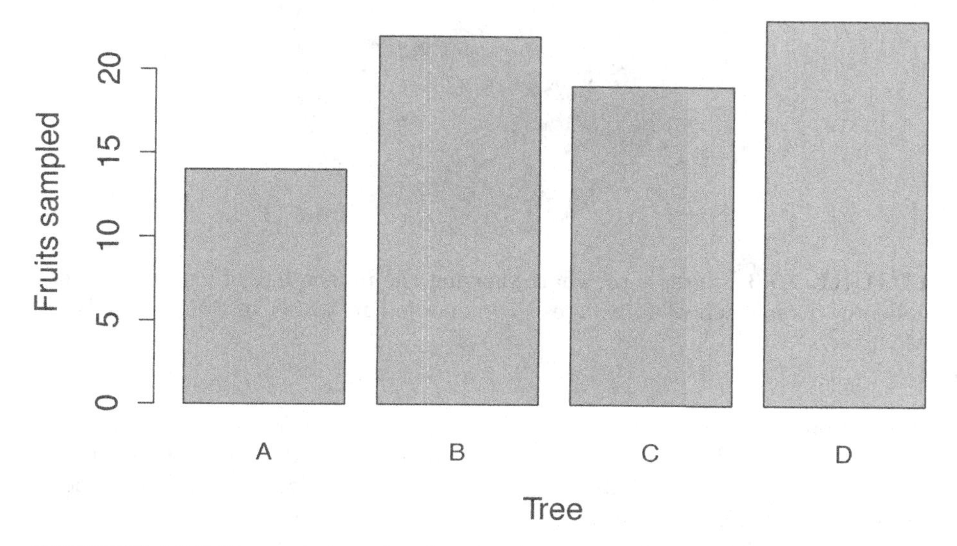

FIGURE 10.2 Example barplot showing how many fruits were collected from each of four trees (78 collected in total) in 2010 from Baja, Mexico.

In Figure 10.2, each tree is represented by a separate bar on the x-axis. Unlike a histogram, the bars do not touch each other, which reinforces the idea that different categories of data are being shown (in this case, different trees). The height of a bar indicates how many fruits were sampled from each tree. For example, 14 fruits were sampled from tree A, and 22 fruits were sampled from tree B. At a glance, it is therefore possible to compare different trees and make inferences about how they differ in sampled fruits.

Pie charts are similar to barplots in that both present categorical data, but pie charts are more effective for visualising the relative quantity for each category. That is, pie charts illustrate the percentage of measurements for each category. For example, in the case of the fig fruits, it might be useful to visualise what percentage of fruits were sampled from each tree. A pie chart could be used to evaluate this, with pie slices corresponding to different trees and the size of each slice reflecting the percentage of the total sampled fruits that came from each tree (Figure 10.3).

Pie charts can be useful in some situations, but in the biological and environmental sciences, they are not used as often as barplots. In contrast to pie charts, barplots present absolute quantities (in Figure 10.2, e.g., the actual number of fruits sampled per tree), and it is still possible with barplots to infer

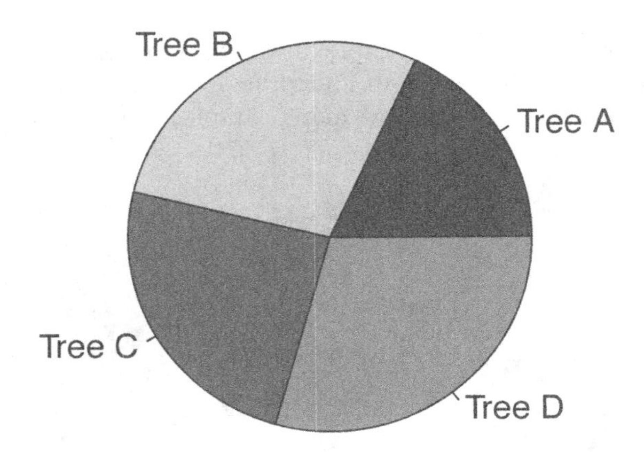

FIGURE 10.3 Example pie chart showing the percentage of fruits that were collected from each of four trees (78 collected in total) in 2010 from Baja, Mexico.

the percentage each category contributes to the total from the relative sizes of the bars. Pie charts, in contrast, only illustrate relative percentages unless numbers are used to indicate absolute quantities. Unless percentage alone is important, barplots are often the preferred way to communicate categorical data.

10.3 Box-whisker plots

Box-whisker plots (also called boxplots) can be used to visualise distributions in a different way than histograms. Instead of presenting the full distribution, as in a histogram, a box-whisker plot shows where summary statistics are located (summary statistics are explained in Chapter 11 and Chapter 12). This allows the distribution of data to be represented in a more compact way, but does not show the full shape of a distribution. Figure 10.4 compares a box-whisker plot of fig fruit widths (10.4A) with a histogram of fig fruit widths (10.4B). In other words, both of the panels (A and B) in Figure 10.4 show the same information in two different ways (note that these are the same data as presented in Figure 10.1).

To show how the panels of Figure 10.4 correspond to one another more clearly, points indicate where the summary statistics shown in the boxplot (Figure 10.4A) are located in the histogram (Figure 10.4B). These summary statistics include the median (circles), quartiles (squares), and the limits of the distribution (i.e., minimum and maximum values; triangles). Note that in boxplots, if outliers exist, they are presented as separate points.

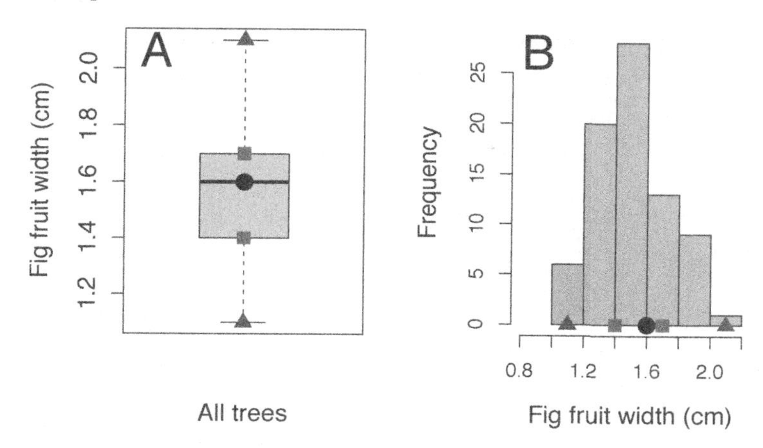

FIGURE 10.4 Boxplot (A) of fig fruit widths (cm) for 78 fig fruits collected in 2010 in Baja, Mexico. Panel A presents the same data as a histogram. Points in the boxplot indicate the median (circle), first and third quartiles (squares), and the limits of the distribution (triangles). Corresponding locations are shown on the histogram in panel B.

One benefit of a boxplot is that it is possible to show the distribution of multiple variables simultaneously. For example, the distribution of fig fruit width can be shown for each of the four trees side by side on the same x-axis of a boxplot (Figure 10.5). While it is possible to show histograms side by side, it will quickly take up a lot of space.

The boxplot in Figure 10.5 can be used to quickly compare the distribution of Trees A-D. The point at the bottom of the distribution of Tree A shows an outlier. This outlier is an especially low value of fig fruit width compared to the other fruits of Tree A.

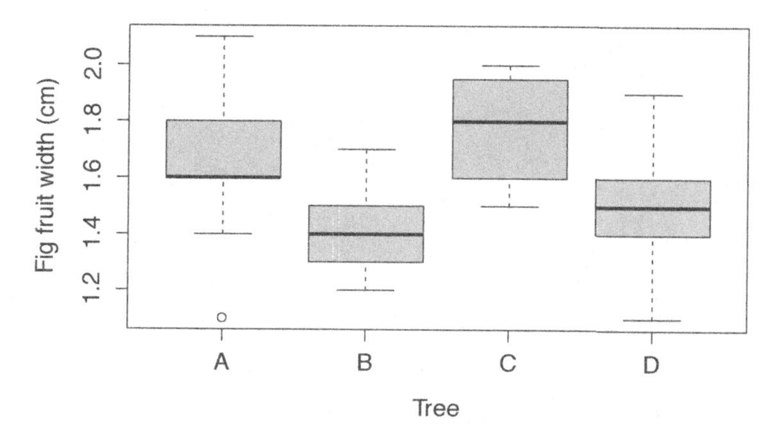

FIGURE 10.5 Boxplot of fig fruit widths (cm) collected from four separate trees sampled in 2010 from Baja, Mexico.

11

Measures of central tendency

Summary statistics describe properties of data in a single number (e.g., mean), or a set of numbers (e.g., quartiles). This chapter focuses on summary statistics that describe the centre of a distribution. It also introduces quantiles, which divide a distribution into different percentages of the data (e.g., lowest 50% or highest 75%). Throughout this section, verbal and mathematical explanations of summary statistics will be presented alongside histograms or boxplots that convey the same information. The point of doing this is to help connect the two ways of summarising the data. All of the summary statistics that follow describe calculations for a *sample* and are therefore estimates of the true values in a *population*. Recall from Chapter 4 the difference between a population and a sample. This book focuses on statistical techniques, not statistical theory, so summary statistics will just focus on how to estimate statistics from sampled data instead of how statistics are defined mathematically[1].

11.1 The mean

The arithmetic mean (hereafter just *the mean*[2]) of a sample is one of the most commonly reported statistics when communicating information about a dataset. The mean is a measure of central tendency, so it is located somewhere in the middle of a distribution. Figure 11.1 shows the same histogram of fig fruit widths shown in Figure 10.1, but with an arrow indicating where the mean of the distribution is located.

The mean is calculated by adding up the values of all of the data and dividing this sum by the total number of data (Sokal & Rohlf, 1995). This is a fairly straightforward calculation, so we can use the mean as an example to demonstrate some new mathematical notation that will be used throughout the book. We will start with a concrete example with actual numbers, then end with a more abstract equation describing how any sample mean is calculated. The

[1] A good textbook for learning about theoretical statistics and the mathematics underlying the techniques in this book is Miller & Miller (2004).

[2] There are other types of means, such as the geometric mean or the harmonic mean, but we will not use these at all.

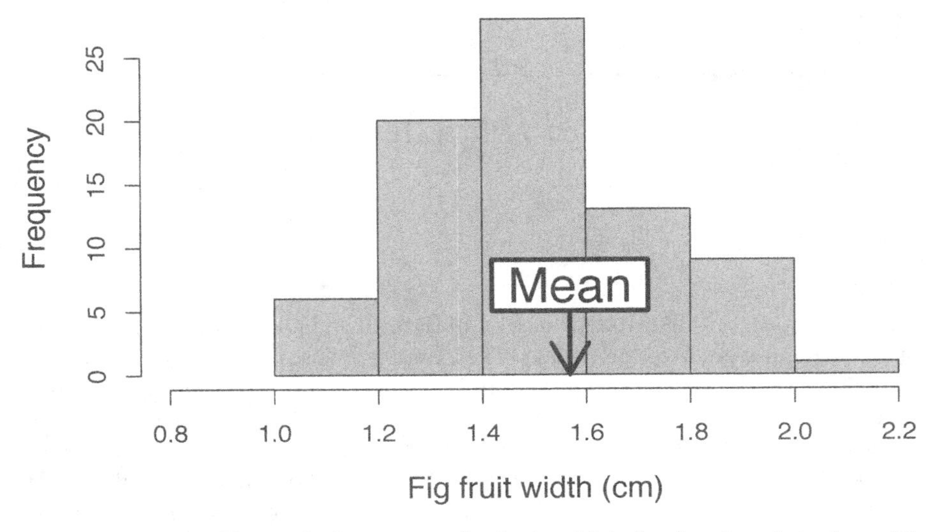

FIGURE 11.1 Example histogram fig fruit width (cm) using data from 78 fig fruits collected in 2010 from Baja, Mexico. The mean of the distribution is indicated with an arrow.

notation might be a bit confusing at first, but learning it will make understanding statistical concepts easier later in the book. There are a lot of equations in what follows, but this is because I want to explain what is happening as clearly as possible, step by step. We start with the following eight values.

```
4.2, 5.0, 3.1, 4.2, 3.8, 4.6, 4.0, 3.5
```

To calculate the mean of a sample, we just need to add up all of the values and divide by 8 (the total number of values),

$$\bar{x} = \frac{4.2 + 5.0 + 3.1 + 4.2 + 3.8 + 4.6 + 4.0 + 3.5}{8}.$$

Note that I have used the symbol $\bar{x}$ to represent the mean of x, which is a common notation (Sokal & Rohlf, 1995). In the example above, $\bar{x} = 4.05$.

Writing the full calculation above is not a problem because we only have 8 points of data. But sample sizes are often much larger than 8. If we had a sample size of 80 or 800, then there is no way that we could write down every number to show how the mean is calculated. One way to get around this is to use ellipses and just show the first and last couple of numbers,

$$\bar{x} = \frac{4.2 + 5.0 + ... + 4.0 + 3.5}{8}.$$

This is a more compact, and perfectly acceptable, way to write the sample mean.

But it is often necessary to have an even more compact way of indicating the sum over a set of values (i.e., top of the fraction above). To do this, each value can be symbolised by an x, with a unique subscript i, so that x_i corresponds to a specific value in the list above. The usefulness of this notation, x_i, will become clear soon. It takes some getting used to, but Table 11.1 shows each symbol with its corresponding value to make it more intuitive.

TABLE 11.1 Sample dataset that includes eight values.

Symbol	x_1	x_2	x_3	x_4	x_5	x_6	x_7	x_8
Value	4.2	5.0	3.1	4.2	3.8	4.6	4.0	3.5

Note that we can first replace the actual values with their corresponding x_i, so the mean can be written as,

$$\bar{x} = \frac{x_1 + x_2 + x_3 + x_4 + x_5 + x_6 + x_7 + x_8}{8}.$$

Next, we can rewrite the top of the equation in a different form using a summation sign,

$$\sum_{i=1}^{8} x_i = x_1 + x_2 + x_3 + x_4 + x_5 + x_6 + x_7 + x_8.$$

Like the use of x_i, the summation sign $\sum$ takes some getting used to, but here it just means 'sum up all of the x_i values'. You can think of it as a big 'S' that just says 'sum up'. The bottom of the S tells you the starting point, and the top of it tells you the ending point, for adding numbers. Verbally, we can read this as saying, 'starting with $i = 1$, add up all of the x_i values until $i = 8$'. We can then replace the long list of x values with a summation,

$$\bar{x} = \frac{\sum_{i=1}^{8} x_i}{8}.$$

This still looks a bit messy, so we can rewrite the above equation. Instead of dividing the summation by 8, we can multiply it by 1/8, which gives us the same answer,

$$\bar{x} = \frac{1}{8} \sum_{i=1}^{8} x_i.$$

There is one more step. We have started with 8 actual values and ended with a compact and abstract equation for calculating the mean. But if we want a general description for calculating *any* mean, then we need to account for sample sizes not equal to 8. To do this, we can use N to represent the sample size. In our example, $N = 8$, but it is possible to have a sample size be any finite value above zero. We can therefore replace 8 with N in the equation for the sample mean,

$$\bar{x} = \frac{1}{N}\sum_{i=1}^{N} x_i.$$

There we have it. Verbally, the above equation tells us to multiply $1/N$ by the sum of all x_i values from 1 to N. This describes the mean for any sample that we might collect.

11.2 The mode

The mode of a dataset is simply the value that appears most often. As a simple example, we can again consider the sample dataset of 8 values.

```
4.2, 5.0, 3.1, 4.2, 3.8, 4.6, 4.0, 3.5
```

In this dataset, the values 5.0, 3.1, 3.8, 4.6, 4.0, and 3.5 are all represented once. But the value 4.2 appears twice, once in the first position and once in the fourth position. Because 4.2 appears most frequently in the dataset, it is the mode of the dataset.

Note that it is possible for a dataset to have more than one mode. Also, somewhat confusingly, distributions that have more than one peak are often described as multimodal, even if the peaks are not of the same height (Sokal & Rohlf, 1995). For example, the histogram in Figure 11.2 might be described as bimodal because it has two distinct peaks (one around 10 and the other around 14), even though these peaks are not the same size.

In very rare cases, data might have a U-shape. The lowest point of the U would then be described as the antimode (Sokal & Rohlf, 1995).

11.3 The median and quantiles

The median of a dataset is the middle value when the data are sorted. More technically, the median is defined as the value that has the same number of lower and higher values than it (Sokal & Rohlf, 1995). If there is an odd number of values in the dataset, then finding the median is often easy. For example, the median of the values {8, 5, 3, 2, 6} is 5. This is because if we

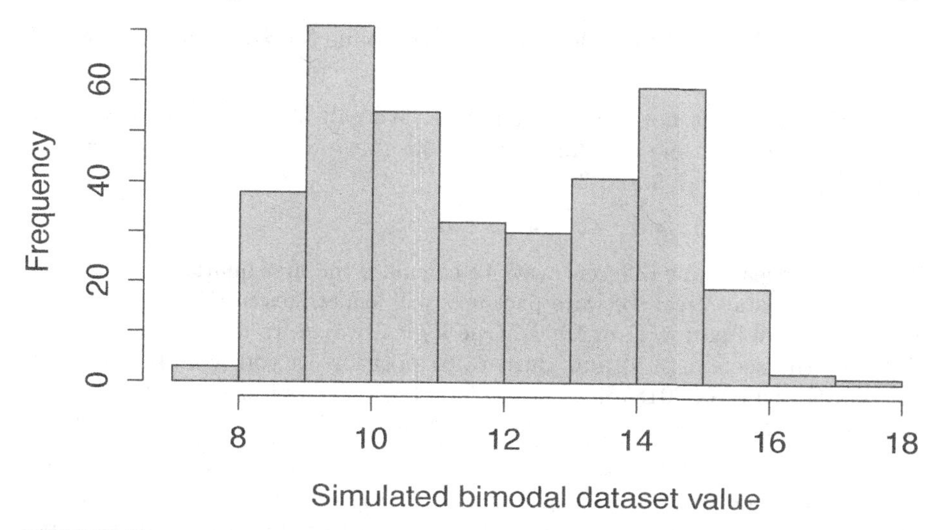

FIGURE 11.2 Example histogram of a hypothetical dataset that has a bimodal distribution.

sort the values from lowest to highest (2, 3, 5, 6, 8), the value 5 is exactly in the middle. It gets more complicated for an even number of values, such as the sample dataset used for explaining the mean and mode.

```
4.2, 5.0, 3.1, 4.2, 3.8, 4.6, 4.0, 3.5
```

We can order these values from lowest to highest.

```
3.1, 3.5, 3.8, 4.0, 4.2, 4.2, 4.6, 5.0
```

Again, there is no middle value here. But we can find a value that has the same number of lower and higher values. To do this, we just need to find the mean of the middle 2 numbers, in this case 4.0 and 4.2, which are in positions 4 and 5, respectively. The mean of 4.0 and 4.2 is $(4.0 + 4.2)/2 = 4.1$, so 4.1 is the median value.

The median is a type of quantile. A quantile divides a sorted dataset into different percentages that are lower or higher than it. Hence, the median could also be called the 50% quantile because 50% of values are lower than the median and 50% of values are higher than it. Two other quantiles besides the median are also noteworthy. The first quartile (also called the 'lower quartile') defines the value for which 25% of values are lower and 75% of values are higher. The third quartile (also called the 'upper quartile') defines the value for which 75% of values are lower and 25% of values are higher. Sometimes this is easy to calculate. For example, if there are only five values in a dataset, then the lower quartile is the number in the second position when the data are sorted because 1 value (25%) is below it and 3 values (75%) are above it.

For example, for the values {1, 3, 4, 8, 9}, the value 3 is the first quartile and 8 is the third quartile.

In some cases, it is not always this clear. We can show how quantiles get more complicated using the same 8 values as above where the first quartile is somewhere between 3.5 and 3.8.

```
3.1, 3.5, 3.8, 4.0, 4.2, 4.2, 4.6, 5.0
```

There are at least nine different ways to calculate the first quartile in this case, and different statistical software packages will sometimes use different default methods (Hyndman & Fan, 1996). One logical way is to calculate the mean between the second (3.5) and third (3.8) position as you would do for the median (Rowntree, 2018), $(3.5 + 3.8)/2 = 3.65$. Jamovi uses a slightly more complex method, which will give a value of 3.725 (The jamovi project, 2024).

It is important to emphasise that no one way of calculating quantiles is the one and only correct way. Statisticians have just proposed different approaches to calculating quantiles from data, and these different approaches sometimes give slightly different results. This can be unsatisfying when first learning statistics because it would be nice to have a single approach that is demonstrably correct, i.e., the *right* answer under all circumstances. Unfortunately, this is not the case here, nor is it the case for a lot of statistical techniques. Often there are different approaches to answering the same statistical question and no simple right answer. For this book, we will always report calculations of quantiles from jamovi. But it is important to recognise that different statistical tools might give different answers (Hyndman & Fan, 1996).

12

Measures of spread

It is often important to know how much a set of numbers is spread out. That is, do all of the data cluster close to the mean, or are most values distant from the mean? For example, all of the numbers below are quite close to the mean of 5.0 (three numbers are exactly 5.0).

```
4.9, 5.3, 5.0, 4.7, 5.1, 5.0, 5.0
```

In contrast, all of the numbers that follow are relatively distant from the same mean of 5.0.

```
3.0, 5.6, 7.8, 1.2, 4.3, 8.2, 4.9
```

This chapter focuses on summary statistics that describe the spread of data. The approach in this chapter is similar to Chapter 11, which provided verbal and mathematical explanations of measures of central tendency. We will start with the most intuitive measures of spread: the range and inter-quartile range. Then, we will move on to some more conceptually challenging measures of spread: the variance, standard deviation, coefficient of variation, and standard error. These more challenging measures can be a bit confusing at first, but they are absolutely critical for doing statistics. The best approach to learning them is to see them and practice using them in different contexts, which we will do throughout this book.

12.1 The range

The range of a set of numbers is probably the most intuitive measure of spread. It is simply the difference between the highest and the lowest value of a dataset (Sokal & Rohlf, 1995). To calculate it, we just need to take the highest value minus the lowest value. If we want to be technical, then we can write a general equation for the range of a random variable X,

$$\text{Range}(X) = \max(X) - \min(X).$$

But really, all we need to worry about is finding the highest and lowest values, then subtracting. Consider again the two sets of numbers introduced at the beginning of the chapter. In examples, it is often helpful to imagine numbers as representing something concrete that has been measured, so suppose that these numbers are the measured masses (in grams) of leaves from two different plants. Below are the masses of plant A, in which leaf masses are very similar and close to the mean of 5.

```
4.9, 5.3, 5.0, 4.7, 5.1, 5.0, 5.0
```

Plant B masses are below, which are more spread out around the same mean of 5.

```
3.0, 5.6, 7.8, 1.2, 4.3, 8.2, 4.9
```

To get the range of plant A, we just need to find the highest (5.3 g) and lowest (4.7 g) mass, then subtract,

$$\text{Range}(plant\ A) = 5.3 - 4.7 = 0.6$$

Plant A therefore has a range of 0.6 g. We can do the same for plant B, which has a highest value of 8.2 g and lowest value of 1.2 g,

$$\text{Range}(plant\ B) = 8.2 - 1.2 = 7.0$$

Plant B therefore has a much higher range than plant A.

It is important to mention that the range is highly sensitive to outliers (Navarro & Foxcroft, 2022). Just adding a single number to either plant A or plant B could dramatically change the range. For example, imagine if we measured a leaf in plant A to have a mass of 19.7 g (i.e., we found a huge leaf!). The range of plant A would then be $19.7 - 4.7 = 15$ instead of 0.6. Just this one massive leaf would then make the range of plant A more than double the range of plant B. This lack of robustness can really limit how useful the range is as a statistical measure of spread.

12.2 The inter-quartile range

The inter-quartile range (usually abbreviated as 'IQR') is conceptually the same as the range. The only difference is that we are calculating the range between quartiles rather than the range between the highest and lowest numbers in the dataset. A general formula subtracting the first quartile (Q_1) from the third quartile (Q_3) is,

$$IQR = Q_3 - Q_1.$$

Recall from Chapter 11 how to calculate first and third quartiles. As a reminder, we can sort the leaf masses for plant A below.

```
4.7, 4.9, 5.0, 5.0, 5.0, 5.1, 5.3
```

The first quartile will be 4.95. The third quartile will be 5.05. The IQR of plant A is therefore,

$$IQR_{\text{plant A}} = 5.05 - 4.95 = 0.1.$$

We can calculate the IQR for plant B in the same way. Here are the masses of plant B leaves sorted.

```
1.2, 3.0, 4.3, 4.9, 5.6, 7.8, 8.2
```

The first quartile of plant B is 3.65, and the third quartile is 6.70. To get the IQR of plant B,

$$IQR_{\text{plant B}} = 6.70 - 3.65 = 3.05.$$

An important point about the IQR is that it is more robust than the range (Dytham, 2011). Recall that if we found an outlier leaf of 19.7 g on plant A, it would change the range of plant leaf mass from 0.6 to 15 g. The IQR is not nearly so sensitive. If we include the outlier, the first quartile for plant A changes from $Q_1 = 4.95$ to $Q_1 = 4.975$. The third quartile changes from $Q_3 = 5.05$ to $Q_3 = 5.150$. The resulting IQR is therefore $5.150 - 4.975 = 0.175$. Hence, the IQR only changes from 0.1 to 0.175, rather than from 0.6 to 15. The one outlier therefore has a huge effect on the range, but only a modest effect on the IQR.

12.3 The variance

The range and inter-quartile range were reasonably intuitive, in the sense that it is not too difficult to think about what a range of 10, e.g., actually means in terms of the data. We now move to measures of spread that are less intuitive. These measures of spread are the variance, standard deviation, coefficient of variation, and standard error. These can be confusing and unintuitive at first, but they are extremely useful. We will start with the variance; this section is long because I want to break the variance down carefully, step by step.

The sample variance of a dataset is a measure of the expected squared distance of data from the mean. To calculate the variance of a sample, we need to know the sample size (N, i.e., how many measurements in total), and the mean of the sample ($\bar{x}$). We can calculate the variance of a sample (s^2) as follows,

$$s^2 = \frac{1}{N-1} \sum_{i=1}^{N} (x_i - \bar{x})^2 .$$

This looks like a lot, but we can break down what the equation is doing verbally. First, we can look inside the summation ($\sum$). Here we are taking an individual measurement x_i, subtracting the mean $\bar{x}$, then squaring. We do this for each x_i, summing up all of the values from $i = 1$ to $i = N$. This part of the equation is called the **sum of squares** (SS),

$$SS = \sum_{i=1}^{N} (x_i - \bar{x})^2 .$$

That is, we need to subtract the mean from each value x_i, square the result, and add everything up. Once we have this sum, SS, then we just need to multiply by $1/(N-1)$ to get the variance.

An example of how to do the actual calculation should help make it easier to understand what is going on. We can use the same values from plant A earlier.

```
4.9, 5.3, 5.0, 4.7, 5.1, 5.0, 5.0
```

To calculate the variance of plant A leaf masses, we start with the sum of squares. That is, take 4.9, subtract the sample mean of 5.0 ($4.9 - 5.0 = -0.1$), then square the result ($(-0.1)^2 = 0.01$). We do the same for 5.3, $(5.3 - 5.0)^2 = 0.09$, and add it to the 0.01, then continue down the list of numbers finishing with 5.0. This is what the sum of squares calculation looks like written out,

$$SS = (4.9-5)^2+(5.3-5)^2+(5-5)^2+(4.7-5)^2+(5.1-5)^2+(5-5)^2+(5-5)^2.$$

Remember that the calculations in parentheses need to be done first, so the next step for calculating the sum of squares would be the following,

$$SS = (-0.1)^2 + (0.3)^2 + (0)^2 + (-0.3)^2 + (0.1)^2 + (0)^2 + (0)^2.$$

Next, we need to square all of the values,

$$SS = 0.01 + 0.09 + 0 + 0.09 + 0.01 + 0 + 0.$$

If we sum the above, we get $SS = 0.2$. We now just need to multiply this by $1/(N-1)$, where $N = 7$ because this is the total number of measurements in the plant A dataset,

$$s^2 = \frac{1}{7-1} (0.2) .$$

From the above, we get a variance of approximately $s^2 = 0.0333$.

Fortunately, it will almost never be necessary to calculate a variance manually in this way. Jamovi will do all of these steps and calculate the variance for us (Chapter 14 explains how). The only reason that I present the step-by-step calculation here is to help explain the equation for s^2. The details can be helpful for understanding how the variance works as a measure of spread. For example, note that what we are really doing here is getting the distance of each value from the mean, $x_i - \bar{x}$. If these distances tend to be large, then it means that most data points (x_i) are far away from the mean ($\bar{x}$), and the variance (s^2) will therefore increase. The differences $x_i - \bar{x}$ are squared because we need all of the values to be positive, so that variance increases regardless of whether a value x_i is higher or lower than the mean. It does not matter if x_i is 0.1 lower than $\bar{x}$ (i.e., $x_i - \bar{x} = -0.1$), or 0.1 higher (i.e., $x_i - \bar{x} = 0.1$). In both cases, the deviation from the mean is the same. Moreover, if we did not square the values, then the sum of $x_i - \bar{x}$ values would always be 0 (you can try this yourself)[1]. Lastly, it turns out that the variance is actually a special case of a more general concept called the *covariance*, which we will look at later in Chapter 30 and which helps the squaring of differences make a bit more sense.

We sum up all of the squared deviations to get the SS, then divide by the sample size minus 1, to get the mean squared deviation from the mean. That is, the whole process gives us the *average* squared deviation from the mean. But wait, why is it the sample size minus 1, $N - 1$? Why would we subtract 1 here? The short answer is that in calculating a *sample* variance, s^2, we are almost always trying to estimate the corresponding *population* variance (σ^2). And if we were to just use N instead of $N - 1$, then our s^2 would be a biased estimate of σ^2 (see Chapter 4 for a reminder on the difference between samples and populations). By subtracting 1, we are correcting for this bias to get a more accurate estimate of the population variance[2]. It is not necessary to do

[1]If you are wondering why we square the difference $x_i - \bar{x}$ instead of just taking its absolute value, this is an excellent question! You have just invented something called the mean absolute deviation. There are some reasons why the mean absolute deviation is not as good of a measure of spread as the variance. Navarro & Foxcroft (2022) explain the mean absolute deviation, and how it relates to the variance, very well in section 4.2.3 of their textbook. We will not get into these points here, but it would be good to check out Navarro & Foxcroft (2022) for more explanation.

[2]To get the true population variance σ^2, we would also need to know the true mean μ. But we can only estimate μ from the sample, $\bar{x}$. That is, what we would really want to calculate is $x_i - \mu$, but the best we can do is $x_i - \bar{x}$. The consequence of this is that there will be some error that underestimates the true distance of x_i values from the population mean, μ. Here is the really cool part: to determine the extent to which our estimate of the variance is biased by using $\bar{x}$ instead of μ, we just need to know the expected squared difference between the two values, $(\bar{x} - \mu)^2$. It turns out that this difference (i.e., the bias of our estimate s^2) is just σ^2/N, that is, the true variance of the population divided by the sample size. If we subtract this value from σ^2, so $\sigma^2 - \sigma^2/N$, then we can get the expected difference between the true variance and the estimate from the sample size. We can rearrange $\sigma^2 - \sigma^2/N$ to get $\sigma^2 \times (N - 1)/N$, which means that we need to correct our sample variance by $N/(N - 1)$ to get an unbiased estimate of σ^2. If all of this is confusing, that is okay! This is really only relevant for those interested in statistical theory, which is not the focus of this book.

this ourselves; jamovi will do it automatically (The jamovi project, 2024). The subtraction is required due to a reduction in the *degrees of freedom* that occurs when using the sample mean in the equation for variance[3].

Degrees of freedom is a difficult concept to understand, but we can broadly define the degrees of freedom as the number of independent pieces of information that we have when calculating a statistic (Grafen & Hails, 2022; Upton & Cook, 2014). When we need to estimate a parameter from a dataset in the process of calculating a statistic, we lose an independent piece of information, and therefore a degree of freedom (Pandey & Bright, 2008). For example, if we collect $N = 10$ samples, then there are 10 independent pieces of information that we can use to calculate the mean. To calculate the variance, we use all of these 10 samples *and* $\bar{x}$ (note that $\bar{x}$ appears in the equation for variance above). This suggests that we actually have 11 independent pieces of information when making our calculation (all of the samples and the sample mean). But not all of these values are actually free to vary. When we know what the 10 sample values are, then the sample mean is fixed. And if we know what 9 sample values are and what the mean is, then we can work out what the last sample value must be. We therefore lose a degree of freedom by including $\bar{x}$ in the calculation of variance, so we need to divide by $N - 1$ rather than N to avoid bias (Fowler et al., 1998; Wardlaw, 1985). Another way of thinking about degrees of freedom is as the number of differences that can arise between what we sample versus what we expect just based on the mathematics (Fryer, 1966). In other words, how much is our calculation free to vary just due to chance? In calculating the variance, the mathematics introduces a constraint by including $\bar{x}$, thereby reducing degrees of freedom.

This was a lot of information. The variance is not an intuitive concept. In addition to being a challenge to calculate, the calculation of a variance leaves us with a value in units squared. That is, for the example of plant leaf mass in grams, the variance is measured in grams squared, g^2, which is not particularly easy to interpret. For more on this, Navarro & Foxcroft (2022) have a really good section on the variance. Despite its challenges as a descriptive statistic, the variance has some mathematical properties that are very useful (Navarro & Foxcroft, 2022), especially in the biological and environmental sciences.

For example, variances are additive, meaning that if we are measuring two separate characteristics of a sample, A and B, then the variance of A+B equals the variance of A plus the variance of B, i.e., $\text{Var}(A + B) = \text{Var}(A) +$

[3]In the case of sample variance, note that we needed to use all the values x_i in the dataset and the sample mean $\bar{x}$. But if we know what all of the x_i values are, then we also know $\bar{x}$. And if we know all but one value of x_i and $\bar{x}$, then we could figure out the last x_i. Hence, while we are using N values in the calculation of s^2, the use of $\bar{x}$ reduces the degree to which these values are free to vary. We have lost 1 degree of freedom in the calculation of $\bar{x}$, so we need to account for this in our calculation of s^2 by dividing by $N - 1$. This is another way to think about the $N - 1$ correction factor (Sokal & Rohlf, 1995) explained in the previous footnote.

$\text{Var}(B)$.[4] This is relevant to genetics when measuring heritability. Here, the total variance in the phenotype of a population (e.g., body mass of animals) can be partitioned into variance attributable to genetics plus variance attributable to the environment,

$$\text{Var}(Phenotype) = \text{Var}(Genotype) + \text{Var}(Environment).$$

This is also sometimes written as $V_P = V_G + V_E$. Applying this equation to calculate heritability ($H^2 = V_G/V_P$) can be used to predict how a population will respond to natural selection. This is just one place where variance reveals itself to be a highly useful statistic in practice. Nevertheless, as a descriptive statistic to communicate the spread of a variable, it usually makes more sense to calculate the standard deviation of the mean.

12.4 The standard deviation

The standard deviation of the mean (s) is just the square root of the variance,

$$s = \sqrt{\frac{1}{N-1}\sum_{i=1}^{N}(x_i - \bar{x})^2}.$$

This is a simple step, mathematically, but it also is easier to understand conceptually as a measure of spread (Navarro & Foxcroft, 2022). By taking the square root of the variance, our units are no longer squared, so we can interpret the standard deviation in the same terms as our original data. For example, the leaf masses of plant A and plant B in the example above were measured in grams. While the variance of these masses was in grams squared, the standard deviation is in grams, just like the original measurements. For plant A, we calculated a leaf mass variance of $s^2 = 0.0333\,\text{g}^2$, which means that the standard deviation of leaf masses is $s = \sqrt{0.0333\,\text{g}^2} = 0.1825\,\text{g}$. Because we are reporting s in the original units, it is a very useful measure of spread to report, and it is an important one to be able to interpret. To help with the interpretation, an interactive tool can more effectively show how the heights of trees in a forest change across different standard deviation values[5]. Another interactive tool can help show how the shape of a histogram changes when the

[4] This has one caveat, which is not important for now. Values of A and B must be uncorrelated. That is, A and B cannot covary. If A and B covary, i.e., $\text{Cov}(A, B) \neq 0$, then $\text{Var}(A + B) = \text{Var}(A) + \text{Var}(B) + \text{Cov}(A, B)$. That is, we need to account for the covariance when calculating $\text{Var}(A + B)$.

[5] https://bradduthie.github.io/stats/app/forest/

standard deviation of a distribution is changed[6]. Chapter 14 explains how to calculate the standard deviation in jamovi.

12.5 The coefficient of variation

The coefficient of variation (CV) is just the standard deviation divided by the mean,

$$CV = \frac{s}{\bar{x}}.$$

Dividing by the mean seems a bit arbitrary at first, but this can often be useful for comparing variables with different means or different units. The reason for this is that the units cancel out when dividing the standard deviation by the mean. For example, for the leaf masses of plant A, we calculated a standard deviation of 0.1825 g and a mean of 5 g. We can see the units cancel below,

$$CV = \frac{0.1825\ g}{5\ g} = 0.0365.$$

The resulting CV of 0.0365 has no units; it is *dimensionless* (Lande, 1977). Because it has no units, it is often used to compare measurements with much different means or with different measurement units. For example, Sokal & Rohlf (1995) suggest that biologists might want to compare tail length variation between animals with much different body sizes, such as elephants and mice. The standard deviation of tail lengths between these two species will likely be much different just because of their difference in size, so by standardising by mean tail length, it can be easier to compare relative standard deviation. This is a common application of the CV in biology, but it needs to be interpreted carefully (Pélabon et al., 2020).

Often, we will want to express the coefficient of variation as a percentage of the mean. To do this, we just need to multiply the CV above by 100%. For example, to express the CV as a percentage, we would multiply the 0.0365 above by 100%, which would give us a final answer of $CV = 3.65\%$.

[6]https://bradduthie.github.io/stats/app/normal_pos_neg/

12.6 The standard error

The standard error of the mean is the last measure of spread that I will introduce. It is slightly different than the previous measurements in that it is a measure of the variation in the *mean* of a sample rather than the sample itself. That is, the standard error tells us how far our sample mean $\bar{x}$ is expected to deviate from the true mean μ. Technically, the standard error of the mean is the standard deviation *of sample means* rather than the standard deviation *of samples.* What does that even mean? It is easier to explain with a concrete example.

Imagine that we want to measure nitrogen in the water of Airthrey Loch (note that 'loch' is the Scottish word for 'lake'). We collect 12 water samples and record the nitrate levels in milligrams per litre (mg/l). The measurements are reported below.

```
0.63, 0.60, 0.53, 0.72, 0.61, 0.48, 0.67, 0.59, 0.67, 0.54, 0.47, 0.87
```

We can calculate the mean of the above sample to be $\bar{x} = 0.615$, and we can calculate the standard deviation of the sample to be $s = 0.111$. We do not know what the *true* mean μ is, but our best guess is the sample mean $\bar{x}$. Suppose, however, that we then went back to the loch to collect another 12 measurements (assume that the nitrogen level of the lake has not changed in the meantime). We would expect to get values similar to our first 12 measurements, but certainly not the *exact* same measurements, right? The sample mean of these new measurements would also be a bit different. Maybe we actually go out and do this and get the following new sample.

```
0.47, 0.56, 0.72, 0.61, 0.54, 0.64, 0.68, 0.54, 0.48, 0.59, 0.62, 0.78
```

The mean of our new sample is 0.603, which is a bit different from our first. In other words, the sample means vary. We can therefore ask what is the variance and standard deviation of the sample means. In other words, suppose that we kept going back out to the loch, collecting 12 new samples, and recording the sample mean each time? The standard deviation of those sample means would be the standard error. **The standard error is the standard deviation of $\bar{x}$ values around the true mean μ.** But we do not actually need to go through the repetitive resampling process to estimate the standard error. We can estimate it with just the standard deviation and the sample size. To do this, we just need to take the standard deviation of the sample (s) and divide by the square root of the sample size ($\sqrt{N}$),

$$SE = \frac{s}{\sqrt{N}}.$$

In the case of the first 12 samples from the loch in the example above,

$$SE = \frac{0.111}{\sqrt{12}} = 0.032.$$

The standard error is important because it can be used to evaluate the uncertainty of the sample mean in comparison with the true mean. We can use the standard error to place confidence intervals around our sample mean to express this uncertainty. We will calculate confidence intervals in Chapter 18, so it is important to understand what the standard error is measuring.

If the concept of standard error is still a bit unclear, we can work through one more hypothetical example. Suppose again that we want to measure the nitrogen concentration of a loch. This time, however, assume that we somehow *know* that the true mean nitrogen concentration is $\mu = 0.7$, and that the true standard deviation of water sample nitrogen concentration is $\sigma = 0.1$. Of course, we can never actually know the *true* parameter values, but we can use a computer to simulate sampling from a population in which the true parameter values are known. In Table 12.1, we simulate the process of going out and collecting 10 water samples from Airthrey Loch. This collecting of 10 water samples is repeated 20 different times. Each row is a different sampling effort, and columns report the 10 samples from each effort.

TABLE 12.1 Simulated samples (S1-S20) of nitrogen content from water samples of Airthrey Loch. Values are sampled from a normal distribution with a mean of 0.7 and a standard deviation of 0.1.

S1	0.72	0.82	0.62	0.75	0.62	0.68	0.61	0.59	0.65	0.80
S2	0.63	0.77	0.58	0.71	0.60	0.74	0.64	0.61	0.86	0.80
S3	0.62	0.70	0.68	0.50	0.89	0.72	0.83	0.64	0.79	0.69
S4	0.77	0.68	0.84	0.62	0.79	0.60	0.63	0.80	0.56	0.81
S5	0.72	0.68	0.67	0.68	0.94	0.67	0.58	0.71	0.58	0.69
S6	0.71	0.66	0.69	0.59	0.71	0.77	0.71	0.84	0.75	0.70
S7	0.83	0.54	0.75	0.58	0.61	0.68	0.61	0.65	0.69	0.79
S8	0.80	0.73	0.56	0.64	0.75	0.86	0.78	0.70	0.83	0.81
S9	0.64	0.72	1.07	0.58	0.79	0.64	0.66	0.64	0.56	0.65
S10	0.68	0.71	0.86	0.88	0.64	0.84	0.73	0.73	0.56	0.64
S11	0.77	0.62	0.82	0.82	0.74	0.78	0.90	0.62	0.68	0.76
S12	0.84	0.66	0.71	0.85	0.56	0.82	0.76	0.69	0.63	0.84
S13	0.70	0.54	0.77	0.77	0.58	0.72	0.52	0.59	0.65	0.74
S14	0.78	0.67	0.72	0.59	0.77	0.66	0.68	0.69	0.71	0.47
S15	0.71	0.71	0.71	0.73	0.80	0.62	0.63	0.86	0.55	0.64
S16	0.68	0.61	0.56	0.84	0.67	0.75	0.80	0.76	0.74	0.70
S17	0.69	0.81	0.66	0.59	0.90	0.82	0.79	0.65	0.83	0.76
S18	0.80	0.80	0.58	0.60	0.77	0.74	0.74	0.65	0.61	0.73
S19	0.58	0.69	0.63	0.69	0.75	0.82	0.67	0.55	0.62	0.74
S20	0.68	0.73	0.81	0.62	0.75	0.69	0.70	0.70	0.65	0.76

We can calculate the mean of each sample by calculating the mean of each row. These 20 means are reported below.

```
      [,1]  [,2]  [,3]  [,4]  [,5]  [,6]  [,7]  [,8]  [,9] [,10]
[1,] 0.686 0.694 0.706 0.710 0.692 0.713 0.673 0.746 0.695 0.727
[2,] 0.751 0.736 0.658 0.674 0.696 0.711 0.750 0.702 0.674 0.709
```

The standard deviation of the 20 sample means reported above is 0.0265613. Now suppose that we only had Sample 1 (i.e., top row of data). The standard deviation of Sample 1 is $s = 0.0824891$. We can calculate the standard error from these sample values below,

$$s = \frac{0.0824891}{\sqrt{10}} = 0.0260853.$$

The estimate of the standard error from calculating the standard deviation of the sample means is therefore 0.0265613, and the estimate from just using the standard error formula and data from only Sample 1 is 0.0260853. These are reasonably close, and they would be even closer if we had either a larger sample size in each sample (i.e., higher N) or a larger number of samples.

13

Skew and kurtosis

Following measures of central tendency in Chapter 11 and measures of spread in Chapter 12, there are two more concepts that are worth briefly mentioning. Both of these concepts concern the shape of a distribution. The first concept is skew, and the second is kurtosis. This is also a good place to introduce what the word 'moment' refers to in the statistics.

13.1 Skew

The **skew** of a distribution refers to its asymmetry (Dytham, 2011; Sokal & Rohlf, 1995), which can be observed in a histogram (Figure 13.1). What we are looking for here are the sides of the distribution, also known as the 'tails'. When the left tail sticks out as it does in Figure 13.1A, we can describe the distribution as 'left-skewed' or 'negatively skewed'. When the right tail sticks out as it does in Figure 13.1B, we can describe the distribution as 'right-skewed' or 'positively skewed'.

When a distribution is skewed, the mean and median of the distribution will be different (Sokal & Rohlf, 1995). This is because the mean of the distribution will be pulled towards the tail that sticks out due to the extreme values in the tail. In contrast, the median is robust to these extreme values. This can be important when interpreting the median versus mean as a measure of central tendency (Reichmann, 1970). For example, household income data are typically highly right-skewed because a small proportion of households have a much higher income than the typical household (McDonald et al., 2013). Consequently, when data are strongly skewed, the mean and median give us different information, and it often makes sense to use the median as an indication of what is typical (Chiripanhura, 2011).

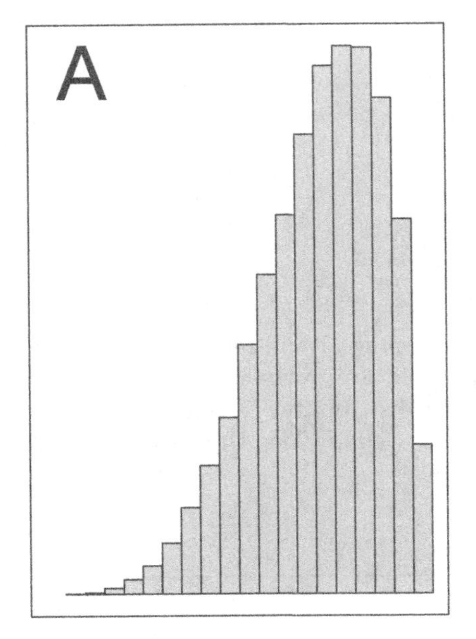

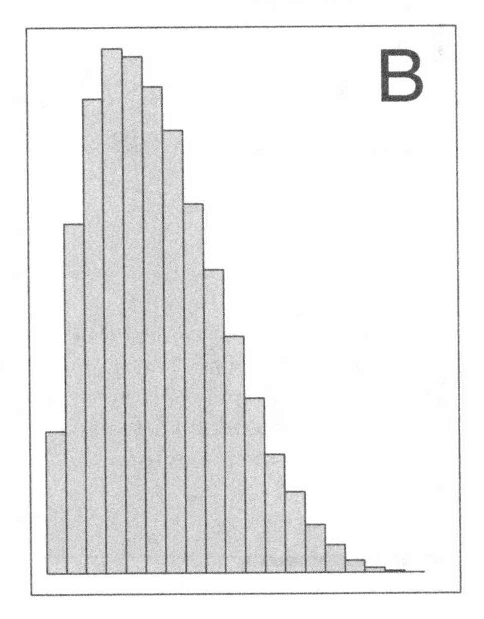

FIGURE 13.1 Histograms showing a (A) distribution that has a negative (i.e., 'left') skew and (B) distribution that has a positive (i.e., 'right') skew.

13.2 Kurtosis

The **kurtosis** describes the extent to which a distribution is flat. In other words, whether data tend to be in the centre and at the tails, or whether data tend to be between the centre and tails (Sokal & Rohlf, 1995). If there are more values in the centre and tails of the distribution, then the distribution is described as 'leptokurtic' (Figure 13.2A). If there are more values between the centre and the tails, then the distribution is described as 'platykurtic' (Figure 13.2B).

The extent to which any given distribution is leptokurtic versus platykurtic is defined relative to the normal distribution, which will be introduced in Chapter 15.

13.3 Moments

When actually doing statistics in the biological and environmental sciences, just being able to recognise skew and kurtosis on a histogram is usually sufficient.

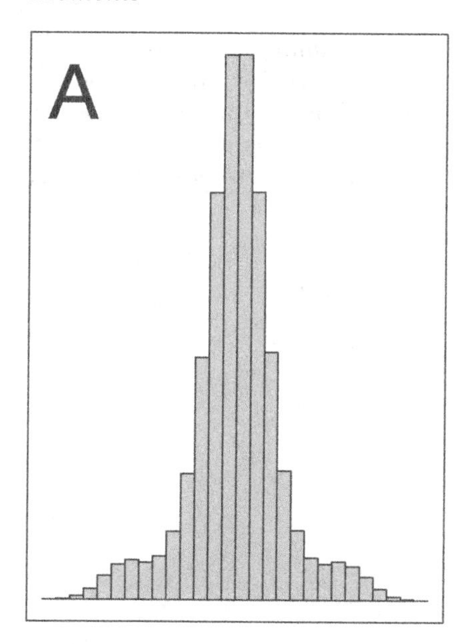

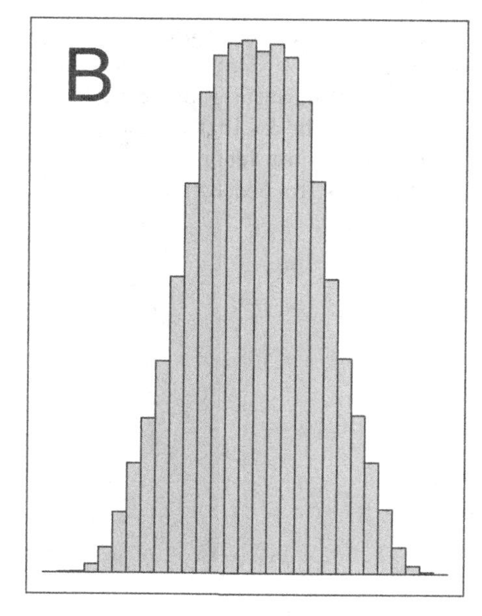

FIGURE 13.2 Histograms showing a (A) leptokurtic distribution and (B) platykurtic distribution.

It is possible to actually calculate skew and kurtosis as we would measures of central tendency and spread (Groeneveld & Meeden, 1984; Rahman, 1968; Sokal & Rohlf, 1995), but this is very rarely necessary (but see Doane & Seward, 2011). Nevertheless, this is a good place to introduce the concept of what a moment is in statistics.

The word 'moment' usually refers to a brief period in time. But this is not what the word means in statistics. In statistics, a **moment** is related to how much some value (more technically, a random variable) is expected to deviate from the mean (Upton & Cook, 2014). The term has its origins in physics (Miller & Miller, 2004; Sokal & Rohlf, 1995). A moment statistic describes the expected deviation from the mean, raised to some power. We have already looked at the second moment about the mean (i.e., the expected deviation raised to the power 2). This was the variance,

$$s^2 = \frac{1}{N-1}\sum_{i=1}^{N}(x_i - \bar{x})^2 .$$

A different exponent would indicate a different moment. For example, a 3 would give us the third moment (skew), and a 4 would give us the fourth moment (kurtosis). The equations for calculating these sample moments are a bit more complicated than simply replacing the 2 with a 3 or 4 in the equation for variance. This is because the corrections for the third and fourth moments are no longer $N - 1$. The corrections now also need to be scaled by the standard deviation (for details, see Sokal & Rohlf, 1995). For the purpose of doing statistics in the biological and environmental sciences, it is usually sufficient to recognise that second, third, and fourth moments refer to variance, skew, and kurtosis, respectively.

14

Practical. *Plotting and statistical summaries in jamovi*

This chapter focuses on applying the concepts from Chapters 9–13 in jamovi (The jamovi project, 2024). The data that we will work with in this chapter were inspired by Law et al. (2014). Law et al. (2014) focused on beaver reintroduction in Scottish habitats and investigated its consequences for the white water lily, *Nymphaea alba*, which beavers regularly consume.

This chapter will analyse data on the petiole diameter (mm) from *N. alba* collected from seven different sites on the west coast of Scotland (the petiole is the structure that attaches the plant stem to the blade of the leaf). The *N. alba* dataset is available online[1]. The data are not in a tidy format, so it is important to first reorganise the data so that they can be analysed in jamovi. Once the data are properly organised, we will use jamovi to plot them, calculate summary statistics, apply appropriate decimals, significant figures, and rounding, and compare petiole diameters across sites.

14.1 Reorganise the dataset into a tidy format

The *N. alba* dataset is not in a tidy format. All of the numbers from this dataset are measurements of petiole diameter in millimetres from *N. alba*, but each row contains seven samples because each column shows a different site.

Remember that to make these data tidy and usable in jamovi, we need each row to be a unique observation. What we really want then is a dataset with two columns of data. The first column should indicate the site, and the second column should indicate the petiole diameter. This can be done in two ways. First, we could use a spreadsheet programme like LibreOffice or MS Excel to create a new dataset with two columns, one column with the site information and the other column with the petiole diameters. Second, we could use the 'Data' tab in jamovi to create two new columns of data (one for site and the other for petiole diameter). Either way, we need to copy and paste site names

[1] `https://bradduthie.github.io/stats/data/Nymphaea_alba.csv`

into the first column and petiole diameters in the second column. This is a bit tedious, but it is an important step in the process of data analysis. See Figure 14.1 for how this would look in jamovi.

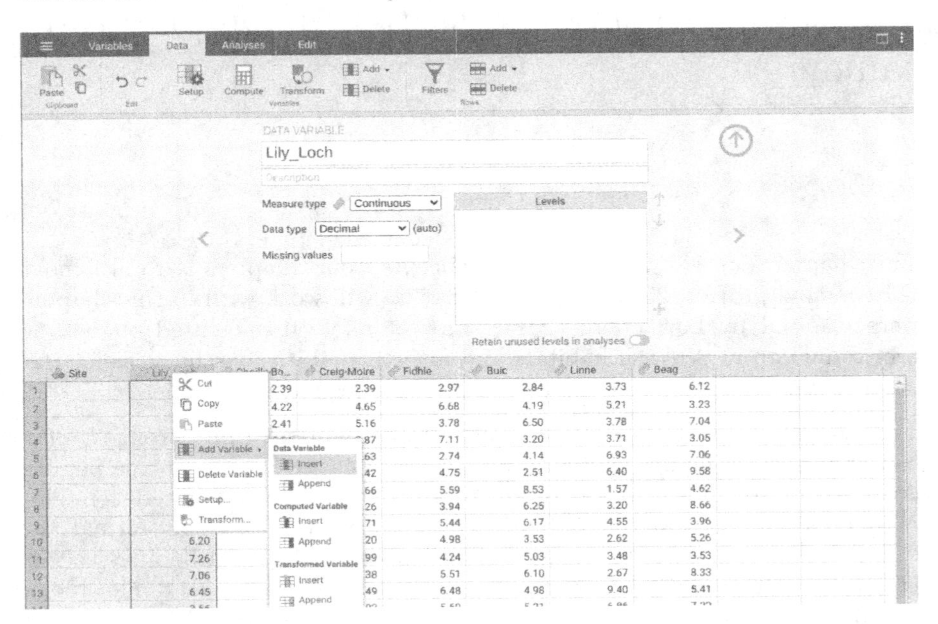

FIGURE 14.1 Tidying the raw data of petiole diameters from lily pad measurements across seven sites in Scotland. A new column of data is created by right-clicking on an existing column and choosing 'Add Variable'.

Note that to insert a new column in jamovi, we need to right-click on an existing column and select 'Add Variable' → 'Insert'. A new column will then pop up in jamovi, and we can give this an informative name. Make sure to specify that the 'Site' column should be a nominal measure type, and the 'petiole_diameter_mm' column should be a continuous measure type. The first six rows of the dataset should look like the below.

```
       Site petiole_diameter_mm
1 Lily_Loch                7.42
2 Lily_Loch                3.58
3 Lily_Loch                7.47
4 Lily_Loch                6.07
5 Lily_Loch                6.81
6 Lily_Loch                8.05
```

With the reorganised dataset, we are now ready to do some analysis in jamovi. We will start with some plotting.

14.2 Histograms and box-whisker plots

We will start by making a histogram of the full dataset of petiole diameter. To do this, we need to go to the 'Analyses' tab of the jamovi toolbar, then select the 'Exploration' button. Next, select the 'Descriptives' option (Figure 14.2). This will open a new window where it is possible to create plots and calculate summary statistics. The white box on the left of the Descriptive interface lists all of the variables in the dataset. Below this box, there are options for selecting different summary statistics ('Statistics') and building different graphs ('Plots'). To get started, select the petiole diameter variable in the box to the left, then move it to the 'Variables' box (top right) using the → arrow. Next, open the Plots option at the bottom of the interface. Choose the 'Histogram' option by clicking the checkbox. A histogram will open up in the window on the right (you might need to scroll down).

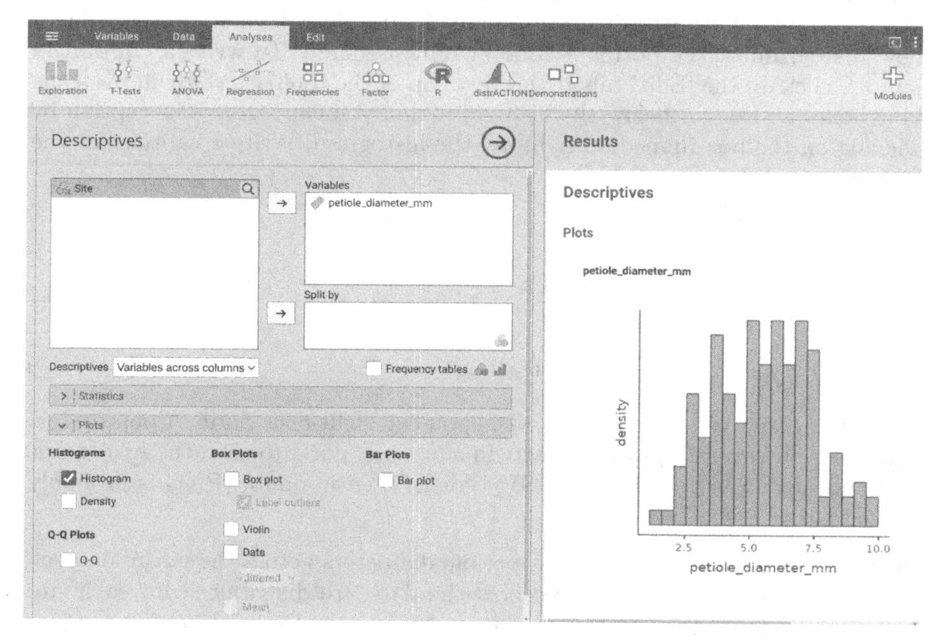

FIGURE 14.2 Jamovi Descriptives toolbar with petiole diameter selected and a histogram produced in the plotting window.

Take a look at the histogram to the right (Figure 14.2). Just looking at the histogram, write down what you think the following summary statistics will be.

Mean: ____

Median: ____

Standard deviation: ____

Based on the histogram, do you think that the mean and median are the same? Why or why not?

The histogram needs better-labelled axes and an informative caption. To label the axes better, go back to the data tab and double-click on the column heading 'petiole_diameter_mm'. Change the name of the data variable to 'Petiole diameter (mm)'. The newly named variable will then appear when a new histogram of the petiole diameter data is made. To write a caption in jamovi, click on the 'Edit' tab at the very top of the toolbar. You will see some blue boxes above and below the histogram, and you can write your caption by clicking on the box immediately below the histogram. Write a caption for the histogram below.

If you want to save the histogram, then you can right-click on it. A pop-up box will give you several options; select 'Image → Export' to save the histogram. You can save it as a PDF, PNG, SVG, or EPS (if in doubt, PNG is probably the easiest to use).

In the first example, we looked at petiole diameters across the entire dataset, but suppose that we want to see how the data are distributed for each site individually. To do this, we just need to go back to the Descriptives box (Figure 14.2) and put the 'Site' variable into the box on the lower right called 'Split by'. Do this by selecting 'Site' then using the lower → arrow to bring it to the 'Split by' box. Instead of one histogram of petiole diameters, you will now see seven different histograms, one for each site, all stacked on top of each other. This might be useful, but all of these histograms together is a bit busy. Instead, we can use a box-whisker plot to compare the distributions of petiole diameters across different sites.

To create a boxplot, simply check 'Box plot' from the Plots options (you might want to uncheck 'Histogram', but it is not necessary). You should now see all of the different sites on the x-axis of the newly created boxplot and a summary of the petiole diameters on the y-axis. Based on the boxplot, which site appears to have the highest and lowest median petiole diameter?

Highest: ______________________________

Lowest: ______________________________

There is one more trick that is useful with box-whisker plots in jamovi. The current plots show a summary of each site, but it might also be useful to plot the actual data points to give some more information about the distribution of petiole diameters. You can do this by checking the option 'Data', which places the petiole diameter of each sample over the box and whiskers for each site. The y-axis shows the petiole diameter of each data point. By default, the points are jittered on the x-axis, which just means that they are placed randomly on the x-axis within a site. This is just to ensure that points will not be placed directly on top of each other if they are the same value. If you prefer, you can use the pull-down menu right below the Data checkbox to select 'Stacked' instead of 'Jittered'. The stacked option will place points side by side. Think about where the points are in relation to the box and whiskers of the plot; this should help you develop an intuitive understanding of how to read box-whisker plots.

14.3 Calculate summary statistics

We can calculate the summary statistics using the 'Descriptives' option in jamovi, just as we did with the histogram and box-whisker plots. Before doing anything else, again place the petiole diameter variable in the box of variables, but do not split the dataset by site just yet because we first want summary statistics across the entire dataset. Below the box of variables, but above the 'Plots' options, there are options for selecting different summary statistics. Open up this new box and have a look at the different summary statistics that can be calculated. To calculate all of the variables explained in Chapter 11 and Chapter 12, check the following 11 boxes:

- N: ______________________________
- Std. deviation: ______________________________
- Variance: ______________________________
- Minimum: ______________________________
- Maximum: ______________________________
- Range: ______________________________
- IQR: ______________________________

- Mean: ______________________________
- Median: ______________________________
- Mode: ______________________________
- Std. error of mean: ______________________________

When you do this, the Statistics option in jamovi should look like it does in Figure 14.3.

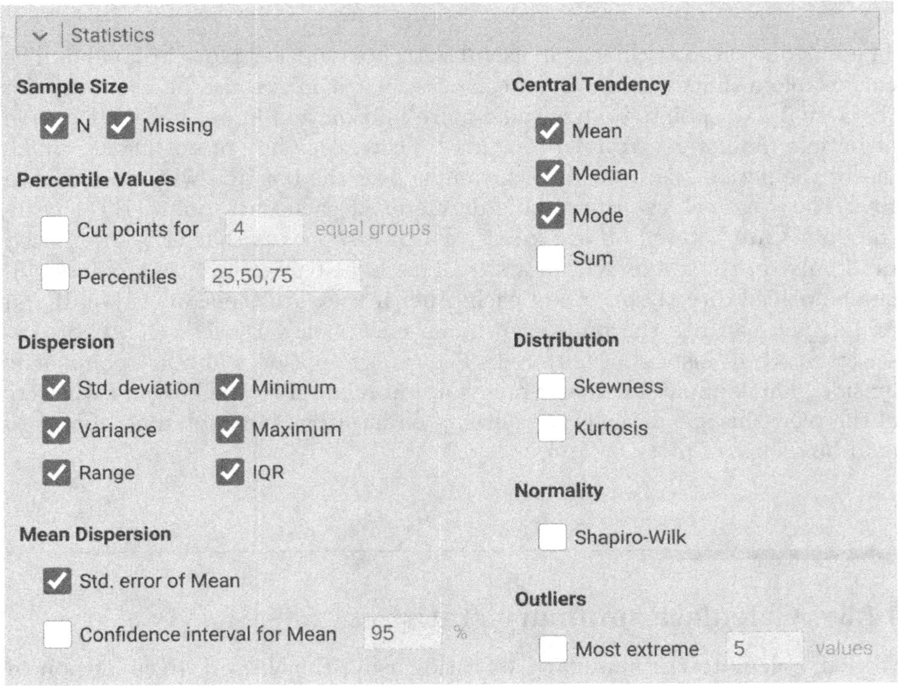

FIGURE 14.3 Jamovi Descriptives toolbar showing the summary statistics available to report.

Once you check these boxes, you will see a 'Descriptives' table open on the right-hand side of jamovi. This table will report all of the summary statistics that you have checked. Write down the values for the summary statistics next to the corresponding bullet points above.

Next, split these summary statistics up by site. Notice the very large table that is now produced on the right-hand side of jamovi. Which of the seven sites in the dataset has the highest mean petiole diameter, and what is its mean?

Site: ______________________________

Mean: ______________________________

Which of the seven sites has the lowest variance in petiole diameter, and what is its variance?

Site: ______________________________

Variance: ______________________________

Make sure that you are able to find and interpret these summary statistics in jamovi. Explore different options to get more comfortable using jamovi for building plots and reporting summary statistics. Can you find the first and third quartiles for each site? Report the third quartiles for each site below.

Beag: ______________________________

Buic: ______________________________

Choille-Bharr: ______________________________

Creig-Moire: ______________________________

Fidhle: ______________________________

Lily_Loch: ______________________________

Linne: ______________________________

Next, we will look at reporting summary statistics to different significant figures.

14.4 Reporting decimals and significant figures

Using the same values that you reported above for the whole dataset (i.e., not broken down by site), report each summary statistic to two significant figures. Remember to round accurately if you need to reduce the number of significant figures from the original values to the new values below.

- N: ______________________________
- Std. deviation: ______________________________
- Variance: ______________________________
- Minimum: ______________________________
- Maximum: ______________________________
- Range: ______________________________
- IQR: ______________________________
- Mean: ______________________________
- Median: ______________________________
- Mode: ______________________________
- Std. error of mean: ______________________________

Remember from Section 14.2 that you were asked to write down what you thought the mean, median, and standard deviation were just by inspecting the histogram. Compare your answers in that section with the rounded statistics listed above. Were you able to get a similar value from the histogram as calculated in jamovi from the data? What can you learn from the histogram that you cannot from the summary statistics, and what can you learn from the summary statistics that you cannot from the histogram? Write your reflections in the space below.

Next, we will produce barplots to show the mean petiole diameter for each site.

14.5 Comparing across sites

To make a barplot that compares the mean petiole diameters across sites, we again use the Descriptives option in jamovi. Place petiole diameter as the variable, and split this by site. Next, go down to the plotting options and check 'Bar plot'. You will see a barplot produced in the window to the right with different sites on the x-axis. Bar heights show the mean petiole diameter for each site. Notice the intervals shown for each bar (i.e., the vertical lines in the centre of the bars that go up and down different lengths). These error bars are centred on the mean petiole diameter (bar height) and show one standard error above and below the site mean. Recall back from Chapter 12; what information do these error bars convey about the estimated mean petiole diameter?

What can you say about the mean petiole diameters across the different sites? Do these sites appear to have very different mean petiole diameters?

There were 20 total petiole diameters sampled from each site. If we were to go back out to these seven sites and sample another 20 petiole diameters, could we **really** expect to get the exact same site means? Assuming the site means would be at least a bit different for our new sample, is it possible that the sites with the highest or lowest petiole diameters might also differ in our new sample? If so, then what does this say about our ability to make conclusions about the differences in petiole diameter among sites?

15

Introduction to probability models

Suppose that we flip a fair coin over a flat surface. There are two possibilities for how the coin lands on the surface. Either the coin lands on one side (heads) or the other side (tails), but we do not know the outcome in advance. If these two events (heads or tails) are equally likely, then we could reason that there is a 50% chance that a flipped coin will land heads up and a 50% chance that it will land heads down. What do we actually mean when we say this? For example, when we say that there is a 50% chance of the coin landing heads up, are we making a claim about our own uncertainty, how coins work, or how the world works? We might mean that we simply do not know whether or not the coin will land heads up, so a 50-50 chance just reflects our own ignorance about what will actually happen when the coin is flipped. Alternatively, we might reason that if a fair coin were to be flipped many times, all else being equal, then about half of flips should end heads up, so a 50% chance is a reasonable prediction for what will happen in any given flip. Or, perhaps we reason that events such as coin flips really are guided by chance on some deeper fundamental level, such that our 50% chance reflects some real causal metaphysical process in the world. These are questions concerning the philosophy of probability. The philosophy of probability is an interesting sub-discipline in its own right, with implications that can and do affect how researchers do statistics (Edwards, 1972; Gelman & Shalizi, 2013; Mayo, 1996; Mayo, 2021; Navarro & Foxcroft, 2022; Suárez, 2020).

In this chapter, we will not worry about the philosophy of probability[1] and instead focus on the mathematical rules of probability as applied to statistics. These rules are important for predicting real-world events in the biological and environmental sciences. For example, we might need to make predictions concerning the risk of disease spreading in a population, or the risk of extreme events such as droughts occurring given increasing global temperatures. Probability is also important for testing scientific hypotheses. For example, if we sample two different groups and calculate that they have different means (e.g., two different fields have different mean soil nitrogen concentrations), we might

[1] In the interest of transparency, this book presents a *frequentist* interpretation of probability (Mayo, 1996). While this approach does reflect the philosophical inclinations of the author, the reason for working from this interpretation has more to do with the statistical tests that are most appropriate for an introductory statistics book, which are also the tests most widely used in the biological and environmental sciences.

want to know the probability that this difference between means could have arisen by chance. Here I will introduce practical examples of probability, then introduce some common probability distributions.

15.1 Instructive example

Probability focuses on the outcomes of trials, such as the **outcome** (heads or tails) of the **trial** of a coin flip. The probability of a specific outcome is the relative number of times it is expected to happen given a large number of trials,

$$P(outcome) = \frac{\text{Number of times outcome occurs}}{\text{Total number of trials}}.$$

For the outcome of a flipped coin landing on heads,

$$P(heads) = \frac{\text{Flips landing on heads}}{\text{Total number of flips}}.$$

As the total number of flips becomes very large, the number of flips that land on heads should get closer and closer to half the total, $1/2$ or 0.5 (more on this later). The above equations use the notation $P(E)$ to define the probability (P) of some event (E) happening. Note that the number of times an outcome occurs cannot be less than 0, so $P(E) \geq 0$ must always be true. Similarly, the number of times an outcome occurs cannot be greater than the number of trials; the most frequently it can happen is in *every* trial, in which case the top and bottom of the fraction are the same value. Hence, $P(E) \leq 1$ must also always be true. Probabilities therefore range from 0 (an outcome *never* happens) to 1 (an outcome *always* happens).

It might be more familiar and intuitive at first to think in terms of percentages (i.e., from 0 to 100% chance of an outcome, rather than from 0 to 1), but there are good mathematical reasons for thinking about probability on a 0–1 scale (it makes calculations easier). For example, suppose we have two coins, and we want to calculate the probability that they will both land on heads if we flip them at the same time. That is, we want to know the probability that coin 1 lands on heads **and** coin 2 lands on heads. We can assume that the coins do not affect each other in any way, so each coin flip is **independent** of the other (i.e., the outcome of coin 1 does not affect the outcome of coin 2, and vice versa – this kind of assumption is often very important in statistics). Each coin, by itself, is expected to land on heads with a probability of 0.5, $P(heads) = 0.5$. When we want to know the probability that two or more independent events will happen, we *multiply* their probabilities. In the case of both coins landing on heads, the probability is therefore,

$$P(Coin_1 = heads \cap Coin_2 = heads) = 0.5 \times 0.5 = 0.25.$$

Note that the symbol $\cap$ is basically just a fancy way of writing 'and' (technically, the intersection between sets; see set theory for details). Verbally, all this is saying is that the probability of coin 1 landing on heads *and* the probability of coin 2 landing on heads equals 0.5 times 0.5, which is 0.25.

But why are we *multiplying* to get the joint probability of both coins landing on heads? Why not add, for example? We could just take it as a given that multiplication is the correct operation to use when calculating the probability that multiple events will occur. Or we could do a simple experiment to confirm that 0.25 really is about right (e.g., by flipping two coins 100 times and recording how many times both coins land on heads). But neither of these options would likely be particularly satisfying. Let us first recognise that adding the probabilities cannot be the correct answer. If the probability of each coin landing on heads is 0.5, then adding probabilities would imply that the probability of both landing on heads is $0.5 + 0.5 = 1$. This does not make any sense because we know that there are other possibilities, such as both coins landing on tails, or one coin landing on heads and the other landing on tails. Adding probabilities cannot be the answer, but why multiply?

We can think about probabilities visually, as a kind of probability space. When we have only one trial, then we can express the probability of an event along a line (Figure 15.1).

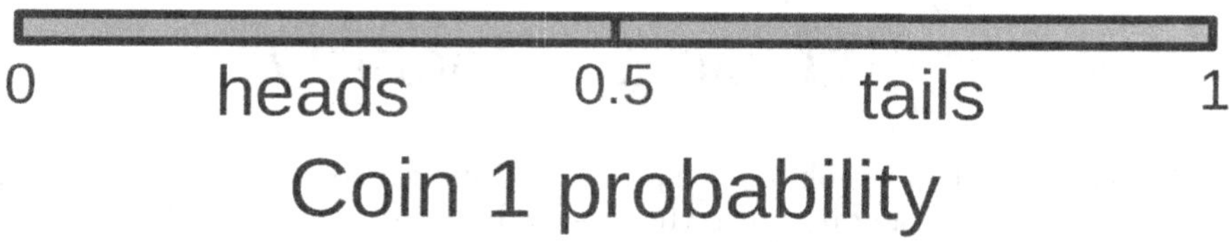

FIGURE 15.1 Total probability space for flipping a single coin and observing its outcome (heads or tails). Given a fair coin, the probability of heads equals a proportion 0.5 of the total probability space, while the probability of tails equals the remaining 0.5 proportion.

The total probability space is 1, and 'heads' occupies a density of 0.5 of the total space. The remaining space, also 0.5, is allocated to 'tails'. When we add a second independent trial, we now need two dimensions of probability space (Figure 15.2). The probability of heads or tails for coin 1 (the horizontal axis of Figure 15.2) remains unchanged, but we add another axis (vertical this time) to think about the equivalent probability space of coin 2.

Now we can see that the area in which both coin 1 and coin 2 land on heads has a proportion of 0.25 of the total area. This is a geometric representation of what

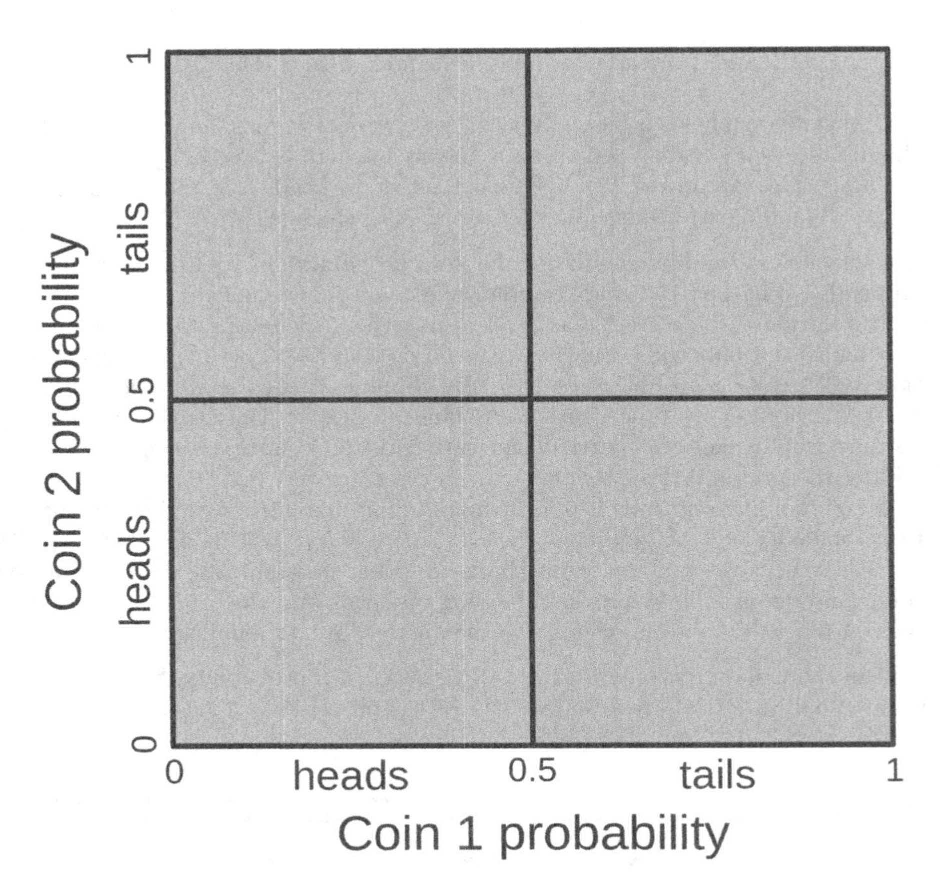

FIGURE 15.2 Total probability space for flipping two coins and observing their different possible outcomes (heads-heads, heads-tails, tails-heads, and tails-tails). Given two fair coins, the probability of flipping heads for each equals 0.25, which corresponds to the lower left square of the probability space.

we did when calculating $P(Coin_1 = heads \cap Coin_2 = heads) = 0.5 \times 0.5 = 0.25$. The multiplication works because multiplying probabilities carves out more specific regions of probability space. Note that the same pattern would apply if we flipped a third coin. In this case, the probability of all three coins landing on heads would be $0.5 \times 0.5 \times 0.5 = 0.125$, or $0.5^3 = 0.125$.

What about when we want to know the probability of one outcome **or** another outcome happening? Here is where we add. Note that the probability of a coin flip landing on heads or tails must be 1 (there are only two possibilities!). What about the probability of both coins landing on the same outcome, that is, either both coins landing on heads or both landing on tails? We know that the probability of both coins landing on heads is 0.25. The probability of both coins landing on tails is also 0.25, so the probability that both coins land on either heads **or** tails is $0.25 + 0.25 = 0.5$. The visual representation in Figure 15.2 works for this example too. Note that heads-heads and tails-tails outcomes are represented by the lower left and upper right areas of probability space, respectively. This is 0.5 (i.e., 50%) of the total probability space.

15.2 Biological applications

Coin flips are instructive, but the relevance for biological and environmental sciences might not be immediately clear. In fact, probability is extremely relevant in nearly all areas of the natural sciences. The following are just two hypothetical examples where the calculations in the previous section might be usefully applied:

1. From a recent report online, suppose you learn that 1 in 40 people in your local area is testing positive for COVID-19. You find yourself in a small shop with 6 other people. What is the probability that at least 1 of these 6 other people would test positive for COVID-19? To calculate this, note that the probability that any given person has COVID-19 is $1/40 = 0.025$, which means that the probability that a person does **not** must be $1 - 0.025 = 0.975$ (they either do or do not, and the probabilities must sum to 1). The probability that **all** 6 people *do not* have COVID-19 is therefore $(0.975)^6 = 0.859$. Consequently, the probability that at least 1 of the 6 people **does** have COVID-19 is $1 - 0.859 = 0.141$, or 14.1%.

2. Imagine you are studying a population of sexually reproducing, diploid (i.e., 2 sets of chromosomes), animals, and you find that a particular genetic locus has 3 alleles with frequencies $P(A_1) = 0.40$, $P(A_2) = 0.45$, and $P(A_3) = 0.15$. What is the probability that a randomly sampled animal will be heterozygous with 1 copy of the A_1 allele and 1 copy of the A_3 allele? Note that there are 2 ways for A_1 and A_3 to arise in an individual, just like there were 2 ways to get a heads and tails coin in the Section 14.1 example (see Figure 14.2). The individual could either get an A_1 in the first position and A_3 in the second position, or an A_3 in the first position and A_1 in the second position. We can therefore calculate the probability as, $P(A_1) \times P(A_3) + P(A_3) \times P(A_1)$, which is $(0.40 \times 0.15) + (0.15 \times 0.4) = 0.12$, or 12% (in population genetics, we might use the notation $p = P(A_1)$ and $r = P(A_3)$, then note that $2pr = 0.12$).

In both of these examples, we made some assumptions, which might or might not be problematic. In the first example, we assumed that the 6 people in our shop were a random and independent sample from the local area (i.e., people with COVID-19 are not more or less likely to be in the shop, and the 6 people in the shop were not associated in a way that would affect their individual probabilities of having COVID-19). In the second example, we assumed that individuals mate randomly, and that there is no mutation, migration, or selection on genotypes (Hardy, 1908). It is important to recognise these assumptions when we are making them, because violations of assumptions could affect the probabilities of events!

15.3 Sampling with and without replacement

It is often important to make a distinction between sampling with or without replacement. Sampling with replacement just means that whatever has been sampled once gets put back into the population before sampling again. Sampling without replacement means that whatever has been sampled does not get put back into the population before sampling again. An example makes the distinction between sampling with and without replacement clearer.

FIGURE 15.3 Playing cards can be useful for illustrating concepts in probability. Here we have 5 hearts (left) and 5 spades (right).

Figure 15.3 shows 10 playing cards: 5 hearts and 5 spades. If we shuffle these cards thoroughly and randomly select 1 card, what is the probability of selecting a heart? This is simply,

$$P(heart) = \frac{\text{5 hearts}}{\text{10 total cards}} = 0.5.$$

What is the probability of randomly selecting 2 hearts? This depends if we are sampling with or without replacement. If we sample 1 card, then put it back into the deck before sampling the second card, then the probability of sampling a heart does not change (in both samples, we have 5 hearts and 10 cards). Hence, the probability of sampling 2 hearts with replacement is $P(heart) \times P(heart) = 0.5 \times 0.5 = 0.25$. If we do not put the first card back into the deck before sampling again, then we have changed the total number of cards. After sampling the first heart, we have one fewer heart in the deck and one fewer card, so the new probability for sampling a heart becomes,

$$P(heart) = \frac{\text{4 hearts}}{\text{9 total cards}} = 0.444.$$

Since the probability has changed after the first heart is sampled, we need to use this adjusted probability when sampling without replacement. In this case, the probability of sampling two hearts is $0.5 \times 0.444 = 0.222$. This is a bit lower than the probability of sampling with replacement because we have decreased the number of hearts that can be sampled. When sampling from a set, it is important to consider whether the sampling is done with or without replacement.

15.4 Probability distributions

Up until this point, we have been considering the probabilities of specific outcomes. That is, we have considered the probability that a coin flip will be heads, that an animal will have a particular combination of alleles, or that we will randomly select a particular suit of card from a deck. Here we will move from specific outcomes and consider the *distribution* of outcomes. For example, instead of finding the probability that a flipped coin lands on heads, we might want to consider the distribution of the number of times that it does (in this case, 0 times or 1 time; Figure 15.4).

This is an extremely simple distribution. There are only two discrete possibilities for the number of times the coin will land on heads, 0 or 1. And the probability of both outcomes is 0.5, so the bars in Figure 15.4 are the same height. Next, we will consider some more interesting distributions.

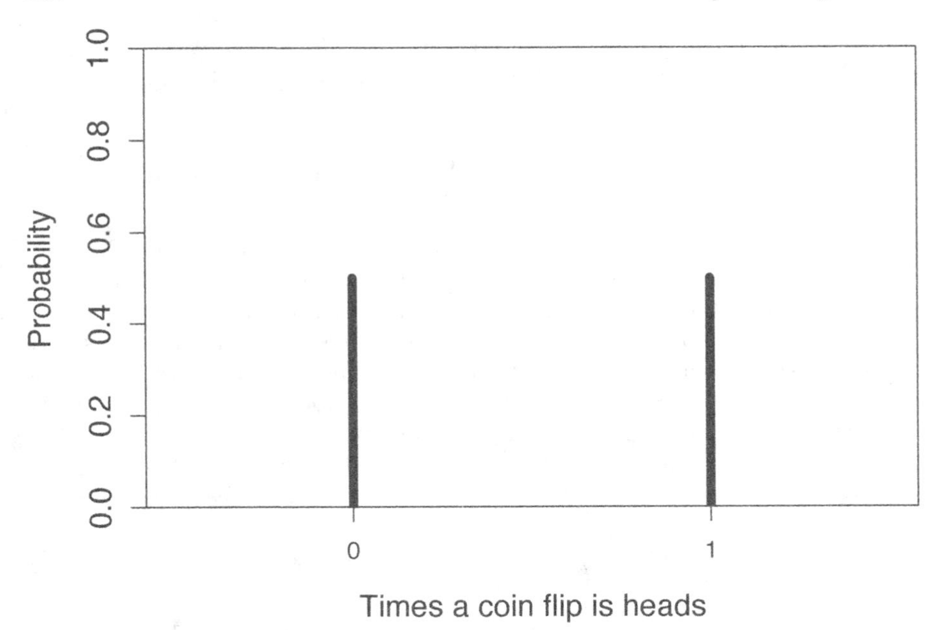

FIGURE 15.4 Probability distribution for the number of times that a flipped coin lands on heads in 1 trial.

15.4.1 Binomial distribution

The simple distribution with a single trial of a coin flip was actually an example of a binomial distribution. More generally, a binomial distribution describes the number of successes in some number of trials (Miller & Miller, 2004). The word 'success' should not be taken too literally here; it does not necessarily indicate a good outcome, or an accomplishment of some kind. A success in the context of a binomial distribution just means that an outcome *did* happen as opposed to it *not* happening. If we define a coin flip landing on heads as a success, we could consider the probability distribution of the number of successes over 10 trials (Figure 15.5).

Figure 15.5 shows that the most probable outcome is that 5 of the 10 coins flipped will land on heads. This makes some sense because the probability that any one flip lands on heads is 0.5, and 5 is 1/2 of 10. But 5 out of 10 heads happens only with a probability of about 0.25. There is also about a 0.2 probability that the outcome is 4 heads, and the same probability that the outcome is 6 heads. Hence, the probability that we get an outcome of between 4 and 6 heads is about $0.25 + 0.2 + 0.2 = 0.65$. In contrast, the probability of getting all heads is very low (about 0.00098).

More generally, we can define the number of successes using the random variable X. We can then use the notation $P(X = 5) = 0.25$ to indicate the

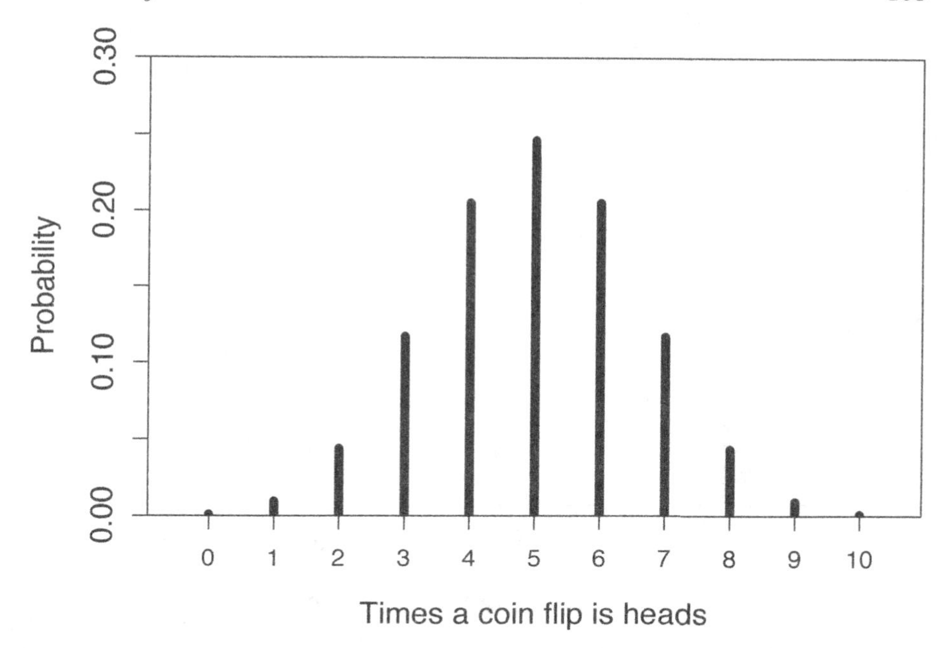

FIGURE 15.5 Probability distribution for the number of times that a flipped coin lands on heads in 10 trials.

probability of 5 successes, or $P(4 \leq X \leq 6) = 0.65$ as the probability that the number of successes is greater than or equal to 4 and less than or equal to 6.

Imagine that you were told a coin was fair, then flipped it 10 times. Imagine that 9 flips out of the 10 came up heads. Given the probability distribution shown in Figure 15.5, the probability of getting 9 or more heads in 10 flips given a fair coin is very low ($P(X \geq 9) \approx 0.011$). Would you still believe that the coin is fair after these 10 trials? How many, or how few, heads would it take to convince you that the coin was not fair? This question gets to the heart of a lot of hypothesis-testing in statistics, and we will see it more in Chapter 21.

Note that a binomial distribution does not need to involve a fair coin with equal probability of success and failure. We can consider again the first example in Section 15.2, in which 1 in 40 people in an area are testing positive for COVID-19, then ask what the probability is that 0–6 people in a small shop would test positive (Figure 15.6).

Note that the shape of this binomial distribution is different from the coin flipping trials in Figure 15.5. The distribution is skewed, with a high probability of 0 successes and a diminishing probability of 1 or more successes.

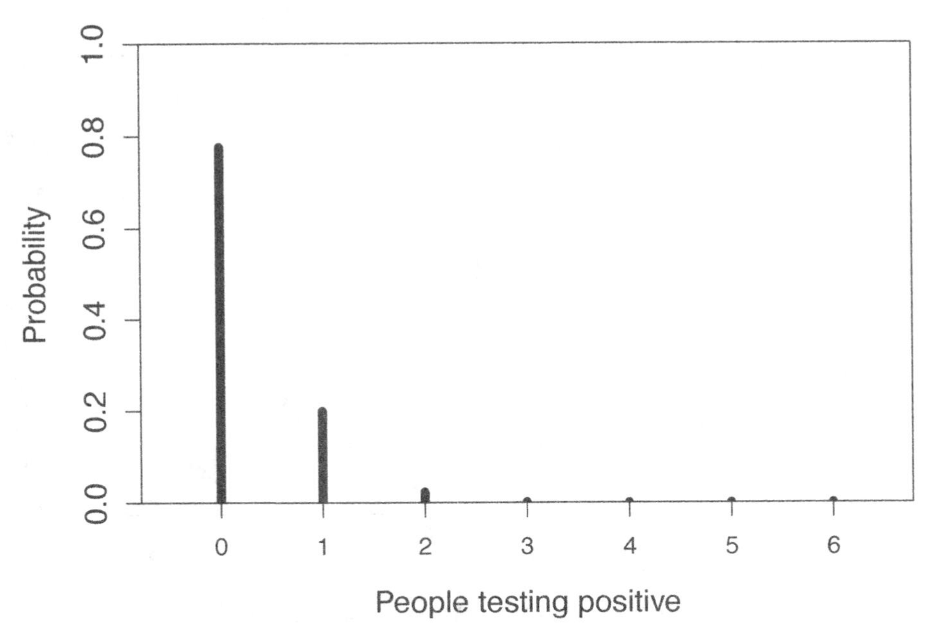

FIGURE 15.6 Probability distribution for the number of people who have COVID-19 in a shop of 6 when the probability of testing positive is 0.025.

The shape of a statistical probability distribution can be defined mathematically. Depending on the details (more on this later), we call the equation defining the distribution either a probability mass function or a probability density function. This book is about statistical techniques, not statistical theory, so we will relegate these equations to footnotes.[2] What is important to know is that the shape of a distribution is modulated by **parameters**. The shape of a binomial distribution is determined by two parameters: the number of trials

[2]For those interested, more technically, we can say that a random variable X has a binomial distribution if and only if its probability mass function is defined by (Miller & Miller, 2004),

$$b\left(x; n, \theta\right) = \binom{n}{x} \theta^{x} \left(1 - \theta\right)^{n-x}.$$

In this binomial probability mass function, $x = 0, 1, 2, ..., n$ (i.e., x can take any integer value from 0 to n). Note that the n over the x in the first parentheses on the right-hand side of the equation is a binomial coefficient, which can be read 'n choose x'. This can be written out as,

$$\binom{n}{x} = \frac{n!}{x!(n-x)!}.$$

The exclamation mark indicates a factorial, such that $n! = n \times (n-1) \times (n-2) \times ... \times 2 \times 1$. That is, the factorial multiplies every decreasing integer down to 1. For example, $4! = 4 \times 3 \times 2 \times 1 = 24$. None of this is critical to know for applying statistical techniques to biological and environmental science data, but it demonstrates just a bit of the theory underlying statistical tools.

(n) and the probability of success (θ). In Figure 15.5, there were 10 trials each with a success probability of 0.5 (i.e., $n = 10$ and $\theta = 0.5$). In Figure 15.6, there were 6 trials each with a success probability of 0.025 (i.e., $n = 6$ and $\theta = 0.025$). This difference in parameter values is why the two probability distributions have a different shape.

15.4.2 Poisson distribution

Imagine sitting outside on a park bench along a path that is a popular route for runners. On this particular day, runners pass by the bench at a steady rate of about 4 per minute, on average. We might then want to know the *distribution* of the number of runners passing by per minute. That is, given that we see 4 runners per minute on average, what is the probability that we will see just 2 runners pass in any given minute. What is the probability that we will see 8 runners pass in a minute? This hypothetical example is modelled with a Poisson distribution. A Poisson distribution describes events happening over some interval (e.g., happening over time or space). There are a lot of situations where a Poisson distribution is relevant in biological and environmental sciences:

- Number of times a particular species will be encountered while walking a given distance
- Number of animals a camera trap will record during a day
- Number of floods or earthquakes that will occur in a given year

The shape of a Poisson distribution is described by just one parameter, λ. This parameter is both the mean and the variance of the Poisson distribution. We can therefore get the probability that some number of events (x) will occur just by knowing λ (Figure 15.7).

Like the binomial distribution, the Poisson distribution can also be defined mathematically[3]. Also like the binomial distribution, probabilities in the Poisson distribution focus on **discrete** observations. This is, probabilities are assigned to a specific number of successes in a set of trials (binomial distribution) or the number of events over time (Poisson distribution). In both cases, the probability distribution focuses on countable numbers. In other words, it does not make any sense to talk about the probability of a coin landing on heads 3.75 times after 10 flips, nor the probability of 2.21 runners passing by a park bench in a given minute. The probability of either of these events

[3]A random variable X has a Poisson distribution if and only if its probability mass function is defined by (Miller & Miller, 2004),

$$p(x; \lambda) = \frac{\lambda^x e^x}{x!}.$$

Recall from Chapter 1 Euler's number, $e \approx 2.718282$, and from the previous footnote that the exclamation mark indicates a factorial. In the Poisson probability mass function, x can take any integer value greater than or equal to 0.

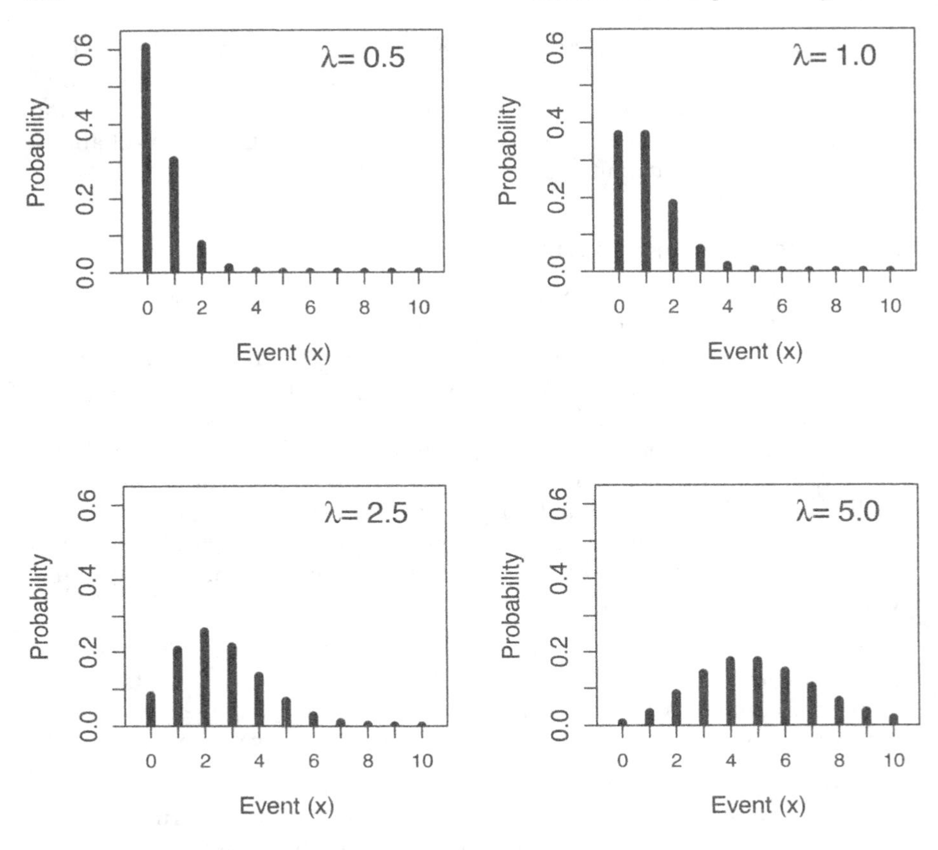

FIGURE 15.7 Poisson probability distributions given different rate parameter values.

happening is zero, which is why Figures 15.5–15.7 all have spaces between the vertical bars. These spaces indicate that values between the integers are impossible. When observations are discrete like this, they are defined by a *probability mass function.* In the next section, we consider distributions with a continuous range of possible sample values; these distributions are defined by a *probability density function.*

15.4.3 Uniform distribution

We now move on to continuous distributions, starting with the continuous uniform distribution. I introduce this distribution mainly to clarify the difference between a discrete and continuous distribution. While the uniform distribution is very important for a lot of statistical tools (notably, simulating pseudorandom numbers), it is not something that we come across much in biological or environmental science data. The continuous uniform distribution has two parameters, α and β (Miller & Miller, 2004)[4]. Values of α and β can be any real number (not just integers). For example, suppose that $\alpha = 1$ and $\beta = 2.5$. In this case, Figure 15.8 shows the probability distribution for sampling some value x.

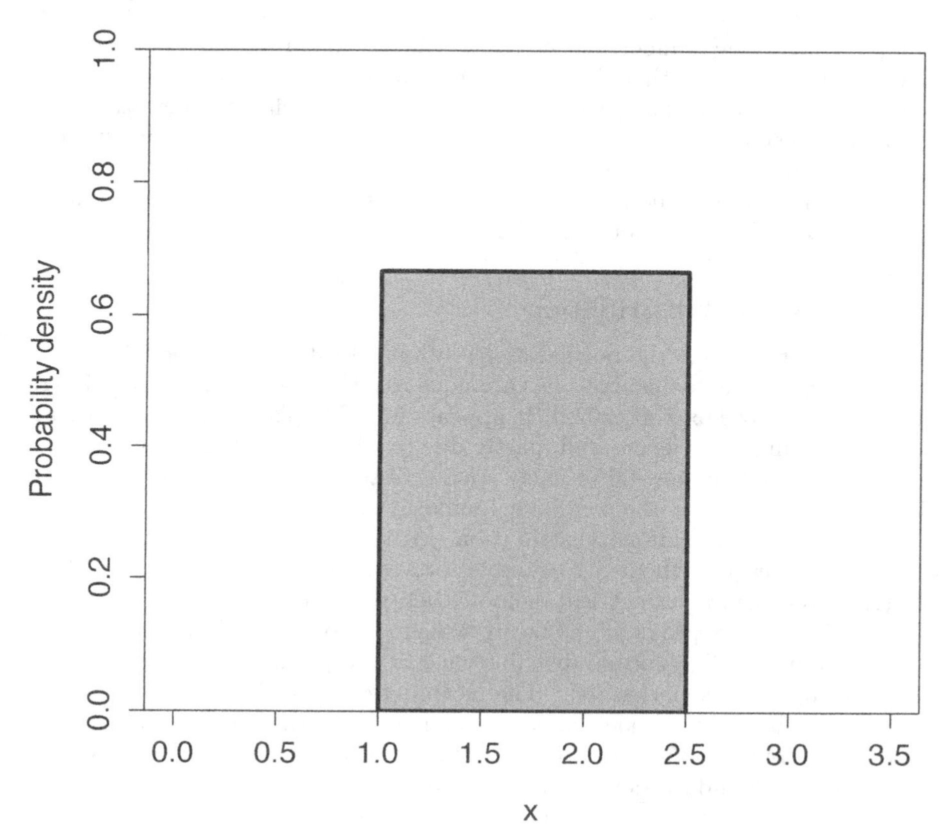

FIGURE 15.8 A continuous uniform distribution in which a random variable X takes a value between 1 and 2.5.

The height of the distribution in Figure 15.8 is $1/(\beta - \alpha) = 1/(2.5 - 1) \approx 0.667$. All values between 1 and 2.5 have equal probability of being sampled.

[4]A random variable X has a continuous uniform distribution if and only if its probability density function is defined by (Miller & Miller, 2004),

$$u(x; \alpha, \beta) = \frac{1}{\beta - \alpha},$$

where $\alpha < x < \beta$, and $u(x; \alpha, \beta) = 0$ everywhere else. The value x can take any real number.

Here is a good place to point out the difference between the continuous distribution versus the discrete binomial and Poisson distributions. From the uniform distribution of Figure 15.8, we can, theoretically, sample *any* real value between 1 and 2.5 (e.g., 1.34532 or 2.21194; the sampled value can have as many decimals as our measuring device allows). There are uncountably infinite real numbers, so it no longer makes sense to ask what is the probability of sampling a specific number. For example, what is the probability of sampling a value of *exactly* 2, rather than, say, 1.999999 or 2.000001, or something else arbitrarily close to 2? The probability of sampling a specific number exactly is negligible. Instead, we need to think about the probability of sampling within intervals. For example, what is the probability of sampling a value between 1.9 and 2.1, or any value greater than 2.2? This is the nature of probability when we consider continuous distributions.

15.4.4 Normal distribution

The last distribution, the normal distribution (also known as the 'Gaussian distribution' or the 'bell curve') has a unique role in statistics (Miller & Miller, 2004; Navarro & Foxcroft, 2022). It appears in many places in the biological and environmental sciences and, partly due to the central limit theorem (see Chapter 16), is fundamental to many statistical tools. The normal distribution is continuous, just like the continuous uniform distribution from the previous section. Unlike the uniform distribution, with the normal distribution, it is possible (at least in theory) to sample *any* real value, $-\infty < x < \infty$. The distribution has a symmetrical, smooth bell shape (Figure 15.8), in which probability density peaks at the mean, which is also the median and mode of the distribution. The normal distribution has two parameters: the mean (μ) and the standard deviation (σ).[5] The mean determines where the peak of the distribution is, and the standard deviation determines the width or narrowness of the distribution. Note that we are using μ for the mean here instead of $\bar{x}$, and σ for the standard deviation instead of s, to differentiate the *population* parameters from the *sample* estimates of Chapter 11 and Chapter 12.

The normal distribution shown in Figure 15.9 is called the **standard normal distribution**, which means that it has a mean of 0 ($\mu = 0$) and a standard deviation of 1 ($\sigma = 1$). Note that because the standard deviation of a distribution is the square-root of the variance (see Chapter 12), and $\sqrt{1} = 1$, the variance of the standard normal distribution is also 1. We will look at the standard normal distribution more closely in Chapter 16.

[5]A random variable X has a normal distribution if and only if its probability density function is defined by (Miller & Miller, 2004),

$$n(x; \mu, \sigma) = \frac{1}{\sigma\sqrt{2\pi}} e^{-\frac{1}{2}\left(\frac{x-\mu}{\sigma}\right)^2}.$$

In the normal distribution, $-\infty < x < \infty$. Note the appearance of two irrational numbers introduced back in Chapter 1: π and e.

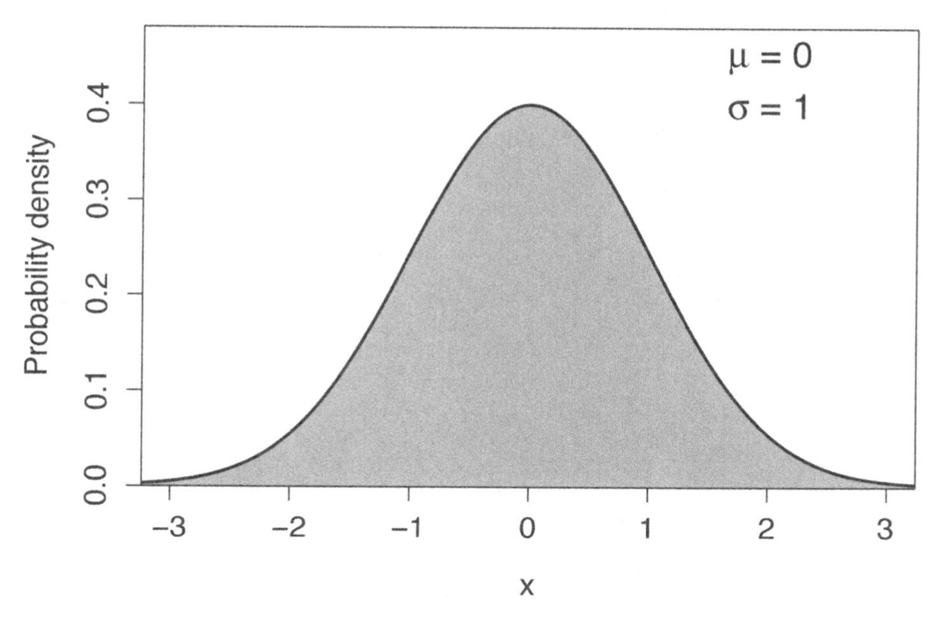

FIGURE 15.9 A standard normal probability distribution, which is defined by a mean value of 0 and a standard deviation of 1.

15.5 Summary

This chapter has introduced probability models and different types of distributions. It has focused on the key points that are especially important for understanding and implementing statistical techniques. As such, a lot of details have been left out. For example, the probability distributions considered in Section 15.4 comprise only a small number of example distributions that are relevant for biological and environmental sciences. In Chapter 16, we will get an even closer look at the normal distribution and why it is especially important.

16

Central Limit Theorem

The previous chapter finished by introducing the normal distribution. This chapter focuses on the normal distribution in more detail and explains why it is so important in statistics.

16.1 The distribution of means is normal

The central limit theorem (CLT) is one of the most important theorems in statistics. It states that if we sample values from **any** distribution and calculate the mean, as we increase our sample size N, the distribution *of the mean* gets closer and closer to a normal distribution (Miller & Miller, 2004; Sokal & Rohlf, 1995; Spiegelhalter, 2019)[1]. This statement is busy and potentially confusing at first, partly because it refers to two separate distributions: the sampling distribution and the distribution of the sample mean. We can take this step by step, starting with the sampling distribution.

The sampling distribution could be any of the four distributions introduced in Chapter 15 (binomial, Poisson, uniform, or normal). Suppose that we sample the binomial distribution from Figure 15.6, the one showing the number of people out of 6 who would test positive for COVID-19 if the probability of testing positive was 0.025. Assume that we sample a value from this distribution (i.e., a number from 0 to 6) 100 times (i.e., $N = 100$). If it helps, we can imagine going to 100 different shops, all of which are occupied by 6 people. From these 100 samples, we can calculate the sample mean $\bar{x}$. This would be the mean number of people in a shop who would test positive for COVID-19. If we were just collecting data to try to estimate the mean number of people with COVID-19 in shops of 6, this is where our calculations might stop. But here is where the second distribution becomes relevant.

[1]For those interested, a mathematical proof of the CLT can be found in Miller & Miller (2004). Here we will demonstrate the CLT by simulation. As an aside, the CLT also applies to the sum of sample values, which will also have a distribution that approaches normality as $N \to \infty$.

Suppose that we could somehow go back out to collect *another* 100 samples from 100 completely different shops. We could then get the mean of this new sample of $N = 100$ shops. To differentiate, we can call the first sample mean $\bar{x}_1$ and this new sample mean $\bar{x}_2$. Will $\bar{x}_1$ and $\bar{x}_2$ be the exact same value? Probably not! Since our samples are independent and random from the binomial distribution (Figure 15.6), it is almost certain that the two sample means will be at least a bit different. We can therefore ask about the *distribution* of sample means. That is, what if we kept going back out to get more samples of 100, calculating additional sample means $\bar{x}_3$, $\bar{x}_4$, $\bar{x}_5$, and so forth? What would this distribution look like? It turns out, it would be a normal distribution!

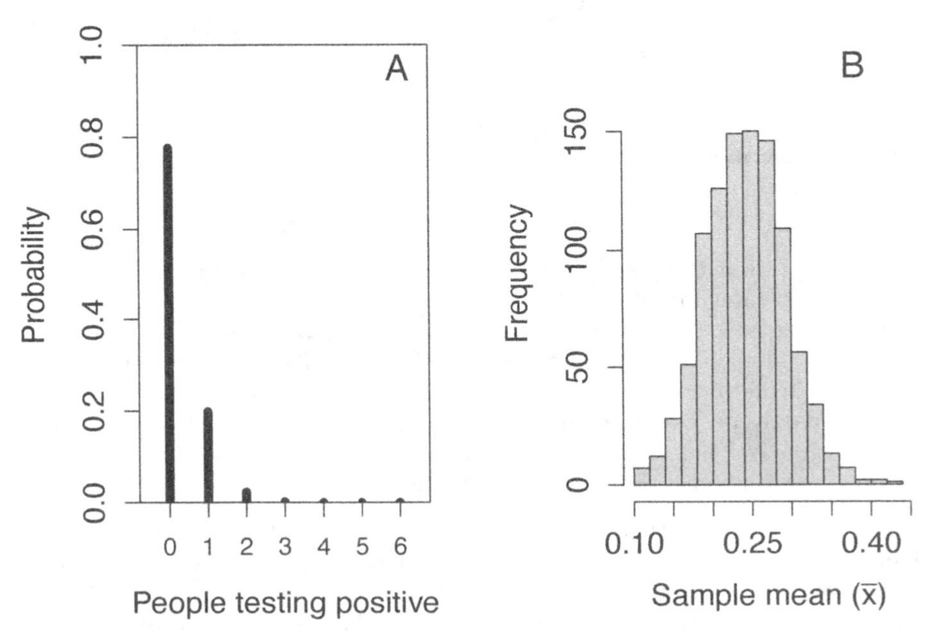

FIGURE 16.1 Simulated demonstration of the central limit theorem. (A) Recreation of Figure 15.6 showing the probability distribution for the number of people who have COVID-19 in a shop of 6 when the probability of testing positive is 0.025. (B) The distribution of 1000 means sampled from panel (A), where the sample size is 100.

To demonstrate the CLT in action, Figure 16.1 shows the two distributions side by side. The first (Figure 16.1A) shows the original distribution from Figure 15.6, from which samples are collected and sample means are calculated. The second (Figure 16.1B) shows the distribution of 1000 sample means (i.e., $\bar{x}_1, \bar{x}_2, ..., \bar{x}_{999}, \bar{x}_{1000}$). Each mean $\bar{x}_i$ is calculated from a sample of $N = 100$ from the distribution in Figure 16.1A. Sampling is simulated using a random number generator on a computer (Chapter 17 shows an example of how to do this in jamovi).

The distribution of sample means shown in Figure 16.1B is not perfectly normal. We can try again with an even bigger sample size of $N = 1000$, this time with a Poisson distribution where $\lambda = 1$ in Figure 15.7. Figure 16.2 shows this result, with the original Poisson distribution shown in Figure 16.2A, and the corresponding distribution built from 1000 sample means shown in Figure 16.2B.

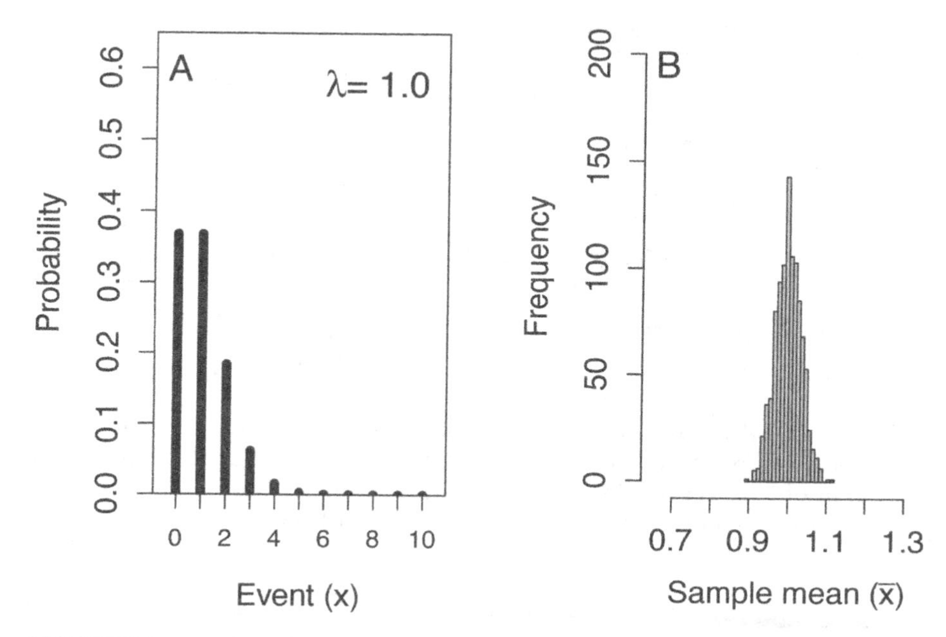

FIGURE 16.2 Simulated demonstration of the central limit theorem. (A) Recreation of Figure 15.7 showing the probability distribution for the number of events occurring in a Poisson distribution with a rate parameter of 1. (B) The distribution of 1000 means sampled from panel (A), where the sample size is 1000.

We can try the same approach with the continuous uniform distribution shown in Figure 15.8. This time, we will use an even larger sample size of $N = 10000$ to get our 1000 sample means. The simulated result is shown in Figure 16.3.

In all cases, regardless of the original sampling distribution (binomial, Poisson, or uniform), the distribution of sample *means* has the shape of a normal distribution. This normal distribution of sample means has important implications for statistical hypothesis testing. The CLT allows us to make inferences about the means of non-normally distributed distributions (Sokal & Rohlf, 1995), to create confidence intervals around sample means, and to apply statistical hypothesis tests that would otherwise not be possible. We will look at these statistical tools in future chapters.

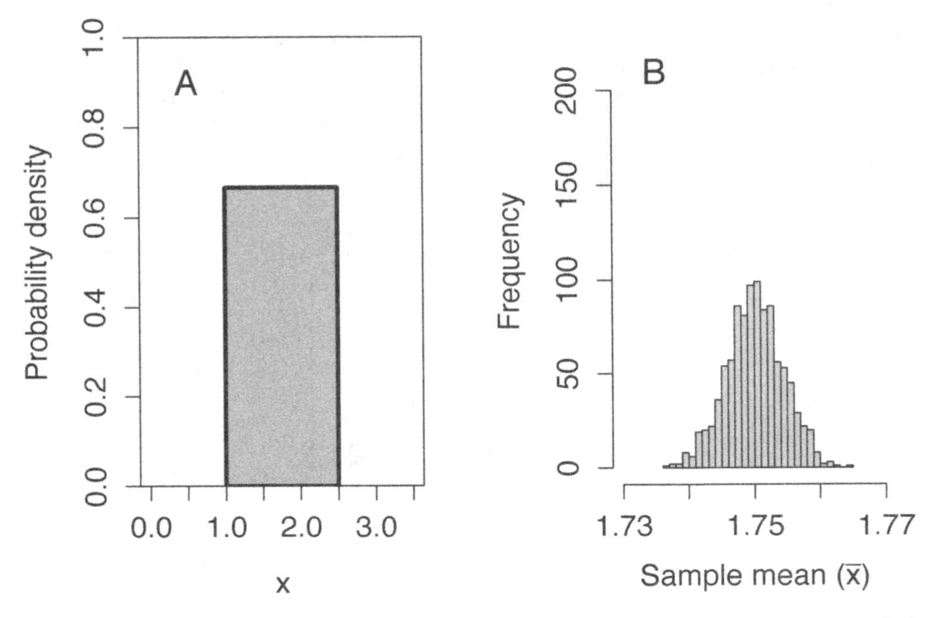

FIGURE 16.3 Simulated demonstration of the central limit theorem. (A) Recreation of Figure 15.8 showing a continuous uniform distribution with a minimum of 1 and a maximum of 2.5. (B) The distribution of 1000 means sampled from panel (A), where the sample size is 10000.

16.2 Probability and z-scores

We can calculate the probability of sampling some range of values from the normal distribution if we know the distribution's mean (μ) and standard deviation (σ). For example, because the normal distribution is symmetric around the mean, the probability of sampling a value greater than the mean will be 0.5 (i.e., $P(x > \mu) = 0.5$), and so will the probability of sampling a value less than the mean (i.e., $P(x < \mu) = 0.5$). Similarly, about 68.2% of the normal distribution's probability density lies within 1 standard deviation of the mean (shaded region of Figure 16.4), which means that the probability of randomly sampling a value x that is greater than $\mu - \sigma$ but less than $\mu + \sigma$ is $P(\mu - \sigma < x < \mu + \sigma) = 0.682$.

Remember that total probability always needs to equal 1. This remains true whether it is the binomial distribution that we saw with the coin flipping example in Chapter 15, or any other distribution. Consequently, the area under the curve of the normal distribution (i.e., under the curved line of Figure 16.4) must equal 1. When we say that the probability of sampling a value within 1

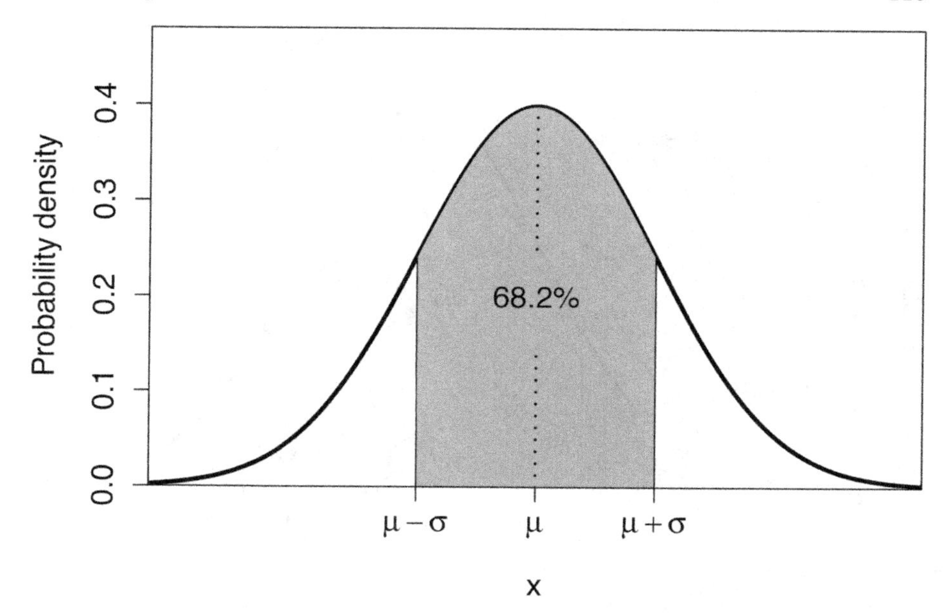

FIGURE 16.4 Normal distribution in which the shaded region shows the area within one standard deviation of the mean (dotted line), that is, the shaded region starts on the left at the mean minus one standard deviation, then ends at the right at the mean plus one standard deviation. This shaded area encompasses 68.2% of the total area under the curve.

standard deviation of the mean is 0.682, this also means that the *area* of this region under the curve equals 0.682 (i.e., the shaded area in Figure 16.4). And, again, because the whole area under the curve sums to 1, that must mean that the unshaded area of Figure 16.4 (where $x < \mu - \sigma$ or $x > \mu + \sigma$) has an area equal to $1 - 0.682 = 0.318$. That is, the probability of randomly sampling a value x in this region is $P(x < \mu - \sigma \cup x > \mu + \sigma) = 0.318$, or 31.8% (note that the $\cup$, is just a fancy way of saying 'or', in this case; technically, the *union* of two sets).

We can calculate other percentages using standard deviations too (Sokal & Rohlf, 1995). For example, about 95.4% of the probability density in a normal distribution lies between 2 standard deviations of the mean, i.e., $P(\mu - 2\sigma < x < \mu + 2\sigma) = 0.954$. And about 99.6% of the probability density in a normal distribution lies between 3 standard deviations of the mean, i.e., $P(\mu - 3\sigma < x < \mu + 3\sigma) = 0.996$. We could go on mapping percentages to standard deviations like this; for example, about 93.3% of the probability density in a normal distribution is less than $\mu + 1.5\sigma$ (i.e., less than 1.5 standard deviations greater than the mean; see Figure 16.5).

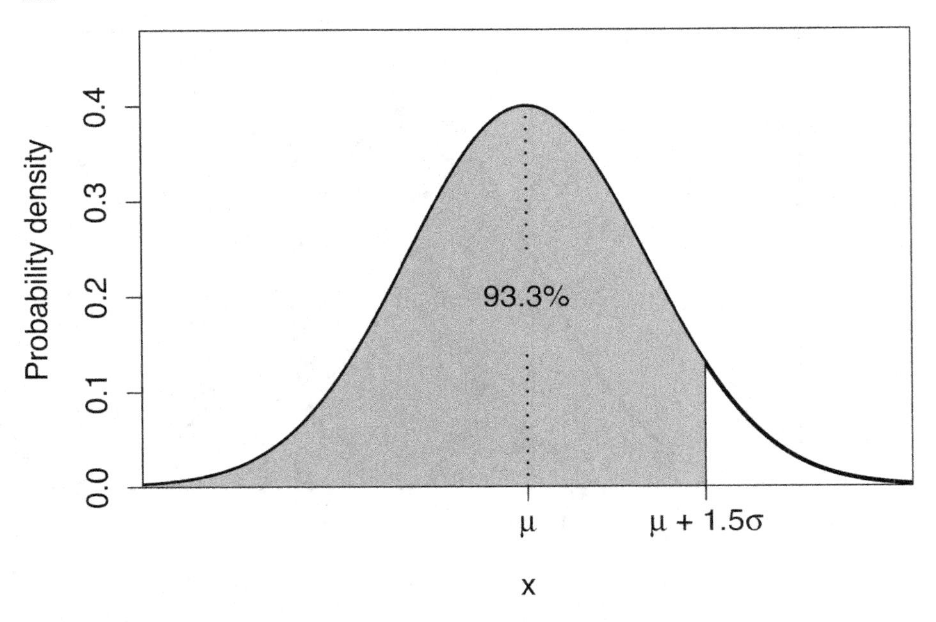

FIGURE 16.5 Normal distribution in which the shaded region shows the area under 1.5 standard deviations of the mean (dotted line). This shaded area encompasses about 93.3% of the total area under the curve.

Notice that there are no numbers on the x-axes of Figure 16.4 or 16.5. This is deliberate; the relationship between standard deviations and probability density applies regardless of the scale. We could have a mean of $\mu = 100$ and standard deviation of $\sigma = 4$, or $\mu = -12$ and $\sigma = 0.34$. It does not matter. Nevertheless, it would be very useful if we could work with some standard values of x when working out probabilities. This is where the standard normal distribution, first introduced in Chapter 15, becomes relevant. Recall that the standard normal distribution has a mean of $\mu = 0$ and a standard deviation (and variance) of $\sigma = 1$. With these standard values of μ and σ, we can start actually putting numbers on the x-axis and relating them to probabilities. We call these numbers **standard normal deviates**, or **z-scores** (Figure 16.6).

What z-scores allow us to do is map probabilities to deviations from the mean of a standard normal distribution (hence 'standard normal deviates'). We can say, e.g., that about 95% of the probability density lies between $z = -1.96$ and $z = 1.96$, or that about 99% lies between $z = -2.58$ and $z = 2.58$ (this will become relevant later). An interactive application[2] can help show how probability density changes with changing z-score.

[2] https://bradduthie.github.io/stats/app/zandp/

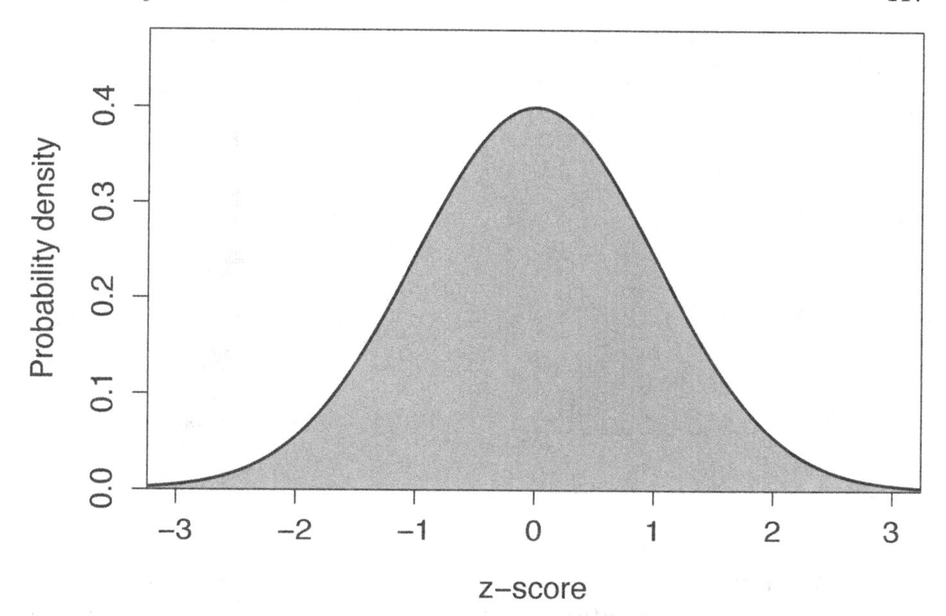

FIGURE 16.6 Standard normal probability distribution with z-scores shown on the x-axis.

Of course, most variables that we measure in the biological and environmental sciences will not fit the standard normal distribution. Almost all variables will have a different mean and standard deviation, even if they are normally distributed. Nevertheless, we can translate any normally distributed variable into a standard normal distribution by subtracting its mean and dividing by its standard deviation. We can see what this looks like visually in Figure 16.7.

In Figure 16.7A, we see the standard normal distribution curve represented by the dotted line at $\mu = 0$ and with a standard deviation of $\sigma = 1$. To the right of this normal distribution we have 10000 values randomly sampled from a normal distribution with a mean of 5 and a standard deviation of 2 (note that the histogram peaks around 5 and is wider than the standard normal distribution because the standard deviation is higher). After subtracting 5 from all of the values in the histogram of Figure 16.7A, then dividing by 2, the data fit nicely within the standard normal curve, as shown in Figure 16.7B. By doing this transformation on the original dataset, z-scores can now be used with the data. Mathematically, here is how the calculation is made,

$$z = \frac{x - \mu}{\sigma}.$$

For example, if we had a value of $x = 9.1$ in our simulated dataset, in which $\mu = 5$ and $\sigma = 2$, then we would calculate $z = (9.1 - 5)/2 = 2.05$. Since we

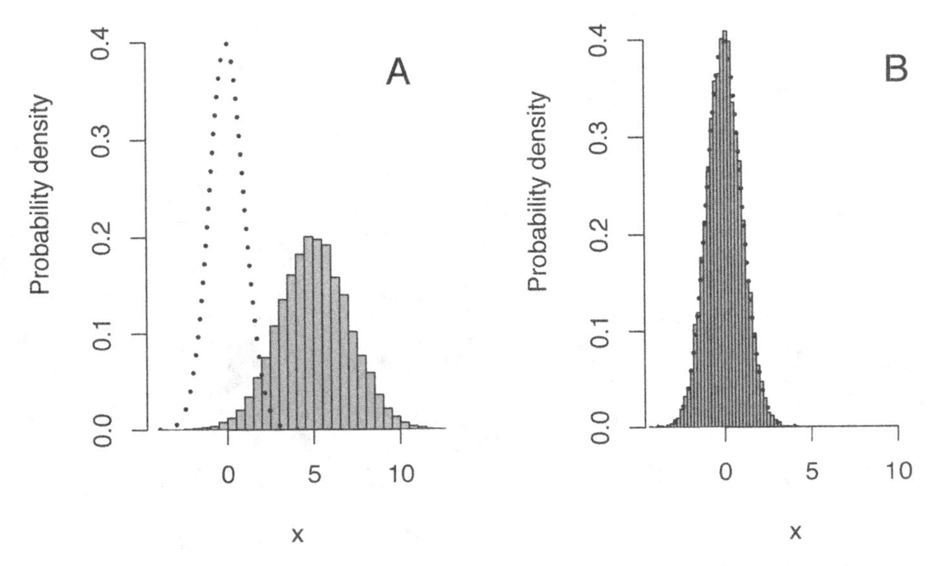

FIGURE 16.7 Visual representation of what happens when we subtract the sample mean from a dataset, then divide by its standard deviation. (A) Histogram (grey bars) show 10000 normally distributed values with a mean of 5 and a standard deviation of 2; the curved dotted line shows the standard normal distribution with a mean of 0 and standard deviation of 1. (B) Histogram after subtracting 5, then dividing by 2, from all values shown in panel (A).

almost never know what the true population mean (μ) and standard deviation (σ) are, we usually need to use the estimates made from our sample,

$$z = \frac{x - \bar{x}}{s}.$$

We could then use jamovi (The jamovi project, 2024), an interactive application, or an old-fashioned z-table[3] to find that only about 2% of values are expected to be higher than $x = 9.1$ in our original dataset. These z-scores will become especially useful for calculating confidence intervals in Chapter 18. They can also be useful for comparing values from variables or statistics measured on different scales (Adams & Collyer, 2016; Cheadle et al., 2003; Sokal & Rohlf, 1995).

[3]Before the widespread availability of computers, which can easily be used to calculate probability densities on a normal distribution, the way to map z-scores to probabilities was using a z-table (`https://www.z-table.com/`). The table would have rows and columns mapping to different z-values, which could be used to find the appropriate probability densities. Such tables would be used for many different distributions, not just the normal distribution. The textbook by Sokal & Rohlf (1995) comes with a nearly 200-page supplemental book that is just statistical tables. These tables are more or less obsolete nowadays, but some people still use them.

17

Practical. *Probability and simulation*

This practical focuses on applying the concepts from Chapter 15 and Chapter 16 in jamovi. There will be three exercises:

1. Calculating probabilities from a dataset
2. Calculating probabilities from a normal distribution
3. Demonstrating the central limit theorem (CLT)

To complete exercises 2 and 3, you will need to download and install two new jamovi modules. Jamovi modules are add-ons that make it possible to run specialised statistical tools inside jamovi. These tools are written by a community of statisticians, scientists, and educators and listed in the jamovi library[1]. Like jamovi, these tools are open source and free to use.

The dataset for this chapter is a bit different. It comes from the Beacon Project (`https://www.thebeaconproject.net/`), which is an interdisciplinary scientific research programme led by Dr Isabel Jones at the University of Stirling. This project focuses on large hydropower dams as a way to understand the trade-offs between different United Nations Sustainable Development Goals. It addresses challenging questions about environmental justice, biodiversity, and sustainable development. The project works with people affected, and sometimes displaced, by dam construction in Brazil, Kazakhstan, India, USA, and UK. Part of this project involves the use of mobile games to investigate how people make decisions about sustainable development.

The game 'Power Up!' is freely available as an Android[2] and iPhone[3] app. Data are collected from players' decisions and used to investigate social-ecological questions. We will use the 'Power Up!' dataset[4] in exercises 1 and 2. To get started, first download this dataset and open it in jamovi. Note that these data are already in a tidy format, so we do not need to do any reorganising. The dataset includes columns for each player's ID, the OS that they use, the dam size that they decided to build in the game, their in-game investment in Biodiversity, Community, and Energy, and their final Score.

[1] `https://www.jamovi.org/library.html`

[2] `https://play.google.com/store/apps/details?id=com.hyperluminal.stirlinguniversity.sustainabledevelopmentgame`

[3] `https://apps.apple.com/gb/app/power-up/id1585634888`

[4] `https://bradduthie.github.io/stats/data/power_up.csv`

17.1 Probabilities from a dataset

Suppose that we want to estimate the probability that a new Power Up! game player will be an Android user. To estimate this probability, we can use the proportion of players in the dataset who are Android users. To get this proportion, we need to divide the number of Android users by the total number of players,

$$P(Android) = \frac{\text{Number of Android users}}{\text{Number of players}}.$$

In jamovi, you could figure this out the long way by counting up the number of rows with 'Android' in the second column, then dividing by the total number of rows. But there is an easier way, which is faster and less prone to human error than manually tallying up items. To do this, go to the 'Analyses' tab in jamovi and navigate to 'Exploration', then 'Descriptives'. Place the 'OS' variable into the 'Variables' box. Next, find the check box called 'Frequency tables' just under the 'Split by' box and above the 'Statistics' drop-down tab. Check this box to get a table of frequencies for Android versus iPhone users.

FIGURE 17.1 Jamovi Descriptives toolbar showing the OS column from the Power Up! dataset selected. The 'Frequency tables' checkbox builds a table of counts and percentages.

The table of frequencies shown in Figure 17.1 includes counts of Android versus iPhone users. We can see that 56 of the 74 total game players use Android, while 18 players use iPhone. To get the proportion of Android users, we could divide 56 by 74 to get 0.7567568. Similarly, for the proportion of iPhone users, we could calculate 18 / 74 = 0.2432432. But jamovi already does this for us, with a bit of rounding. The second column of the Frequencies table gives us these proportions, but expressed as a percentage. The percentage of Android users is 75.7%, and the percentage of iPhone users is 24.3%. Percentages are out of a total of 100, so to get back to the proportions, we can just divide by 100%, 75.7% / 100% = 0.757 for Android and 24.3% / 100% = 0.243 for iPhone. To answer the original question, our best estimate of the probability that a new Power Up! game player will be an Android user is therefore 0.757.

Next, use the same procedure to find the probability that a game player will make a small-, medium-, and large-size dam. Now, fill in Table 17.1 with counts, percentage, and the estimated probability of a player selecting a small, medium, or large dam.

TABLE 17.1 Statistics of Power Up! decisions for dam size.

Dam Size	Counts	Percentage	Estimated Probability
Small			
Medium			
Large			

We can use these estimated probabilities of small, medium, and large dam size selection to predict what will happen in future games. Suppose that a new player decides to play the game. What is the probability that this player chooses a small **or** a large dam?

$P(small \text{ or } large) =$ ______________________________________

Now suppose that 3 new players arrive and decide to play the game. What is the probability that all 3 of these new players choose a large dam?

$P(3\, large) =$ ______________________________________

What is the probability that the first player chooses a small dam, the second player chooses a medium dam, and the third player chooses a large dam?

$P(Player\ 1 = small, Player\ 2 = medium, Player\ 3 = large) =$ __________

Now consider a slightly different type of question. Instead of trying to predict the probability of new player decisions, we will focus on sampling from the existing dataset. Imagine that you randomly choose one of the 74 players with equal probability (i.e., every player is equally likely to be chosen). What is the probability that you choose player 20?

$P(Player\ 20) =$ ______________________________________

What is the probability that you choose player 20, *then* choose a different player with a large dam? As a hint, remember that you are now sampling *without replacement.* The second choice cannot be player 20 again, so the probability of choosing a player with a large dam has changed from the estimated probability in Table 17.1.

$P(Player\ 20,\ Large) =$ ____________________________________

Now we can use the Descriptives tool in jamovi to ask a slightly different question with the data. Suppose that we wanted to estimate the probability that an Android user will choose a large dam. We could multiply the proportion of Android users times the proportion of players who choose a large dam (i.e., find the probability of Android *and* large dam). But this assumes that the two characteristics are independent (i.e., that Android users are not more or less likely than iPhone users to build large dams). To estimate the probability that a player chooses a large dam *given* that they are using Android, we can keep Dam_size in the Variables box, but now put OS in the 'Split by' box. Figure 17.2 shows the output of jamovi. A new frequency table breaks down dam choice for each OS.

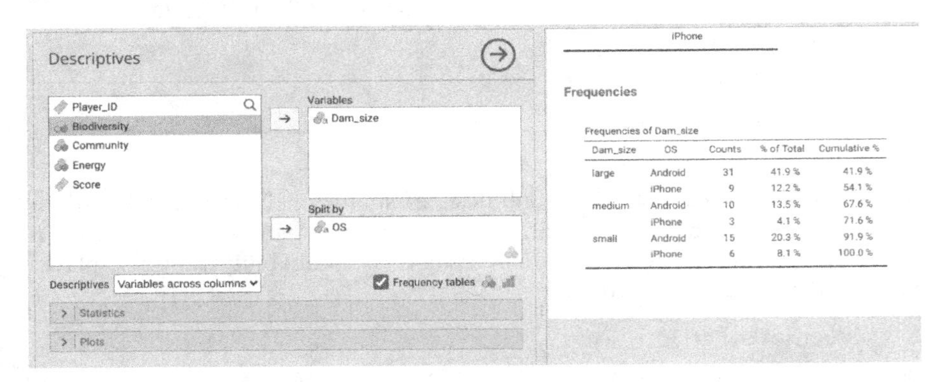

FIGURE 17.2 Jamovi Descriptives toolbar showing the dam size column from the Power Up! dataset selected as a variable split by OS. The 'Frequency tables' checkbox builds a table of counts for small, medium, and large dam size broken down by Android versus iPhone OS.

To get the proportion of Android users who choose to build a large dam, we just need to divide the number of Android users who chose the large dam size by the total number of Android users (i.e., sum of the first column in the Frequencies table; Figure 17.2). Note that the vertical bar, $|$, in the equation below just means 'given' (or, rather, 'conditional up', so the number of players that chose a large dam *given* that they are Android users),

$$P(Large|Android) = \frac{\text{Number of Android users choosing large dam}}{\text{Number of Android users}}.$$

Now, recreate the table in Figure 17.2 and estimate the probability that an Android user will choose to build a large dam,

$P(Large|Android) =$ ________________________________

Is $P(Large|Android)$ much different from the probability that *any* player chooses a large dam, as calculated in Table 17.1? Do you think that the difference is significant?

Next, we will move on to calculating probabilities from a normal distribution.

17.2 Probabilities from a normal distribution

In the example of the first exercise, we looked at OS and dam size choice. Players only use Android or iPhone, and they could only choose one of three sizes of dam. For these nominal variables, estimating the probability of a particular discrete outcome (e.g., Android versus iPhone) was just a matter of dividing counts. But we cannot use the same approach for calculating probabilities from continuous data. Consider, for example, the final score for each player in the column 'Score'. Because of how the game was designed, Score can potentially be any real number, although most scores are somewhere around 100. We can use a histogram to see the distribution of player scores (Figure 17.3).

In this case, it does not really make sense to ask what the probability is of a particular score. If the score can take *any* real value, out to as many decimals as we want, then what is the probability of a score being *exactly* 94.97 (i.e., 94.97 with infinite zeros after it, $94.9700000\bar{0}$)? The probability is infinitesimal, i.e., basically zero, because there are an infinite number of real numbers. Consequently, we are not really interested in the probabilities of specific values of continuous data. Instead, we want to focus on intervals. For example, what is the probability that a player scores higher than 120? What is the probability that a player scores lower than 100? What is the probability that a player scores between 100 and 120?

Take another look at Figure 17.3, then take a guess at each of these probabilities. As a hint, the y-axis of this histogram is showing density instead of frequency. What this means is that the total grey area (i.e., the histogram bars) sums to 1. Guessing the probability that a player scores higher than 120 is the same as

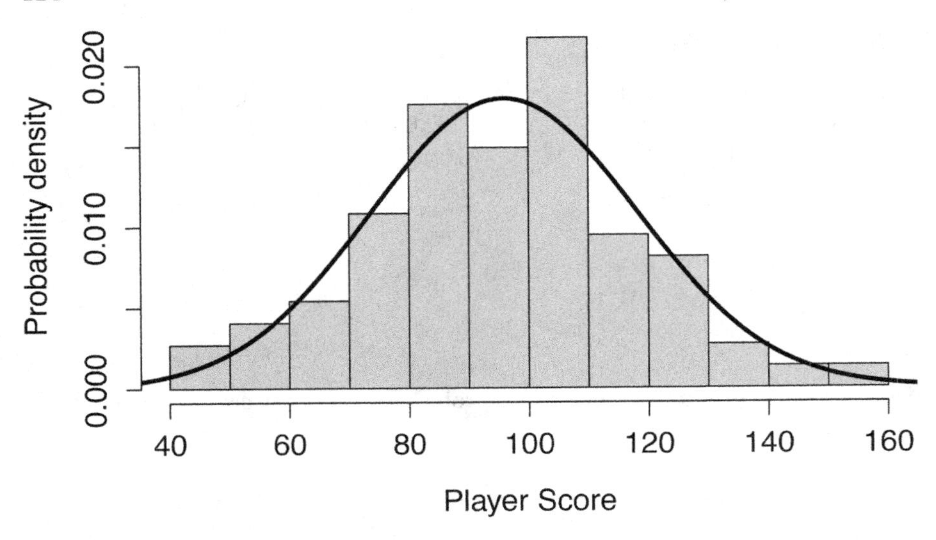

FIGURE 17.3 Distribution of player scores in the game Power Up! shown in histogram bars. The overlaid curve shows the probability density function for a normal distribution that has the same mean and standard deviation as the sample described by the histogram.

guessing the proportion of grey space in the highest four bars of Figure 17.3 (i.e., grey space >120).

$P(Score > 120) =$ ____________________________

$P(Score < 100) =$ ____________________________

$P(100 < Score < 120) =$ ____________________________

Trying to do this by looking at a histogram is not easy, and it is really not the best way to get the above probabilities. We can get much better estimates using jamovi, but we need to make an assumption about the distribution of Player Score. Specifically, we need to assume that the distribution of Player Score has a specific shape. More technically, we must assume a specific probability density function that we can use to mathematically calculate probabilities of different ranges of player scores. Inspecting Figure 17.3, Player Score appears to be normally distributed. In other words, the shape of Player Score distribution appears to be normal, or 'Gaussian'. If we are willing to assume this, then we can calculate probabilities using its mean and standard deviation. Use jamovi to find the mean and the standard deviation of Player Score (note, we can just say that score is unitless, so no need to include units).

Mean score: ____________________________

Standard deviation score: ____________________________

We will assume that the *sample* of scores shown in Figure 17.3 came from a *population* that is normally distributed with the mean and standard deviation that you wrote above (recall sample versus population from Chapter 4). We can overlay this distribution on the histogram above using a curved line (Figure 17.3).

We can interpret the area under the curve in the same way that we interpret the area in the grey bars. As mentioned earlier, the total area of the histogram bars must sum to 1. The total area under the curve must also sum to 1. Both represent the probability of different ranges of player scores. Notice that the normal distribution is not a perfect match for the histogram bars. For example, the middle bar of values illustrating scores between 90 and 100 appears to be a bit low compared to a perfect normal distribution, and there are more scores between 40 and 50 than we might expect. Nevertheless, the two distributions broadly overlap, so we might be willing to assume that the player scores represented in the histogram bars are sampled from the population described by the curve.

Because the curve relating player score to probability density is described by an equation (see Chapter 15), we can use that equation to make inferences about the probabilities of different ranges of scores. The simplest example is the mean of the distribution. Because the normal distribution is symmetric, the area to the left of the mean must be the same as the area to the right of the mean. And since the whole area under the curve must sum to 1, we can conclude that the probability of sampling a player score that is less than the mean is 1/2, and the probability of sampling a player score greater than the mean is also 1/2. Traditionally, we would need to do some maths to get other player score probabilities, but jamovi can do this much more easily.

To get jamovi to calculate probabilities from a normal distribution, we need to go to the Modules option and download a new module. Click on the 'Modules' button, and select the first option called 'jamovi library' from the pull-down menu. From the 'Available' tab, scroll down until you find the Module called 'distrACTION - Quantiles and Probabilities of Continuous and Discrete Distributions' (Rihs & Mayer, 2018). Click the 'Install' button to install it into jamovi. A new button in the toolbar called 'distrACTION' should become visible (Figure 17.4).

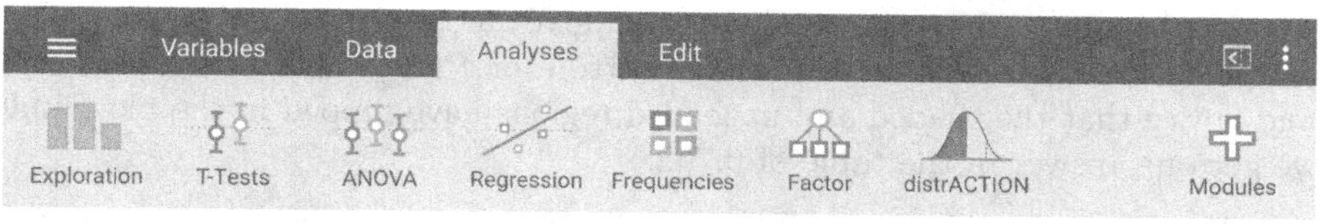

FIGURE 17.4 Jamovi tool bar, which includes an added module called distrACTION.

If the module is not there after installation, then it should be possible to find by again going to Modules and selecting distrACTION from the pull-down menu. Click on the module and choose 'Normal Distribution' from the pull-down menu. Next, we can see a box for the mean and standard deviation (SD) under the 'Parameters' subtitle in bold. Put the mean and the standard deviation calculated from above into these boxes. In the panel on the right, jamovi will produce the same normal distribution that is in Figure 17.3 (note that the axes might be scaled a bit differently).

Given this normal distribution, we can compute the probability that a player scores less than x1 = 80 by checking the box 'Compute probability', which is located just under 'Function' (Figure 17.5). We can then select the first radio button to find the probability that a randomly sampled value X from this distribution is less than x1, $P(X \leq x1)$. Notice in the panel on the right that the probability is given as $P = 0.238$. This is also represented in the plot of the normal distribution, with the same proportion in the lower part of the distribution shaded ($P = 0.238$, i.e., about 23.8%).

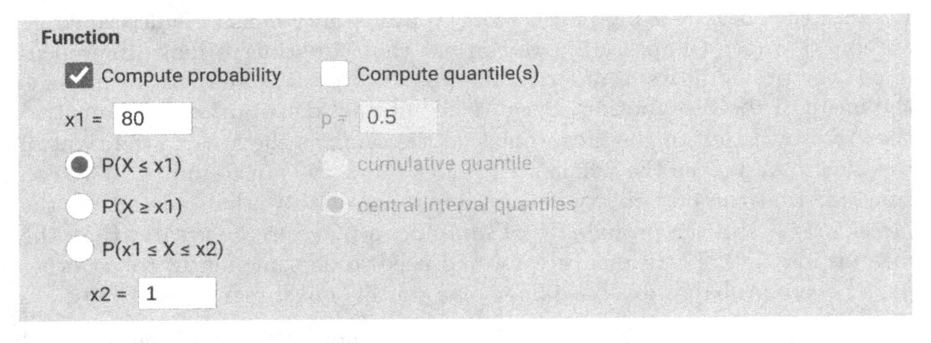

FIGURE 17.5 Jamovi options for the distrACTION module for computing probability for a given normal distribution. The example shown here calculates the probability that a value sampled from the normal distribution of interest is less than 80.

To find the probability that a value is greater than 80, we could subtract our answer of 0.238 from 1, $1 - 0.238 = 0.762$ (remember that the total area under the normal curve equals 1, so the shaded plus the unshaded region must also equal 1; hence, 1 minus the shaded region gives us the unshaded region). We could also just select the second radio button for $P(X \geq x1)$. Give this a try, and notice that the shaded and unshaded regions have flipped in the plot, and we get our answer in the table of 0.762.

Finally, to compute the probability of an interval, we can check the third radio button and set x2 in the bottom box (Figure 17.5). For example, to see the probability of a score between 80 and 120, we can choose select $P(x1 \leq X \leq x2)$, then set $x2 = 120$ in the bottom box. Notice where the shaded area is in the newly drawn plot. What is the probability of a player getting a score between 80 and 120?

$P(80 \leq X \leq 120) =$ ____________________

What is the probability of a player getting a score greater than 130?

$P(X \geq 130) =$ ____________________

Now try the following probabilities for different scores.

$P(X \geq 120) =$ ____________________

$P(X \leq 100) =$ ____________________

$P(100 \leq X \leq 120) =$ ____________________

Note, these last three were the same intervals that you guessed using the histogram. How close was your original guess to the calculations above?

One last question. What is the probability of a player getting a score lower than 70 or higher than 130?

$P(X \leq 70 \cup X \geq 130) =$ ____________________

There is more than one way to figure this last one out. How did you do it, and what was your reasoning?

We will now move on to the central limit theorem.

17.3 Central limit theorem

To demonstrate the central limit theorem, we need to download and install another module in jamovi. This time, go to 'Modules', and from the 'Available' tab, scroll down until you find 'Rj' in the jamovi library. Install 'Rj', then a new button 'R' should become available in the toolbar. This will allow us to run a bit of script using the coding language R. Click on the new 'R' button in the toolbar and select 'Rj Editor' from the pull-down menu. You will see an open editor; this is where the code will go. If it has some code in it already (e.g., `# summary(data[1:3])`), just delete it so that we can start with a clean slate. Copy and paste the following lines into the Rj Editor.

```
v1  <- runif(n = 200, min = 0, max = 100);
v2  <- runif(n = 200, min = 0, max = 100);
v3  <- runif(n = 200, min = 0, max = 100);
v4  <- runif(n = 200, min = 0, max = 100);
v5  <- runif(n = 200, min = 0, max = 100);
v6  <- runif(n = 200, min = 0, max = 100);
v7  <- runif(n = 200, min = 0, max = 100);
v8  <- runif(n = 200, min = 0, max = 100);
v9  <- runif(n = 200, min = 0, max = 100);
v10 <- runif(n = 200, min = 0, max = 100);
v11 <- runif(n = 200, min = 0, max = 100);
v12 <- runif(n = 200, min = 0, max = 100);
v13 <- runif(n = 200, min = 0, max = 100);
v14 <- runif(n = 200, min = 0, max = 100);
v15 <- runif(n = 200, min = 0, max = 100);
v16 <- runif(n = 200, min = 0, max = 100);
v17 <- runif(n = 200, min = 0, max = 100);
v18 <- runif(n = 200, min = 0, max = 100);
v19 <- runif(n = 200, min = 0, max = 100);
v20 <- runif(n = 200, min = 0, max = 100);
v21 <- runif(n = 200, min = 0, max = 100);
v22 <- runif(n = 200, min = 0, max = 100);
v23 <- runif(n = 200, min = 0, max = 100);
v24 <- runif(n = 200, min = 0, max = 100);
v25 <- runif(n = 200, min = 0, max = 100);
v26 <- runif(n = 200, min = 0, max = 100);
v27 <- runif(n = 200, min = 0, max = 100);
v28 <- runif(n = 200, min = 0, max = 100);
v29 <- runif(n = 200, min = 0, max = 100);
v30 <- runif(n = 200, min = 0, max = 100);
v31 <- runif(n = 200, min = 0, max = 100);
v32 <- runif(n = 200, min = 0, max = 100);
v33 <- runif(n = 200, min = 0, max = 100);
v34 <- runif(n = 200, min = 0, max = 100);
v35 <- runif(n = 200, min = 0, max = 100);
v36 <- runif(n = 200, min = 0, max = 100);
v37 <- runif(n = 200, min = 0, max = 100);
```

```
v38 <- runif(n = 200, min = 0, max = 100);
v39 <- runif(n = 200, min = 0, max = 100);
v40 <- runif(n = 200, min = 0, max = 100);

hist(x = v1, main = "", xlab = "Random uniform variable");
```

What this code is doing is creating 40 different datasets of 200 random numbers from 0 to 100 (there is a way to do all of this in much fewer lines of code, but it requires a bit more advanced use of R). The `hist` function plots a histogram of the first variable. To run the code, find the green triangle in the upper right (Figure 17.6).

Rj Editor

```
1  v1  <- runif(n = 200, min = 0, max = 100);
2  v2  <- runif(n = 200, min = 0, max = 100);
3  v3  <- runif(n = 200, min = 0, max = 100);
4  v4  <- runif(n = 200, min = 0, max = 100);
5  v5  <- runif(n = 200, min = 0, max = 100);
6  v6  <- runif(n = 200, min = 0, max = 100);
7  v7  <- runif(n = 200, min = 0, max = 100);
8  v8  <- runif(n = 200, min = 0, max = 100);
```

FIGURE 17.6 Jamovi interface for the Rj Editor module. Code can be run by clicking on the green triangle in the upper right.

When you run the code, the 40 new variables will be created, each variable being made up of 200 random numbers. The histogram for `v1` is plotted to the right (to plot other variables, substitute `v1` in the `hist` function for some other variable). How would you describe the shape of the distribution of `v1`?

Next, we are going to get the mean value of each of the 40 variables. To do this, copy the code below and paste it at the bottom of the Rj Editor (somewhere below the `hist` function).

```
m1  <- mean(v1);
m2  <- mean(v2);
m3  <- mean(v3);
```

```
m4  <- mean(v4);
m5  <- mean(v5);
m6  <- mean(v6);
m7  <- mean(v7);
m8  <- mean(v8);
m9  <- mean(v9);
m10 <- mean(v10);
m11 <- mean(v11);
m12 <- mean(v12);
m13 <- mean(v13);
m14 <- mean(v14);
m15 <- mean(v15);
m16 <- mean(v16);
m17 <- mean(v17);
m18 <- mean(v18);
m19 <- mean(v19);
m20 <- mean(v20);
m21 <- mean(v21);
m22 <- mean(v22);
m23 <- mean(v23);
m24 <- mean(v24);
m25 <- mean(v25);
m26 <- mean(v26);
m27 <- mean(v27);
m28 <- mean(v28);
m29 <- mean(v29);
m30 <- mean(v30);
m31 <- mean(v31);
m32 <- mean(v32);
m33 <- mean(v33);
m34 <- mean(v34);
m35 <- mean(v35);
m36 <- mean(v36);
m37 <- mean(v37);
m38 <- mean(v38);
m39 <- mean(v39);
m40 <- mean(v40);

all_means <- c(m1,  m2,  m3,  m4,  m5,  m6,  m7,  m8,  m9,  m10,
               m11, m12, m13, m14, m15, m16, m17, m18, m19, m20,
               m21, m22, m23, m24, m25, m26, m27, m28, m29, m30,
               m31, m32, m33, m34, m35, m36, m37, m38, m39, m40);
```

Now we have calculated the mean for each variable. The last line of code defines `all_means`, which makes a new dataset that includes the mean value of each of our original variables. Think about what the distribution of these mean values will look like. Sketch what you predict the shape of its distribution will be below.

Now, add one more line of code to the very bottom of the Rj Editor.

```
hist(x = all_means, main = "", xlab = "All variable means");
```

This last line will make a histogram of the means of all 40 variables. Click the green button again to run the code. Compare the distribution of the original `v1` to the means of variables 1–40, and to your prediction above. Is this what you expected? As best you can, explain why the shapes of the two distributions differ.

We did all of this the long way to make it easier to see and think about the relationship between the original, uniformly distributed, variables and the distribution of their means. Now, we can repeat this more quickly using one more jamovi module. Go to 'Modules', and from the 'Available' tab, download the 'clt - Demonstrations' module from the jamovi library. Once it is downloaded, go to the 'Demonstrations' button in the jamovi toolbar and select 'Central Limit Theorem' from the pull-down menu.

To replicate what we did in the Rj Editor above, we just need to set the 'Source distribution' to 'uniform' using the pull-down menu, set the sample size to 200, and set the number of trials to 40 (Figure 17.7). Try doing this, then look at the histogram generated to the lower right. It should look similar, but not identical, to the histogram produced with the R code. Now try increasing the number of trials to 200. What happens to the histogram? What about when you increase the number of trials to 2000?

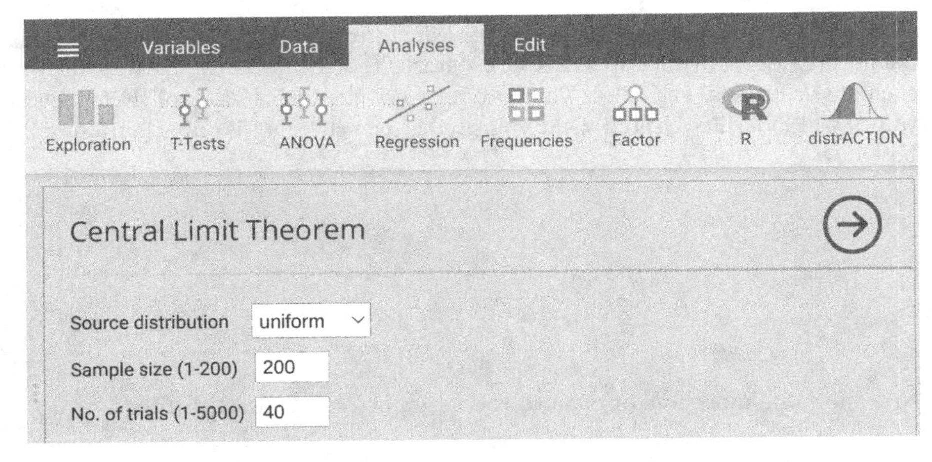

FIGURE 17.7 Jamovi interface for the 'Demonstrations' module, which allows users to randomly generate data from a specific source distribution (normal, uniform, geometric, lognormal, and binary), sample size, and number of trials (i.e., variables)

Try playing around with different source distributions, sample sizes, and numbers of trials. What general conclusion can you make about the distribution of sample means from the different distributions?

18

Confidence intervals

In Chapter 16, we saw how it is possible to calculate the probability of sampling values from a specific interval of the normal distribution (e.g., the probability of sampling a value within 1 standard deviation of the mean). In this chapter, we will see how to apply this knowledge to calculating intervals that express confidence in the mean value of a population.

Remember that we almost never really know the true mean value of a *population*, μ. Our best estimate of μ is the mean that we have calculated from a *sample*, $\bar{x}$ (see Chapter 4 for a review of the difference between populations and samples). But how good of an estimate is $\bar{x}$ of μ, really? Since we cannot know μ, one way of answering this question is to find an interval that expresses a degree of confidence about the value of μ. The idea is to calculate two numbers that we can say with some degree of confidence that μ is between (i.e., a lower confidence interval and an upper confidence interval). The wider this interval is, the more confident that we can be that the true mean μ is somewhere within it. The narrower the interval is, the less confident we can be that our confidence intervals (CIs) contain μ.

Confidence intervals are notoriously easy to misunderstand. I will explain this verbally first, focusing on the general ideas rather than the technical details. Then I will present the calculations before coming back to their interpretation again. The idea follows a similar logic to the standard error from Chapter 12.

Suppose that we want to know the mean body mass of all domestic cats (Figure 18.1). We cannot weigh every living cat in the world, but maybe we can find enough to get a sample of 20. From these 20 cats, we want to find some interval of masses (e.g., 3.9–4.3 kg) within which the *true* mean mass of the population is contained. The only way to be 100% certain that our proposed interval *definitely* contains the true mean would be to make the interval absurdly large. Instead, we might more sensibly ask what the interval would need to be to contain the mean with 95% confidence. What does 'with 95% confidence' actually mean? It means that when we do the calculation to get the interval, the true mean should be somewhere within the interval 95% of the time that a sample is collected.

In other words, if we were to go back out and collect another sample of 20 cats, and then another, and another (and so forth), calculating 95% CIs each time, then in 95% of our samples the true mean will be within our CIs (meaning that

FIGURE 18.1 Two domestic cats sitting side by side with much different body masses.

5% of the time it will be outside the CIs). Note that this is slightly different than saying that there is a 95% probability that the true mean is between our CIs.[1] Instead, the idea is that if we were to repeatedly resample from a population and calculate CIs each time, then 95% of the time the true mean would be within our CIs (Sokal & Rohlf, 1995). If this idea does not make sense at first, that is okay. The calculation is actually relatively straightforward, and we will come back to the statistical concept again afterwards to interpret it. First, we will look at CIs assuming a normal distribution, then the separate case of a binomial distribution.

[1]The reason that these two ideas are different has to do with the way that probability is defined in the frequentist approach to statistics (see Chapter 15). With this approach, there is no way to get the probability of the true mean being within an interval, strictly speaking. Other approaches to probability, such as Bayesian probability, do allow you to build intervals in which the true mean is contained with some probability. These are called 'credible intervals' rather than 'confidence intervals' (e.g., Ellison, 2004). The downside to credible intervals (or not, depending on your philosophy of statistics) is that Bayesian probability is at least partly subjective, i.e., based in some way on the subjective opinion of the individual researcher.

18.1 Normal distribution CIs

Remember from the Central Limit Theorem in Chapter 16 that as our sample size N increases, the distribution of our sample mean $\bar{x}$ will start looking more and more like a normal distribution. Also from Chapter 16, we know that we can calculate the probability associated with any interval of values in a normal distribution. For example, we saw that about 68.2% of the probability density of a normal distribution is contained within one standard deviation of the mean. We can use this knowledge from Chapter 16 to set confidence intervals for any percentage of values around the sample mean ($\bar{x}$) using a standard error (SE) and z-score (z). Confidence intervals include two numbers. The **lower confidence interval** (LCI) is below the mean, and the **upper confidence interval** (UCI) is above the mean. Here is how they are calculated,

$$LCI = \bar{x} - (z \times SE),$$

$$UCI = \bar{x} + (z \times SE).$$

Note that the equations are the same, except that for the LCI, we are subtracting $z \times SE$, and for the UCI we are adding it. The specific value of z determines the confidence interval that we are calculating. For example, about 95% of the probability density of a standard normal distribution lies between $z = -1.96$ and $z = 1.96$ (Figure 18.2). Hence, if we use $z = 1.96$ to calculate LCI and UCI, we would be getting 95% confidence intervals around our mean (an interactive application[2] helps visualise the relationship between probability intervals and z-scores more generally).

Now suppose that we want to calculate 95% CIs around the sample mean of our $N = 20$ domestic cats from earlier. We find that the mean body mass of cats in our sample is $\bar{x} = 4.1$ kg, and that the standard deviation is $s = 0.6$ kg (suppose that we are willing to assume, for now, that $s = \sigma$, that is, we know the true standard deviation of the population). Remember from Chapter 12 that the sample standard error can be calculated as $s/\sqrt{N}$. Our lower 95% confidence interval is therefore,

$$LCI_{95\%} = 4.1 - \left(1.96 \times \frac{0.6}{\sqrt{20}}\right) = 3.837$$

Our upper 95% confidence interval is,

$$UCI_{95\%} = 4.1 + \left(1.96 \times \frac{0.6}{\sqrt{20}}\right) = 4.363$$

[2] https://bradduthie.github.io/stats/app/zandp/

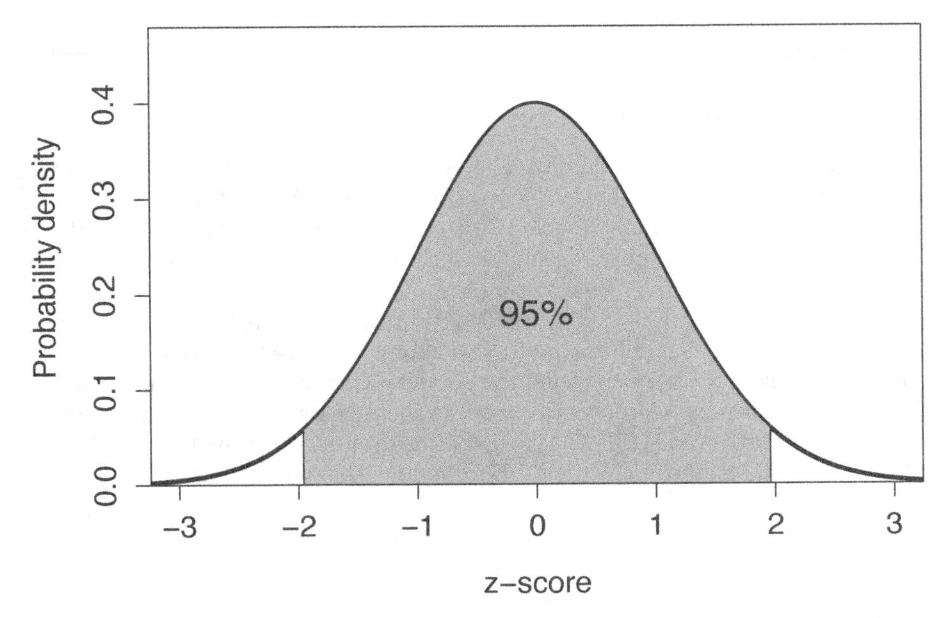

FIGURE 18.2 Standard normal probability distribution showing 95% of probability density surrounding the mean.

Our 95% CIs are therefore $LCI = 3.837$ and $UCI = 4.363$. We can now come back to the statistical concept of what this actually means. If we were to go out and repeatedly collect new samples of 20 cats, and do the above calculations each time, then 95% of the time our true mean cat body mass would be somewhere between the LCI and UCI.

The 95% confidence intervals are the most commonly used in the biological and environmental sciences. In other words, we accept that about 5% of the time (1 in 20 times), our confidence intervals will not contain the true mean that we are trying to estimate. Suppose, however, that we wanted to be a bit more cautious. We could calculate 99% CIs, that is, CIs that contain the true mean in 99% of samples. To do this, we just need to find the z-score that corresponds with 99% of the probability density of the standard normal distribution. This value is about $z = 2.58$, which we could find with an interactive application, a z table[3], some maths, or a quick online search[4]. Consequently, the lower 99% confidence interval for our example of cat body masses would be,

$$LCI_{99\%} = 4.1 - \left(2.58 \times \frac{0.6}{\sqrt{20}}\right) = 3.754$$

[3]https://www.z-table.com/

[4]While it is always important to be careful when searching, typing 'z-score 99% confidence interval' will almost always get the intended result.

Our upper 99% confidence interval is,

$$UCI_{99\%} = 4.1 + \left(2.58 \times \frac{0.6}{\sqrt{20}}\right) = 4.446$$

Notice that the confidence intervals became wider around the sample mean. The 99% CI is now 3.754–4.446, while the 95% CI was 3.837–4.363. This is because if we want to be more confident about our interval containing the true mean, we need to make a bigger interval.

We could make CIs using any percentage that we want, but in practice it is very rare to see anything other than 90% ($z = 1.65$), 95% ($z = 1.96$), or 99% ($z = 2.58$). It is useful to see what these different intervals look like when calculated from actual data (an interactive application[5] illustrates CIs on a histogram). Unfortunately, the CI calculations from this section are a bit of an idealised situation. We assumed that the sample means are normally distributed around the population mean. While we know that this *should* be the case as our sample size increases, it is not quite true when our sample is small. In practice, what this means is that our z-scores are usually not going to be the best values to use when calculating CIs, although they are often good enough when a sample size is large[6]. We will see what to do about this in Chapter 19, but first we turn to the special case of how to calculate CIs from binomial proportions.

18.2 Binomial distribution CIs

For a binomial distribution, our data are counts of successes and failures (see Chapter 15). For example, we might flip a coin 40 times and observe 22 heads and 18 tails. Suppose that we do not know in advance the coin is fair, so we cannot be sure that the probability of it landing on heads is $p = 0.5$. From our collected data, our estimated probability of landing on heads is, $\hat{p} = 22/40 = 0.55$.[7] But how would we calculate the CIs around this estimate? There are multiple ways to calculate CIs around proportions. One common method relies on a normal approximation, that is, approximating the discrete

[5] `https://bradduthie.github.io/stats/app/CI_hist_app/`

[6] What defines a 'small' or a 'large' sample is a bit arbitrary. A popular suggestion (e.g., Sokal & Rohlf, 1995, p. 145) is that any $N < 30$ is too small to use z-scores, but any cut-off N is going to be somewhat arbitrary. Technically, the z-score is not completely accurate until $N \to \infty$, but for all intents and purposes, it is usually only trivially inaccurate for sample sizes in the hundreds. Fortunately, you do not need to worry about any of this when calculating CIs from continuous data in jamovi because jamovi applies a correction for you, which we will look at in Chapter 19.

[7] The hat over the p, ($\hat{p}$) is just being used here to indicate the *estimate* of $P(heads)$, rather than the *true* $P(heads)$.

counts of the binomial distribution using the continuous normal distribution. For example, we can note that the variance of p for a binomial distribution is $\sigma^2 = p(1-p)$ (Box et al., 1978; Sokal & Rohlf, 1995).[8] This means that the standard deviation of p is $\sigma = \sqrt{p(1-p)}$, and p has a standard error,

$$\mathrm{SE}(p) = \sqrt{\frac{p(1-p)}{N}}.$$

We could then use this standard error in the same equation from earlier for calculating CIs. For example, if we wanted to calculate the lower 95% CI for $\hat{p} = 0.55$,

$$LCI_{95\%} = 0.55 - 1.96\sqrt{\frac{0.55(1-0.55)}{40}} = 0.396$$

Similarly, to calculate the upper 95% CI,

$$UCI_{95\%} = 0.55 + 1.96\sqrt{\frac{0.55(1-0.55)}{40}} = 0.704.$$

Our conclusion is that, based on our sample, 95% of the time we flip a coin 40 times, the true mean p will be somewhere between 0.396 and 0.704. These are quite wide CIs, which suggests that our flip of $\hat{p} = 0.55$ would not be particularly remarkable even if the coin was fair ($p = 0.5$).[9] This CI is called the Wald interval. It is easy to calculate, and is instructive for showing another way that CIs can be calculated. But it does not actually do a very good job of accurately producing 95% CIs around a proportion, especially when our proportion is very low or high (Andersson, 2023; Schilling & Doi, 2014).

In fact, there are many different methods proposed to calculate binomial CIs, and a lot of statistical research has been done to compare and contrast these methods (Reed, 2007; Schilling & Doi, 2014; Thulin, 2014). Jamovi uses Clopper-Pearson CIs (Clopper & Pearson E. S., 1934). Instead of relying on a normal approximation, the Clopper-Pearson method uses the binomial probability distribution (see Chapter 15) to build CIs. This is what is known as an 'exact method'. Clopper-Pearson CIs tend to err on the side of caution (they are often made wider than necessary), but they are a much better option than Wald CIs (Reed, 2007).

[8]Note, the variance of total *successes* is simply $np(1-p)$, i.e., just multiply the variance of p by n.

[9]You might ask, why are we doing all of this for the binomial distribution? The central limit theorem is supposed to work for the mean of any distribution, so should that not include the distribution of p too? Can we not just indicate success (heads) with a 1 and failures (tails) with a 0, then estimate the standard error of 22 values of 1 and 18 values of 0? Well, yes! That actually does work and gives an estimate of 0.079663, which is very close to the $\sqrt{\hat{p}(1-\hat{p})/N} = 0.078661$. The problem arises when the sample size is low, or when p is close to 0 or 1, and we are trying to map the z-score to probability density.

19

The t-interval

Chapter 15 introduced the binomial, Poisson, uniform, and normal distributions. In this chapter, I introduce another distribution, the t-distribution. Unlike the distributions of Chapter 15, the t-distribution arises from the need to make accurate statistical inferences, not from any particular kind of data (e.g., successes or failures in a binomial distribution, or events happening over time in a Poisson distribution). In Chapter 18, we calculated confidence intervals (CIs) using the normal distribution and z-scores. In doing so, we made the assumption that the sample standard deviation (s) was the same as the population standard deviation (σ), $s = \sigma$. In other words, we assumed that we knew what σ was, which is almost never true. For large enough sample sizes (i.e., high N), this is not generally a problem, but for lower sample sizes we need to be careful.

If there is a difference between s and σ, then our CIs will also be wrong. More specifically, the uncertainty between our sample estimate (s) and the true standard deviation (σ) is expected to increase the deviation of our sample mean ($\bar{x}$) from the true mean (μ). This means that when we are using the *sample* standard deviation instead of the *population* standard deviation (which is pretty much always), the shape of the standard normal distribution from Chapter 18 (Figure 18.2) will be wrong. The correct shape will be 'wider and flatter' (Sokal & Rohlf, 1995), with more probability density at the extremes and less in the middle of the distribution (Box et al., 1978). What this means is that if we use z-scores when calculating CIs using s, our CIs will not be wide enough, and we will think that we have more confidence in the mean than we really do. Instead of using the standard normal distribution, we need to use a t-distribution[1].

The difference between the standard normal distribution and t-distribution depends on our sample size, N. As N increases, we become more confident that the sample variance will be close to the true population variance (i.e., the deviation of s^2 from σ^2 decreases). At low N, our t-distribution is much wider and flatter than the standard normal distribution. As N becomes large[2], the

[1]This is also called the 'Student's t-distribution'. It was originally discovered by the head brewer of Guinness in Dublin in the early 20th century (Box et al., 1978). The brewer, W. S. Gosset, published under the pseudonym 'A. Student' because Guinness had a policy of not allowing employees to publish (Miller & Miller, 2004).

[2]How large N needs to be for the t-distribution to considered close enough to the normal

t-distribution becomes basically indistinguishable from the standard normal distribution. For calculating CIs from a sample, especially for small sample sizes, it is therefore best to use t-scores instead of z-scores. The idea is the same; we are just multiplying the standard errors by a different constant to get our CIs. For example, in Chapter 18, we multiplied the standard error of 20 cat masses by $z = 1.96$ because 95% of the probability density lies between $z = -1.96$ and $z = 1.96$ in the standard normal distribution. In truth, we should have multiplied by -2.093 because we only had a sample size of $N = 20$. Figure 19.1 shows the difference between the standard normal distribution and the more appropriate t-distribution[3].

Note that in Figure 19.1, a t-distribution with 19 degrees of freedom (*df*) is shown. The t-distribution is parameterised using df, and we lose a degree of freedom when calculating s^2 from a sample size of $N = 20$, so $df = 20 - 1 = 19$ is the correct value (see Chapter 12 for explanation). For calculating CIs, df will always be $N - 1$, and this will be taken care of automatically in jamovi[4] (The jamovi project, 2024).

Recall from Chapter 18 that our body mass measurements of 20 cats had a sample mean of $\bar{x} = 4.1$ kg and sample standard deviation of $s = 0.6$ kg. We calculated the lower 95% CI to be $LCI_{95\%} = 3.837$ and the upper 95% CI to be $UCI_{95\%} = 4.363$. We can now repeat the calculation using the t-score 2.093 instead of the z-score 1.96. Our corrected lower 95% CI is,

$$LCI_{95\%} = 4.1 - \left(2.093 \times \frac{0.6}{\sqrt{20}}\right) = 3.819.$$

---

distribution is subjective. The two distributions get closer and closer as $N \to \infty$, but Sokal & Rohlf (1995) suggest that they are indistinguishable for all intents and purposes once $N > 30$. It is always safe to use the t-distribution when calculating confidence intervals, which is what all statistical programs such as jamovi will do by default, so there is no need to worry about these kinds of arbitrary cutoffs in this case.

[3]We can define the t-distribution mathematically (Miller & Miller, 2004), but it is an absolute beast,

$$f(t) = \frac{\Gamma\left(\frac{v+1}{2}\right)}{\sqrt{\pi v}\Gamma\left(\frac{v}{2}\right)}\left(1 + \frac{t^2}{v}\right)^{-\frac{v+1}{2}}.$$

In this equation, v is the degrees of freedom. The $\Gamma\,()$ is called a 'gamma function', which is basically the equivalent of a factorial function, but for any number z (not just integers), such that $\Gamma(z+1) = \int_0^{\infty} x^z e^{-x} dx$ (where $z > -1$, or, even more technically, the real part of $z > -1$). If z is an integer n, then $\Gamma(n+1) = n!$ (Borowski & Borwein, 2005). What about the rest of the t probability density function? Why is it all so much? The reason is that it is the result of two different probability distributions affecting t independently, a standard normal distribution and a Chi-square distribution (Miller & Miller, 2004). We will look at the Chi-square in Chapter 29. Suffice to say that underlying mathematics of the t-distribution is not important for our purposes in applying statistical techniques.

[4]Another interesting caveat, which jamovi will take care of automatically (so we do not actually have to worry about it), is that when we calculate s^2 to map t-scores to probability densities in the t-distribution, we multiply the sum of squares by $1/N$ instead of $1/(N-1)$ (Sokal & Rohlf, 1995). In other words, we no longer need to correct the sample variance s^2 to account for bias in estimating σ^2 because the t-distribution takes care of this for us.

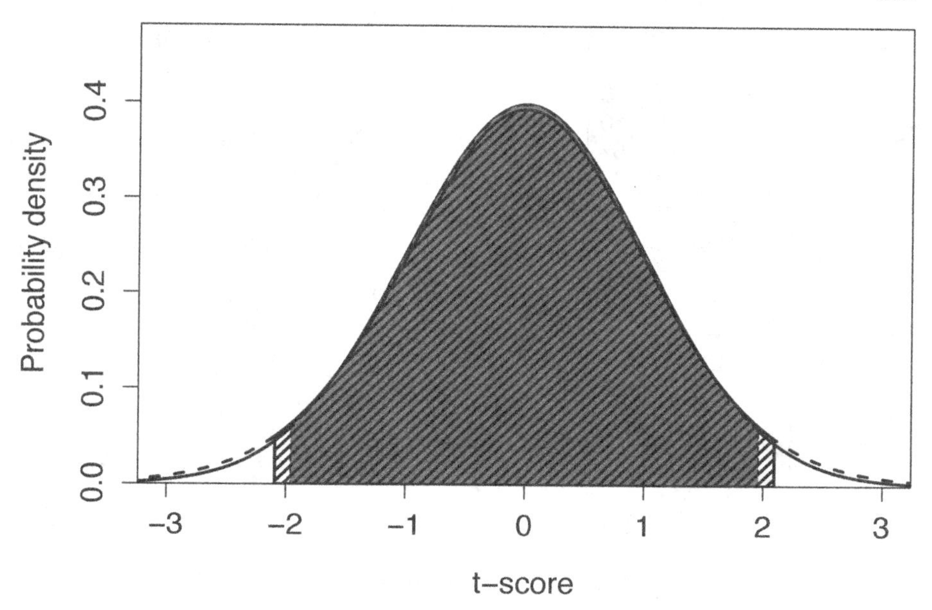

FIGURE 19.1 Standard normal probability distribution showing 95% of probability density surrounding the mean (dark grey shading). On top of the standard normal distribution in dark grey, hatched lines show a t-distribution with 19 degrees of freedom. Hatched lines indicate 95% of the probability density of the t-distribution.

Our upper 95% confidence interval is,

$$UCI_{95\%} = 4.1 + \left(2.093 \times \frac{0.6}{\sqrt{20}}\right) = 4.381.$$

The confidence intervals have not changed too much. By using the t-distribution, the LCI changed from 3.837 to 3.819, and the UCI changed from 4.363 to 4.381. In other words, we only needed our CIs to be a bit wider ($4.381 - 3.819 = 0.562$ for the using t-scores versus $4.363 - 3.837 = 0.526$ using z-scores). This is because a sample size of 20 is already large enough for the t-distribution and standard normal distribution to be very similar (Figure 19.1). But for lower sample sizes and therefore fewer degrees of freedom, the difference between the shapes of these distributions gets more obvious (Figure 19.2).

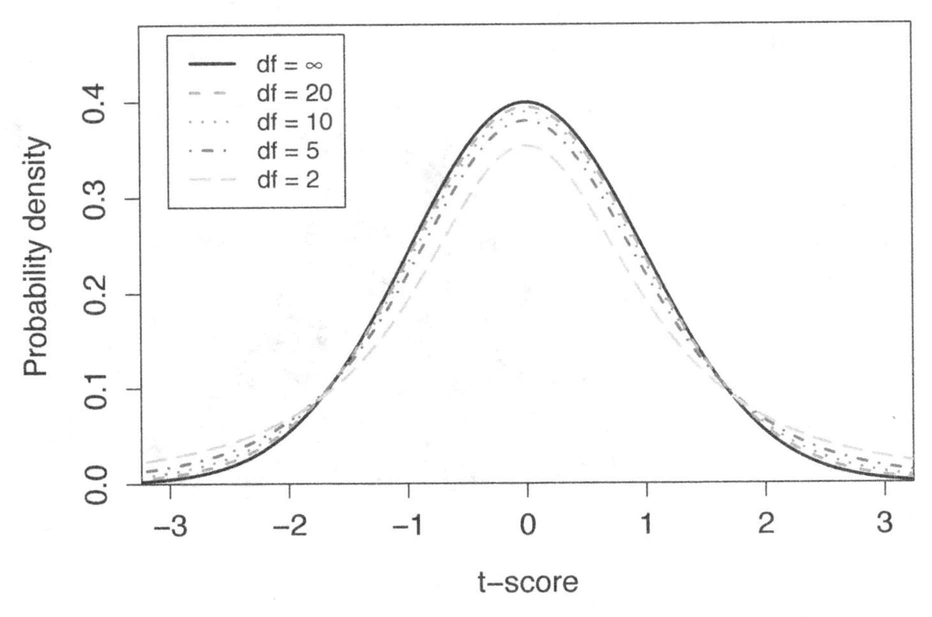

FIGURE 19.2 A t-distribution with infinite degrees of freedom (df) is shown in black; this distribution is identical to the standard normal distribution. Other t-distributions with the same mean and standard deviation, but different degrees of freedom, are indicated by curves of different line types.

The main point of Figure 19.2 is that as degrees of freedom decreases, the t-distribution becomes wider, with more probability density in the tails. Figure 19.2 is quite busy, so an interactive application[5] can make visualising the t-distribution easier. The t-distribution is important throughout most of the rest of this book. It is not just used for calculating confidence intervals. The t-distribution also plays a critical role in hypothesis-testing, which is the subject of Chapter 22 and applied throughout the rest of the book. The t-distribution is therefore very important for understanding many statistical techniques.

[5]https://bradduthie.github.io/stats/app/t_score/

20

Practical. *z- and t-intervals*

This chapter focuses on applying the concepts from Chapter 18 and Chapter 19 in jamovi (The jamovi project, 2024). Specifically, we will practice calculating confidence intervals (CIs). There will be four exercises focused on calculating CIs in jamovi. To complete the first two exercises, you will need the distrACTION module in jamovi. If you need to download it again, the instructions to do this are in the second exercise of Chapter 17 (briefly, go to the Modules option and select 'jamovi library', then scroll down until you find the 'distraACTION' module).

The data for this chapter are inspired by ongoing work in the Woodland Creation and Ecological Networks (WrEN) project (Fuentes-Montemayor, Park, et al., 2022; Fuentes-Montemayor, Watts, et al., 2022). The WrEN project is led by a collaboration between University of Stirling researchers Dr Elisa Fuentes-Montemayor and Prof Kirsty Park, and at Forest Research, Prof Kevin Watts (`https://www.wren-project.com/`). It focuses on questions about what kinds of conservation actions should be prioritised to restore degraded ecological networks. The WrEN project encompasses a huge amount of work and data collection from hundreds of surveyed secondary or ancient woodland sites. Here we will focus on observations of tree diameter at breast height (DBH) and grazing to calculate confidence intervals.

20.1 Confidence intervals with distrACTION

First, it is important to download the distrACTION module if it has not been downloaded already. If the distrACTION module has already been downloaded (see Chapter 17), it should appear in the toolbar of jamovi. Once the distrACTION module has been made available, download the WrEN trees dataset[1] and open it in a spreadsheet. Notice that the dataset is not in a tidy format. There are four different sites represented by different columns in the dataset. The numbers under each column are measurements of tree diameter at breast height (DBH) in centimetres. Before doing anything else, it is therefore

[1] `https://bradduthie.github.io/stats/data/wren_trees.xlsx`

necessary to put the WrEN dataset into a tidy format. The tidy dataset should include two columns: one for site and the other for DBH.

Once the WrEN trees dataset has been reorganised into a tidy format, save it as a CSV file and open it in jamovi. In jamovi, go to Exploration and Descriptives in the toolbar and build a histogram that shows the distribution of DBH. Do these data appear to be roughly normal? Why or why not?

Next, calculate the grand mean and standard deviation of tree DBH (i.e., the mean and standard deviation of trees across all sites).

Grand mean: ________________________________

Grand standard deviation: ________________________________

We will use this mean and standard deviation to compute quantiles and obtain 95% z-scores. First, click on the distrACTION icon in the toolbar. From the distrACTION pull-down menu, select 'Normal Distribution'. To the left, you should see boxes to input parameter values for the mean and standard deviation (SD). Below the 'Parameters' options, you should also see different functions for computing probability or quantiles. To the right, you should see a standard normal distribution (i.e., a normal distribution with a mean of 0 and a standard deviation of 1).

For this exercise, we will assume that the population of DBH from which our sample came is normally distributed. In other words, if we somehow had access to *all possible* DBH measurements in the woodland sites (not just the 120 trees sampled), we assume that DBH would be normally distributed. To find the probability of sampling a tree within a given interval of DBH (e.g., greater than 30), we therefore need to build this distribution with the correct mean and standard deviation. We do not know the *true* mean (μ) and standard deviation (σ) of the population, but our best estimate of these values are the mean ($\bar{x}$) and standard deviation (s) of the sample, as reported above (i.e., the grand mean and standard deviation). Using the Mean and SD parameter input boxes in distrACTION, we can build a normal distribution with the same mean and standard deviation as our sample. Do this now by inputting the calculated Grand mean and Grand standard deviation from above in the appropriate boxes. Note that the normal distribution on the right has the same shape, but the table of parameters has been updated to reflect the mean and standard deviation.

In Chapter 17, we calculated the probability of sampling a value within a given interval of the normal distribution. If we wanted to do the same exercise here, we might find the probability of sampling a DBH < 30 using the Compute probability function (the answer is p = 0.264). Instead, we are now going to do the opposite using the Compute quantile(s) function. We might want to know, for example, what 75% of DBH values will be less than (i.e., what is the cutoff DBH, below which DBH values will be lower than this cutoff with a probability of 0.75). To find this, uncheck the 'Compute probability' box and check the 'Compute quantile(s)' box. Make sure that the 'cumulative quantile' radio button is selected, then set p = 0.75 (Figure 20.1).

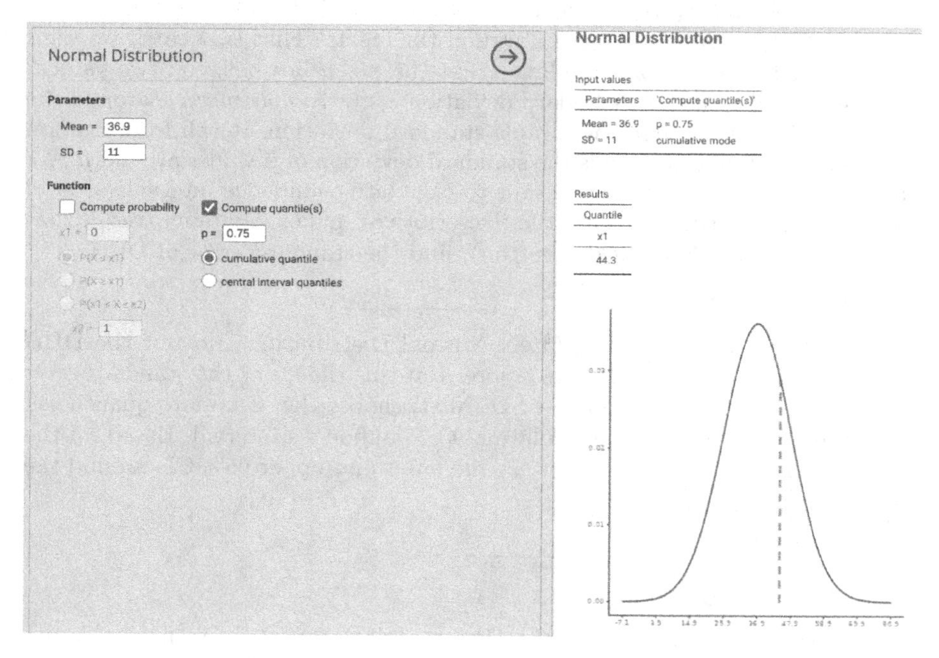

FIGURE 20.1 Jamovi interface for the 'distrACTION' module, in which quantiles have been computed to find the diameter at breast height (DBH) below which 75% of DBHs will be given a normal distribution with a mean of 36.9 and standard deviation of 11. Data for these parameter values were inspired by the Woodland Creation and Ecological Networks (WrEN) project.

From Figure 20.1, we can see that the cumulative 0.75 quantile is 44.3, so if DBH is normally distributed with the mean and standard deviation calculated above, 75% of DBH values in a population will be below 44.3 cm. Using the same principles, what is the cumulative 0.4 quantile for the DBH data?

Quantile: ______________________________

We can also use the Compute quantile(s) option in jamovi to compute interval quantiles. For example, if we want to know the DBH values within which 95% of the probability density is contained, we can set p = 0.95, then select the radio button 'central interval quantiles'. Do this for the DBH data. From the Results table on the right, what interval of DBH values will contain 95% of the probability density around the mean?

Interval: ______________________________

Remember that we are looking at the full sample distribution of DBH. That is, getting intervals for the probability of sampling DBH values around the mean, *not* confidence intervals around the mean as introduced in Chapter 18. How would we get confidence intervals around the mean? That is, what if we want to say that we have 95% confidence that the *mean* lies between two values? We would need to use the standard deviation *of the sample mean* $\bar{x}$ around the true mean μ, rather than the sample standard deviation. Recall from Section 12.6 that the standard error is the standard deviation of $\bar{x}$ values around μ. We can therefore use the standard error to calculate confidence intervals around the mean value of DBH. From the 'Descriptives' panel in jamovi (recall that this is under the 'Exploration' button), find the standard error of DBH,

Std. error of Mean: ______________________________

Now, go back to the distrACTION Normal Distribution and put the DBH mean into the parameters box as before. But this time, put the standard error calculated above into the box for SD. Next, choose the 'Compute quantile(s)' option and set p = 0.95 to calculate a 95% confidence interval. Based on the Results table, what can you infer are the lower and upper 95% CIs around the mean?

Lower 95% CI: ______________________________

Upper 95% CI: ______________________________

Remember that this assumed that the sample means ($\bar{x}$) are normally distributed around the true mean (μ). But as we saw in Chapter 19, when we assume that our sample standard deviation (s) is the same as the population standard deviation (σ), then the shape of the normal distribution will be at least a bit off. Instead, we can get a more accurate estimate of CIs using a t-distribution. Jamovi usually does this automatically when calculating CIs outside of the distrACTION module. To get 95% CIs, go back to the Descriptives panel in jamovi, then choose DBH as a variable. Scroll down to the Statistics options and check 'Confidence interval for Mean' under the **Mean Dispersion** options, and make sure that the number in the box is 95 for 95% confidence. Confidence intervals will appear in the Descriptives table on the right. From this Descriptives table now, write the lower and upper 95% CIs below.

Lower 95% CI: ______________________

Upper 95% CI: ______________________

You might have been expecting a bit more of a difference, but remember, for sufficiently large sample sizes (around N = 30), the normal and t-distributions are very similar (see Chapter 19). We really do not expect much of a difference until sample sizes become small, which we will see in Exercise 20.3.

20.2 Confidence intervals from z- and t-scores

While jamovi can be very useful for calculating CIs from a dataset, you might also need to calculate CIs from just a set of summary statistics (e.g., mean, standard error, and sample size). This activity will demonstrate how to calculate CIs from z- and t-scores. Recall the formula for lower and upper CIs from Section 18.1,

$$LCI = \bar{x} - (z \times SE),$$

$$UCI = \bar{x} + (z \times SE).$$

We could calculate 95% CIs for DBH with just the sample mean ($\bar{x}$), z-score (z), and standard error (SE). We have already calculated $\bar{x}$ and SE for the DBH in Exercise 20.1 above, so we just need to figure out z. Recall that z-scores are *standard normal deviates*, that is, deviations from the mean given a standard normal distribution, in which the mean equals 0 and standard deviation equals 1. For example, $z = -1$ is 1 standard deviation below the mean of a standard normal distribution, and $z = 2$ is 2 standard deviations above the mean of a standard normal distribution. What values of z contain 95% of the probability density of a standard normal distribution? We can use the distrACTION module again to find this out. Select 'Normal Distribution' from the pull-down menu of the distrACTION module. Notice that by default, a standard normal distribution is already set (Mean = 0 and SD = 1). All that we need to do now is compute quantiles for p = 0.95. From these quantiles, what is the proper z-score to use in the equations for LCI and UCI above?

z-score: ______________________

Now, use the values of $\bar{x}$, z, and SE for DBH in the equations above to calculate lower and upper 95% CIs again.

Lower 95% CI: ______________________

Upper 95% CI: ______________________

Are these CIs the same as what you calculated in Exercise 20.1?

Lastly, instead of using the z-score, we can do the same with a t-score. We can find the appropriate t-score from the t-distribution in the distrACTION module. To get the t-score, click on the distrACTION module button and choose 'T-Distribution' from the pull-down menu. To get quantiles with the t-distribution, we need to know the degrees of freedom (*df*) of the sample. Chapter 19 explains how to calculate df from the sample size N. What are the appropriate *df* for DBH?

df: ______________________

Put the df in the Parameters box. Ignore the box for lambda (λ); this is not needed. Under the **Function** options, choose 'Compute quantile(s)' as before to calculate Quantiles. From the Results table, what is the proper t-score to use in the equations for LCI and UCI?

t-score: ______________________

Again, use the values of $\bar{x}$, t, and SE for DBH in the equations above to calculate lower and upper 95% CIs.

Lower 95% CI: ______________________

Upper 95% CI: ______________________

How similar are the estimates for lower and upper CIs when using z- versus t-scores. Reflect on any similarities or differences that you see in all of these different ways of calculating CIs.

20.3 Confidence intervals for different sample sizes

In Exercises 20.1 and 20.2, the sample size of DBH was fairly large ($N = 120$). Now, we will calculate CIs for the mean DBH of each of the four different sites using both z- and t-scores. These sites have much different sample sizes. From the Descriptives tool in jamovi, write the sample sizes for DBH split by site below.

Site 1182: $N =$ ______________

Site 1223: $N =$ ______________

Site 3008: $N =$ ______________

Site 10922: $N =$ ______________

For which of these sites would you predict CIs calculated from z-scores versus t-scores to differ the most?

Site: ____________________

The next part of this exercise is self-guided. In Exercises 20.1 and 20.2, you used different approaches for calculating 95% CIs from the normal and t-distributions. Now, fill in Table 20.1 reporting 95% CIs calculated using each distribution from the four sites using any method you prefer.

TABLE 20.1 95% confidence intervals calculated for tree diameter at breast height (DBH) in centimetres. Data for these parameter values were inspired by the Woodland Creation and Ecological Networks (WrEN) project.

Site	N	95% CIs (Normal)	95% CIs (t-distribution)
1182			
1223			
3008			
10922			

Next, do the same in Table 20.2, but now calculate 99% CIs instead of 95% CIs.

TABLE 20.2 99% confidence intervals calculated for tree diameter at breast height (DBH) in centimetres. Data for these parameter values were inspired by the Woodland Creation and Ecological Networks (WrEN) project.

Site	N	99% CIs (Normal)	99% CIs (t-distribution)
1182			
1223			
3008			
10922			

What do you notice about the difference between CIs calculated from the normal distribution versus the t-distribution across the different sites?

In your own words, what do these CIs *actually mean*?

We will now move on to calculating CIs for proportions.

20.4 Proportion confidence intervals

We will now try calculating CIs for proportional data using the WrEN Sites dataset[2].

Notice that there are more sites included than there were in the dataset used in Exercises 20.1–20.3, and that some of these sites are grazed while others are not (column 'Grazing'). From the Descriptives options, find the number of sites grazed versus not grazed (hint, remember from Chapter 17 to put 'Grazing' in the variable box and click the 'Frequency tables' checkbox).

Grazed: ____________________

Not Grazed: ____________________

From these counts above, what is the estimate (p, or more technically $\hat{p}$, with the hat indicating that it is an estimate) of the proportion of sites that are grazed?

$p =$ ________________

Chapter 18 explained how to calculate lower and upper CIs for binomial distributions (i.e., proportion data). It showed how to calculate Wald CIs, but also noted that Clopper-Pearson CIs are generally more accurate. First, we will calculate Wald CIs by hand, then use jamovi to calculate Clopper-Pearson CIs. To calculate the Wald CIs, we can use equations similar to the ones used for LCI and UCI from Exercise 20.2 above,

$$LCI = p - z \times \mathrm{SE}(p),$$

$$UCI = p + z \times \mathrm{SE}(p),$$

We have already calculated p, and we can find z-scores for CIs in the same way that we did in Exercise 20.2 (i.e., the z-scores associated with 95% CIs do not change just because we are working with proportions). All that is left to

[2] `https://bradduthie.github.io/stats/data/wren_sites.csv`

calculate LCI and UCI are the standard errors of the proportions. Remember from Chapter 18 that these are calculated differently from a standard error of continuous values such as diameter breast height. The formula for standard error of a proportion is,

$$\mathrm{SE}(p) = \sqrt{\frac{p\,(1-p)}{N}}.$$

We can estimate p as the total number of grazed sites divided by N, where N is the total sample size. Using the above equation, what is the standard error of p?

SE(p)= ________________

Using this standard error, what are the Wald lower and upper 95% CIs around p?

Wald $LCI_{95\%}$ = ____________________

Wald $UCI_{95\%}$ = ____________________

Next, find the lower and upper 99% CIs around p and report them below (hint: the only difference here from the calculation of the 95% CIs is the z-score).

Wald $LCI_{99\%}$ = ____________________

Wald $UCI_{99\%}$ = ____________________

Do you notice anything unusual about the lower 99% CI?

Now we can use jamovi to find the Clopper-Pearson 95 and 99% CIs. Jamovi does this for us, so no calculation is required. To calculate Clopper-Pearson CIs, Find the 'Frequencies' button on the toolbar in the 'Analyses' tab. Click on 'Frequencies', then choose '2 Outcomes' from the pull-down menu. You will see a box on the left called 'Proportion Test (2 Outcomes)'. From here, move 'Grazing' from the left box to the right box. Under **Additional Statistics**, check the box for 'Confidence intervals', and make sure that the interval is 95. A table called 'Proportion Test (2 Outcomes)' will appear to the right. Find the row with the Grazing Level 'Yes', then report what you see for p, and the lower and upper CIs below.

p = ______________

Clopper-Pearson $LCI_{95\%}$ = ____________________

Clopper-Pearson $UCI_{95\%}$ = ____________________

To calculate 99% CIs, change the number in the Interval box from 95 to 99. Report the 99% CIs below.

Clopper-Pearson $LCI_{99\%}$ = ____________________

Clopper-Pearson $UCI_{99\%}$ = ____________________

What do you notice about the difference between the Wald CIs and the Clopper-Pearson CIs?

20.5 Another proportion confidence interval

Next, find the 80, 95, and 99% CIs for the proportion of sites that are classified as Ancient woodland using the Clopper-Pearson method for calculating binomial CIs. First consider an 80% CI.

$LCI_{80\%}$ = ____________________

$UCI_{80\%}$ = ____________________

Next, calculate 95% CIs for the proportion of sites classified as Ancient woodland.

$LCI_{95\%}$ = ____________________

$UCI_{95\%}$ = ____________________

Finally, calculate 99% CIs for the proportion of sites classified as Ancient woodland.

$LCI_{99\%}$ = ____________________

$UCI_{99\%}$ = ____________________

Reflect again on what these values actually mean. For example, what does it mean to have 95% confidence that the proportion of sites classified as Ancient woodland are between two values? Are there any situations in which this might be useful, from a scientific or conservation standpoint? There is no right or wrong answer here, but CIs are very challenging to understand conceptually, so having now done the calculations to get them, it is a good idea to think again about what they mean.

21

What is hypothesis testing?

Statistical hypotheses are different from scientific hypotheses. In science, a hypothesis should make some kind of testable statement about the relationship between two or more different concepts or observations (Bouma, 2000). For example, we might hypothesise that in a particular population of sparrows, juveniles that have higher body mass will also have higher survival rates. In contrast, statistical hypotheses compare a sample outcome to the outcome predicted given a relevant statistical distribution (Sokal & Rohlf, 1995). That is, we start with a hypothesis that our data are sampled from some distribution, then work out whether or not we should reject this hypothesis. This concept is counter-intuitive, but it is absolutely fundamental for understanding the logic underlying most modern statistical techniques (Greenland et al., 2016; Mayo, 1996; Sokal & Rohlf, 1995), including all subsequent chapters of this book, so we will focus on it here in-depth. The most instructive way to explain the general idea is with the example of coin flips (Mayo, 1996), as we looked at in Chapter 15.

21.1 How ridiculous is our hypothesis?

Imagine that a coin is flipped 100 times. We are told that the coin is fair, meaning that there is an equal probability of it landing on heads or tails (i.e., the probability is 0.5 for both heads and tails in any given flip). From Section 15.4.1, recall that the number of times out of 100 that the coin flip comes up heads will be described by a binomial distribution. The most probable outcome will be 50 heads and 50 tails, but frequencies that deviate from this perfect 50:50 ratio (e.g., 48 heads and 52 tails) are also expected to be fairly common (Figure 21.1).

The distribution in Figure 21.1 is what we expect to happen if the coin we are flipping 100 times is actually fair. In other words, it is the predicted distribution of outcomes *if our hypothesis that the coin is fair is true* (more on that later). Now, suppose that we actually run the experiment; we flip the coin in question 100 times. Perhaps we observe heads 30 times out of the 100 total flips. From the distribution in Figure 21.1, this result seems *very*

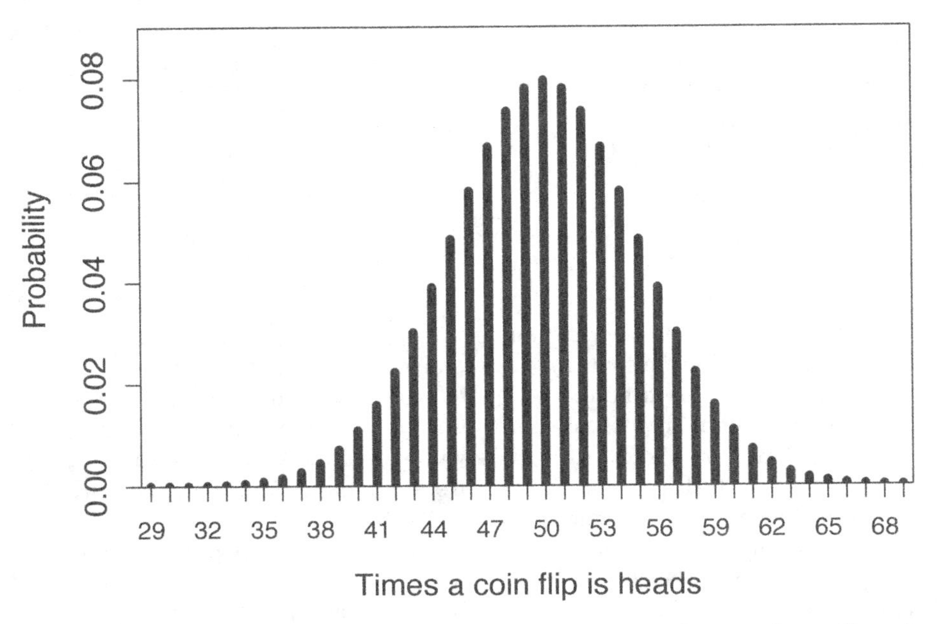

FIGURE 21.1 Probability distribution for the number of times that a flipped coin lands on heads in 100 trials. Note that some areas of parameter space on the x-axis are cut off because the probabilities associated with this number of flips out of 100 being heads are so low.

unlikely if the coin is actually fair. If we do the maths, the probability of observing 30 heads or fewer (i.e., getting anywhere between 0 and 30 heads total) is only $P = 0.0000392507$. And the probability of getting this much of a deviation from 50 heads (i.e., either 20 less than or 20 more than 50) is $P = 0.0000785014$ (two times 0.0000392507, since the binomial distribution is symmetrical around 50). This seems a bit ridiculous! Do we *really* believe that the coin is fair if the probability of getting a result this extreme is so low?

Getting 30 head flips is maybe a bit extreme. What if we flip the coin 100 times and get 45 heads? In this case, if the coin is fair, then we would predict this number of heads or fewer with a probability of about $P = 0.0967$ (i.e., about 9.67% of the time, we would expect to get 45 or fewer heads). And we would predict a deviation as extreme as 5 from the 50:50 ratio of heads to tails with a probability of about $P = 0.193$ (i.e., about 19.3% of the time, we would get 45 heads or fewer, or 55 heads or more). This does not sound nearly so unrealistic. If a fair coin will give us this much of a deviation from the expected 50 heads and 50 tails about 20% of the time, then perhaps our hypothesis is not so ridiculous, and we can conclude the coin is indeed fair.

How improbable does our result need to be to cause us to reject our hypothesis that the coin is fair? There is no definitive answer to this question. In the biological and environmental sciences, we traditionally use a probability of 0.05, but this threshold is completely arbitrary[1]. All it means is that we are willing to reject our hypothesis (i.e., declare the coin to be unfair) when it is actually true (i.e., the coin really *is* fair) about 5% of the time. Note that we do need to decide on some finite threshold for rejecting our hypothesis because even extremely rare events, by definition, can sometimes happen. In the case of 100 coin flips, there is always a small probability of getting *any* number of heads from a fair coin (although getting zero heads would be extraordinarily rare, $P \approx 7.89 \times 10^{-31}$, i.e., a decimal followed by 30 zeros, then a 7). We can therefore never be *certain* about rejecting or not rejecting the hypothesis that we have a fair coin.

This was a very concrete example intended to provide an intuitive way of thinking about hypothesis testing in statistics. In the next section, we will look more generally at what hypothesis testing means in statistics and the terminology associated with it. But everything that follows basically relies on the same general logic as the coin-flipping example here; **if our hypothesis is true, then what is the probability of our result?**

21.2 Statistical hypothesis testing

A statistical test is used to decide if we should reject the hypothesis that some observed value or calculated statistic was sampled from a particular distribution (Sokal & Rohlf, 1995). In the case of the coin example in the previous section, the observed value was the number of heads, and the distribution was the binomial distribution. In other cases, we might, e.g., test the hypothesis that a value was sampled from a normal or t-distribution. In all of these cases, the hypothesis that we are testing is the **null hypothesis**, which we abbreviate as H_0 (e.g., the coin is fair). Typically, H_0 is associated with the lack of an interesting statistical pattern, such as when a coin is fair, when there is no difference between two groups of observations, or when two variables are not associated with each other. This null hypothesis contrasts an **alternative hypothesis**, which we abbreviate as H_A (e.g., the coin is not fair). Alternative hypotheses are always defined by some relationship to H_0 (Sokal & Rohlf, 1995). Typically, H_A is associated with something interesting happening, such as a biased coin, a difference between groups of observations, or an association between two variables. Table 21.1 below presents some null and alternative hypotheses that might be relevant in the biological or environmental sciences.

[1] I have heard many apocryphal stories about how a probability of 0.05 was decided upon, but I have no idea which, if any, of these stories are actually true.

TABLE 21.1 Hypothetical null and alternative hypotheses in the biological and environmental sciences.

Null Hypothesis H_0	Alternative Hypothesis H_A
There is no difference between juvenile and adult sparrow mortality	Mortality differs between juvenile and adult sparrows
Amphibian body size does not change with increasing latitude	Amphibian body size increases with latitude
Soil nitrogen concentration does not differ between agricultural and non-agricultural fields	Soil nitrogen concentration is lower in non-agricultural fields

Notice that alternative hypotheses can indicate direction (e.g., amphibian body size will increase with latitude, or nitrogen content will be lower in non-agricultural fields), or they can be non-directional (e.g., mortality will be different based on life-history stage). When our alternative hypothesis indicates direction, we say that the hypothesis is **one-sided**. This is because we are looking at one side of the null distribution. In the case of our coin example, a one-sided H_A might be that the probability of flipping heads is less than 0.5, meaning that we reject H_0 only given numbers on the left side of the distribution in Figure 21.1 (where the number of times a coin flip is heads is fewer than 50). A different one-sided H_A would be that the probability of flipping heads is greater than 0.5, in which case we would reject H_0 only given numbers on the right side of the distribution. In contrast, when our alternative hypothesis does not indicate direction, we say that the hypothesis is **two-sided**. This is because we are looking at both sides of the null distribution. In the case of our coin example, we might not care in which direction the coin is biased (towards heads or tails), just that the probability of flipping heads does not equal 0.5. In this case, we reject H_0 at both extremes of the distribution of Figure 21.1.

21.3 P-values, false positives, and power

In our hypothetical coin-flipping example, we used P to indicate the probability of getting a particular number of heads out of 100 total flips if our coin was fair. This P (sometimes denoted with a lower-case p) is what we call a 'p-value'.

A p-value is the probability of getting a result as or more extreme than the one observed assuming H_0 is true.[2]

This is separated and in bold because it is a very important concept in statistics, and it is one that is very, very easy to misinterpret[3]. A p-value is *not* the probability that the null hypothesis is true (we actually have no way of knowing this probability). It is also not the probability that an alternative hypothesis is false (we have no way of knowing this probability either). A p-value specifically *assumes that the null hypothesis is true*, then asks what the probability of an observed result would be *conditional upon this assumption.* In the case of our coin flipping example, we cannot really know the probability that the coin is fair or unfair (depending on your philosophy of statistics, this might not even make conceptual sense). But we can say that **if** the coin **is** fair, then an observation of ≤ 45 would occur with a probability of $P = 0.0967$.

Before actually calculating a p-value, we typically set a threshold level (α) below which we will conclude that our p-value is **statistically significant**[4]. As mentioned in Section 21.1, we traditionally set $\alpha = 0.05$ in the biological and environmental sciences (although rarely $\alpha = 0.01$ is used). This means that if $P < 0.05$, then we reject H_0 and conclude that our observation is statistically significant. It also means that even when H_0 really is true (e.g., the coin really is fair), we will mistakenly reject H_0 with a probability of 0.05 (i.e., 5% of the time). This is called a **Type I error** (i.e., false positive), and it typically means that we will infer a pattern of some kind (e.g., difference between groups, or relationship between variables) where none really exists. This is obviously an error that we want to avoid, which is why we set α to a low value.

[2]Technically, it also assumes that all of the assumptions of the model underlying the hypothesis test are true, but we will worry about this later.

[3]In fact, the p-value is so easy to misinterpret and so widely misused, that some scientists have called for them to be abandoned entirely (Wasserstein & Lazar, 2016), but see Stanton-Geddes et al. (2014) and Mayo (2019).

[4]Like p-values, setting thresholds below which we consider P to be significant is at least somewhat controversial (Mayo, 2021; McShane et al., 2019). But the use of statistical significance thresholds is ubiquitous in the biological and environmental sciences, so we will use them throughout this book (it is important to understand them and interpret them).

In contrast, we can also fail to reject H_0 when H_A is actually true. That is, we might mistakenly conclude that there is no evidence to reject the null hypothesis when the null hypothesis really is false. This is called a **Type II error**. The probability that we commit a Type II error, i.e., that we fail to reject the null hypothesis when it is false, is given the symbol β. Since β is the probability that we fail to reject H_0 when it is false, $1 - \beta$ is the probability that we *do* reject H_0 when it is false. This $1 - \beta$ is the **statistical power** of a test. Note that α and β are not necessarily related to each other. Our α is whatever we set it to be (e.g., $\alpha = 0.05$). But statistical power will depend on the size of the effect that we are measuring (e.g., how much bias there is in a coin if we are testing whether or not it is fair), and on the size of our sample. Increasing our sample size will always increase our statistical power, i.e., our ability to reject the null hypothesis when it is really false. Table 21.2 illustrates the relationship between whether or not H_0 is true, and whether or not we reject it.

TABLE 21.2 Summary of Type I and Type II errors in relation to a null hypothesis (H_0).

	Do Not Reject H_0	Reject H_0
H_0 is true	Correct decision	Type I error
H_0 is false	Type II error	Correct decision

Note that we never *accept* a null hypothesis; we just fail to reject it. Statistical tests are not really set up in a way that H_0 can be accepted[5]. The reason for this is subtle, but we can see the logic if we again consider the case of the fair coin. If H_0 is true, then the probability of flipping heads is $P(heads) = 0.5$ (i.e., $H_0 : P(heads) = 0.5$). But even if we fail to reject H_0, this does not mean that we can conclude with any real confidence that our null hypothesis $P(heads) = 0.5$ is true. What if we instead tested the null hypothesis that our coin was *very slightly* biased, such that $H_0 : P(heads) = 0.4999$? If we failed to reject the null hypothesis that $P(heads) = 0.5$, then we would probably also fail to reject a H_0 that $P(heads) = 0.4999$. There is no way to meaningfully distinguish between these two potential null hypotheses by just testing one of them. We therefore cannot conclude that H_0 is correct; we can only find evidence to reject it. In contrast, we can reasonably accept an alternative hypothesis H_A when we reject H_0.

[5]Note that we might, for non-statistical reasons, conclude the absence of a particular phenomenon or relationship between observations. For example, following a statistical test, we might become convinced that a coin really is fair, or that there is no relationship between sparrow body mass and survival. But these are conclusions about scientific hypotheses, not statistical hypotheses.

22

The t-test

A t-test is a simple and widely used statistical hypothesis test that relies on the t-distribution introduced in Chapter 19. In this chapter, we will look at three types of t-tests: (1) the one sample t-test, (2) the independent samples t-test, and (3) the paired samples t-test. We will also look at non-parametric alternatives to t-tests (Wilcoxon and Mann-Whitney tests), which become relevant when the assumptions of t-tests are violated. The use of all of these tests in jamovi will be demonstrated in Chapter 23.

22.1 One sample t-test

Suppose that a biology teacher has created a new approach to teaching and wants to test whether or not their new approach results in student test scores that are higher than the reported national average of 60. This teacher should first define their null and alternative hypotheses.

- H_0: The mean of student test scores equals 60
- H_A: The mean of student test scores is greater than 60

Note that this is a one-sided hypothesis. The teacher is not interested in whether or not the mean test score of their students is below 60. They just want to find out if the mean test scores are greater than 60. Suppose the teacher has 10 students with the following test scores (out of 100).

```
49.3, 62.9, 73.7, 65.5, 69.6, 70.7, 61.5, 73.4, 61.1, 78.1
```

The teacher can use a one sample t-test to test H_0. The one sample t-test will test whether the sample mean of test scores ($\bar{y} = 66.58$) is significantly greater than the reported national average, $\mu_0 = 60$. How does this work? Recall from Chapter 16 that, due to the central limit theorem, the distribution of sample means ($\bar{y}$) will be normally distributed around the true mean μ as sample size N increases. At low N, when we need to estimate the true standard deviation (σ) from the sample standard deviation (s), we need to correct for a bias and use the t-distribution (see Chapter 19). The logic here is to use the t-distribution as the null distribution for $\bar{y}$. If we subtract μ_0 from $\bar{y}$, then we can centre the mean of the null distribution at 0. We can then divide by the standard error of test scores so that we can compare the deviation of $\bar{y}$ from μ_0 in terms of the t-distribution. This is the same idea as calculating a z-score from Section 16.2. In fact, the equations look almost the same,

$$t_{\bar{y}} = \frac{\bar{y} - \mu_0}{\mathrm{SE}(\bar{y})}.$$

In the above equation, $\mathrm{SE}(\bar{y})$ is the standard error of $\bar{y}$.

If the sample mean of test scores is really the same as the population mean $\mu_0 = 60$, then $\bar{y}$ should have a t-distribution. Consequently, values of $t_{\bar{y}}$ far from zero would suggest that the sample mean is improbable given the null distribution predicted if $H_0 : \mu_0 = \bar{y}$ is true. We can calculate $t_{\bar{y}}$ for our above sample (note, $\mathrm{SE}(\bar{y}) = s/\sqrt{N} = 8.334373/\sqrt{10} = 2.63556$),

$$t_{\bar{y}} = \frac{66.58 - 60}{2.63556} = 2.496623.$$

Our t-statistic is therefore 2.496623 (note that a t-statistic can also be negative; this would just mean that our sample mean is less than μ_0, instead of greater than μ_0, but nothing about the t-test changes if this is the case). We can see where this value falls on the t-distribution with 9 degrees of freedom in Figure 22.1.

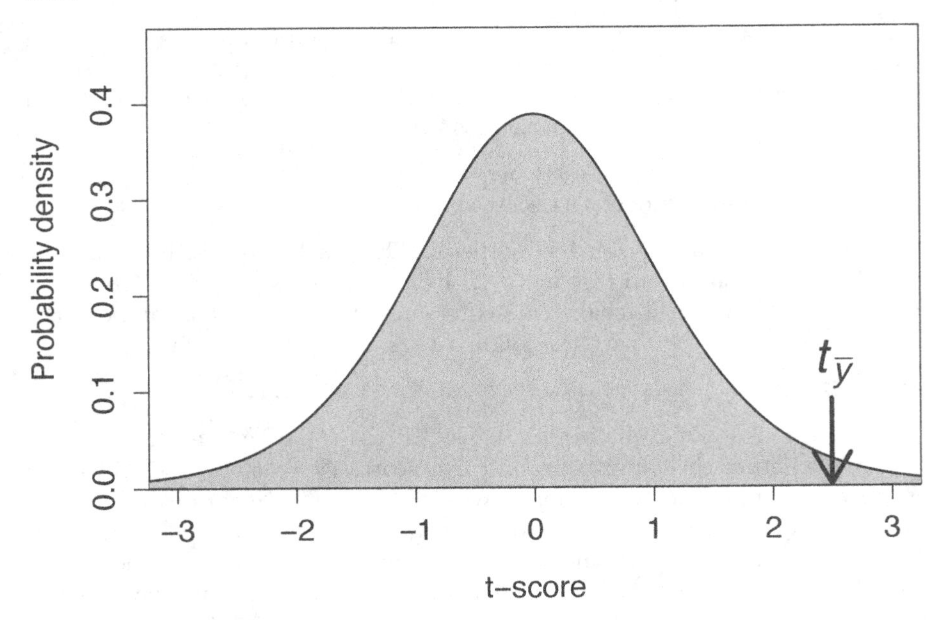

FIGURE 22.1 A t-distribution is shown with a calculated t-statistic of 2.49556 indicated with a downward arrow.

The t-distribution in Figure 22.1 is the probability distribution if H_0 is true (i.e., the student test scores were sampled from a distribution with a mean of $\mu_0 = 60$). The arrow pointing to the calculated $t_{\bar{y}} = 2.496623$ indicates that if H_0 is true, then the sample mean of student test scores $\bar{y} = 66.58$ would be very unlikely. This is because only a small proportion of the probability distribution in Figure 22.1 is greater than or equal to our t-statistic, $t_{\bar{y}} = 2.496623$. In fact, the proportion of t-statistics greater than 2.496623 is only about $P = 0.017$. Hence, if our null hypothesis is true, then the probability of getting a mean student test score of 66.58 or higher is $P = 0.017$ (this is our p-value). It is important to understand the relationship between the t-statistic and the p-value; an interactive application[1] can help visualise. Typically, we set a threshold level of $\alpha = 0.05$, below which we conclude that our p-value is statistically significant (see Chapter 21). Consequently, because our p-value is less than 0.05, we reject our null hypothesis and conclude that student test scores are higher than the reported national average.

22.2 Independent samples t-test

Perhaps the biology teacher is not actually interested in comparing their students' test results with those of the reported national average. After all, there might be many reasons their students score differently from the national average that have nothing to do with their new approach to teaching. To see if their new approach is working, the teacher might instead decide that a better hypothesis to test is whether or not the mean test score from the current year is higher than the mean test score from the class that they taught in the previous year. We can use $\bar{y}_1$ to denote the mean of test scores from the current year, and $\bar{y}_2$ to denote the mean of test scores from the previous year. The test scores of the current year (y_1) therefore remain the same as in the example of the one sample t-test from the previous section.

```
49.3, 62.9, 73.7, 65.5, 69.6, 70.7, 61.5, 73.4, 61.1, 78.1
```

Suppose that in the previous year, there were 9 students in the class (i.e., one fewer than the current year). These 9 students received the following test scores (y_2).

```
57.4, 52.4, 70.5, 71.6, 46.1, 60.4, 70.0, 64.5, 58.8
```

The mean score from last year was $\bar{y}_2 = 61.30$, which does appear to be lower than the mean score of the current year, $\bar{y}_1 = 66.58$. But is the difference between these two means statistically significant? In other words, were the test scores from each year sampled from a population with the same mean,

[1]https://bradduthie.github.io/stats/app/t_score/

such that the population mean of the previous year (μ_2) and the current year (μ_1) are the same? This is the null hypothesis, $H_0 : \mu_1 = \mu_2$.

The general idea for testing this null hypothesis is the same as it was in the one sample t-test. In both cases, we want to calculate a t-statistic, then see where it falls along the t-distribution to decide whether or not to reject H_0. In this case, our t-statistic ($t_{\bar{y}_1 - \bar{y}_2}$) is calculated slightly differently,

$$t_{\bar{y}_1 - \bar{y}_2} = \frac{\bar{y}_1 - \bar{y}_2}{\mathrm{SE}(\bar{y})}$$

The logic is the same as the one sample t-test. If $\mu_1 = \mu_2$, then we also would expect $\bar{y}_1 = \bar{y}_2$ (i.e., $\bar{y}_1 - \bar{y}_2 = 0$). Differences between $\bar{y}_1$ and $\bar{y}_2$ cause the t-statistic to be either above or below 0, and we can map this deviation of $t_{\bar{y}_1 - \bar{y}_2}$ from 0 to the probability density of the t-distribution after standardising by the standard error ($\mathrm{SE}(\bar{y})$).

What is $\mathrm{SE}(\bar{y})$ in this case? After all, there are two different samples y_1 and y_2, so could the two samples not have *different* standard errors? This could indeed be the case, and how we actually conduct the independent samples t-test depends on whether or not we are willing to assume that the two samples came from populations with the same variance (i.e., $\sigma_1 = \sigma_2$). If we are willing to make this assumption, then we can pool the variances (s^2_p) together to get a combined (more accurate) estimate of the standard error $\mathrm{SE}(\bar{y})$ from both samples[2,3]. This version of the independent samples t-test is called the 'Student's t-test'.

If we are unwilling to assume that y_1 and y_2 have the same variance, then we need to use an alternative version of the independent samples t-test. This alternative version is called the Welch's (Welch, 1938), also known as the

[2]This is not a calculation that needs to be done by hand anymore. Statistical software such as jamovi will calculate s_p automatically from a dataset (The jamovi project, 2024). For those interested, the formula that it uses to make this calculation is,

$$s_p = \sqrt{\frac{(n_1 - 1)\, s^2_{y_1} + (n_2 - 1)\, s^2_{y_2}}{n_1 + n_2 - 2}}.$$

This looks like a lot, but really all the equation is doing is adding the two variances ($s^2_{y_1}$ and $s^2_{y_2}$) together, but weighing them by their degrees of freedom ($n_1 - 1$ and $n_2 - 1$), so that the one with a higher sample size has more influence on the pooled standard deviation. To get $\mathrm{SE}(\bar{y})$, we then need to multiply by the square root of $(n_1 + n_2)/(n_1 n_2)$,

$$\mathrm{SE}(\bar{y}) = s_p \sqrt{\frac{n_1 + n_2}{n_1 n_2}}.$$

It might be useful to try this once by hand, but only to convince yourself that it matches the t-statistic produced by jamovi.

[3]The degrees of freedom for the t-statistic is $df = n_1 + n_2 - 2$ (Sokal & Rohlf, 1995). Jamovi handles all of this automatically (The jamovi project, 2024), so we will not dwell on it here.

unequal variances t-test (Dytham, 2011; Ruxton, 2006). In contrast to the Student's t-test, the Welch's t-test does not pool the variances of the samples together[4]. While there are some mathematical differences between the Student's and Welch's independent samples t-tests, the general concept is the same.

This raises the question, when is it acceptable to assume that y_1 and y_2 have the same variance? The sample variance of $s_1^2 = 69.46$ and $s_2^2 = 76.15$. Is this close enough to treat them as the same? Like a lot of choices in statistics, there is no clear right or wrong answer. In theory, if both samples do come from a population with the same variance ($\sigma_1^2 = \sigma_2^2$), then the pooled variance is better because it gives us a bit more statistical power; we can correctly reject the null hypothesis more often when it is actually false (i.e., it decreases the probability of a Type II error). Nevertheless, the increase in statistical power is quite low, and the risk of pooling the variances when they actually are not the same increases the risk that we reject the null hypothesis when it is actually true (i.e., it increases the probability of a Type I error, which we definitely do not want!). For this reason, some researchers advocate using the Welch's by default, unless there is a very good reason to believe y_1 and y_2 are sampled from populations with the same variance (Delacre et al., 2017; Ruxton, 2006).

Here we will adopt the traditional approach of first testing the null hypothesis that $\sigma_1^2 = \sigma_2^2$ using a homogeneity of variances test. If we fail to reject this null hypothesis (i.e., $P > 0.05$), then we will use the Student's t-test, and if we reject it (i.e., $P < 0.05$), then we will use the Welch's t-test. This approach is mostly used for pedagogical reasons; in practice, defaulting to the Welch's t-test is fine (Delacre et al., 2017; Ruxton, 2006). Testing for homogeneity of variances is quite straightforward in most statistical programs, and we will save the conceptual and mathematical details of this for when we look at the F-distribution in Chapter 24. But the general idea is that if $\sigma_1^2 = \sigma_2^2$, then the ratio of variances (σ_1^2/σ_2^2) has its own null distribution (like the normal distribution, or the t-distribution), and we can see the probability of getting a deviation of σ_1^2/σ_2^2 from 1 if $\sigma_1^2 = \sigma_2^2$ is true.

In the case of the test scores from the two samples of students (y_1 and y_2),

---

[4]The equation for calculating the Welch's t-statistic is actually a bit simpler than the Student's t-test that uses the pooled estimate (Ruxton, 2006). Instead of using s_p, we define our t-statistic as,

$$t_s = \frac{\bar{y}_1 - \bar{y}_2}{\sqrt{\frac{s_1^2}{n_1} + \frac{s_2^2}{n_2}}}.$$

The degrees of freedom, however, is quite a bit messier than it is for the independent samples Student's t-test (Sokal & Rohlf, 1995),

$$v = \frac{\left(\frac{1}{n_1} + \frac{(s_2^2/s_1^2)}{n_2}\right)^2}{\frac{1}{n_1^2(n_1-1)} + \frac{(s_2^2/s_1^2)^2}{n_2^2(n_2-1)}}.$$

Of course, jamovi will make these calculations for us, so it is not necessary to do it by hand.

a homogeneity of variance test reveals no evidence that $s_1^2 = 69.46$ and $s_2^2 = 76.15$ are significantly different ($P = 0.834$). We can therefore use the pooled variance and the Student's independent samples t-test. We can calculate $\text{SE}(\bar{y}) = 3.915144$ using the formula for s_p (again, this is not something that ever actually needs to be done by hand), then find $t_{\bar{y}_1 - \bar{y}_2}$,

$$t_{\bar{y}_1 - \bar{y}_2} = \frac{\bar{y}_1 - \bar{y}_2}{\text{SE}(\bar{y})} = \frac{66.58 - 61.3}{3.915144} = 1.348609.$$

As with the one-sample t-test, we can identify the position of $t_{\bar{y}_1 - \bar{y}_2}$ on the t-distribution (Figure 22.2).

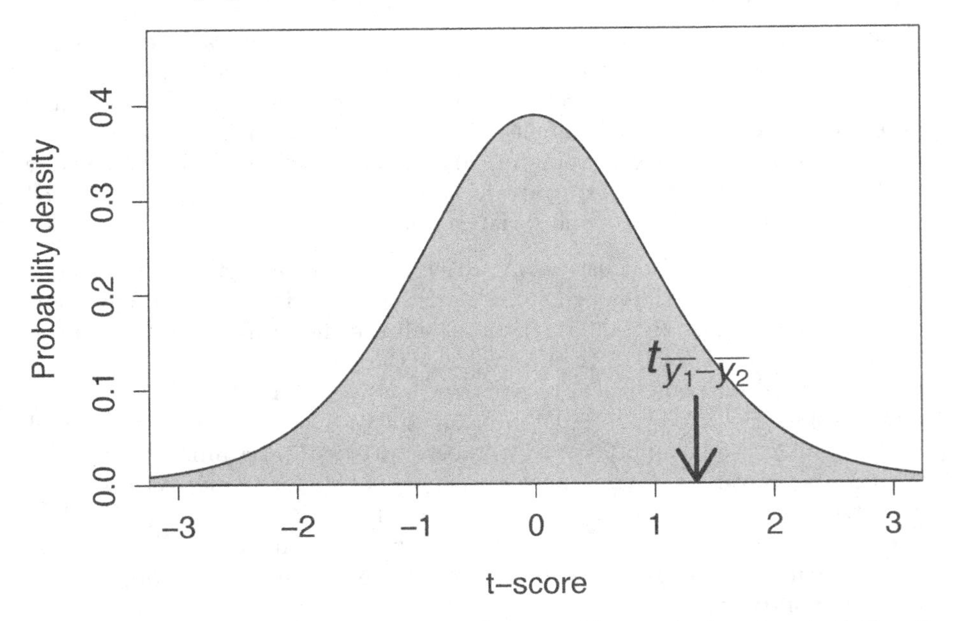

FIGURE 22.2 A t-distribution is shown with a calculated t-statistic of 1.348609 indicated with a downward arrow.

The proportion of t-scores that are higher than $t_{\bar{y}_1 - \bar{y}_2} = 1.348609$ is about 0.098. In other words, given that the null hypothesis is true, the probability of getting a t-statistic this high is $P = 0.098$. Because this p-value exceeds our critical value of $\alpha = 0.05$, we do not reject the null hypothesis. We therefore should conclude that the mean of test scores from the current year ($\bar{y}_1$) is not significantly different from the mean of test scores in the previous year ($\bar{y}_2$). The biology teacher in our example might therefore conclude that mean test results have not improved from the previous year.

22.3 Paired samples t-test

There is one more type of t-test to consider. The paired samples t-test is applied when the data points in one sample can be naturally paired with those in another sample. In this case, data points between samples are not independent. For example, we can consider the student test scores (y_1) yet again.

```
49.3, 62.9, 73.7, 65.5, 69.6, 70.7, 61.5, 73.4, 61.1, 78.1
```

Suppose that the teacher gave these same 10 students (S1–S10) a second test and wanted to see if the mean student score changed from one test to the next (i.e., a two-sided hypothesis).

TABLE 22.1 Test scores from 10 students (S1–S10) for two different tests in a hypothetical biology education example.

	S1	S2	S3	S4	S5	S6	S7	S8	S9	S10
Test 1	49.3	62.9	73.7	65.5	69.6	70.7	61.5	73.4	61.1	78.1
Test 2	46.6	62.7	73.8	58.3	66.8	69.7	64.5	71.3	64.5	78.8
Change	−2.7	−0.2	0.1	−7.2	−2.8	−1	3	−2.1	3.4	0.7

In this case, what we are really interested in is the *change* in scores from Test 1 to Test 2. We want to test the null hypothesis that this change is zero. This is actually the same test as the one-sample t-test. We are just substituting the mean difference in values (i.e., 'Change' in Table 22.1) for $\bar{y}$ and setting $\mu_0 = 0$. We can calculate $\bar{y} = -0.88$ and $\mathrm{SE}(\bar{y}) = 0.9760237$, then set up the t-test as before,

$$t_{\bar{y}} = \frac{-0.88 - 0}{0.9760237} = -0.9016175.$$

Again, we can find the location of our t-statistic $t_{\bar{y}} = -0.9016175$ on the t-distribution (Figure 22.3).

Since this is a two-sided hypothesis, we want to know the probability of getting a t-statistic as extreme as -0.9016175 (i.e., either ± 0.9016175) given that the null distribution is true. In the above t-distribution, 95% of the probability density lies between $t = -2.26$ and $t = 2.26$. Consequently, our calculated $t_{\bar{y}} = -0.9016175$ is not sufficiently extreme to reject the null hypothesis. The p-value associated with $t_{\bar{y}} = -0.9016175$ is $P = 0.391$. We therefore fail to reject H_0 and conclude that there is no significant difference in student test scores from Test 1 to Test 2.

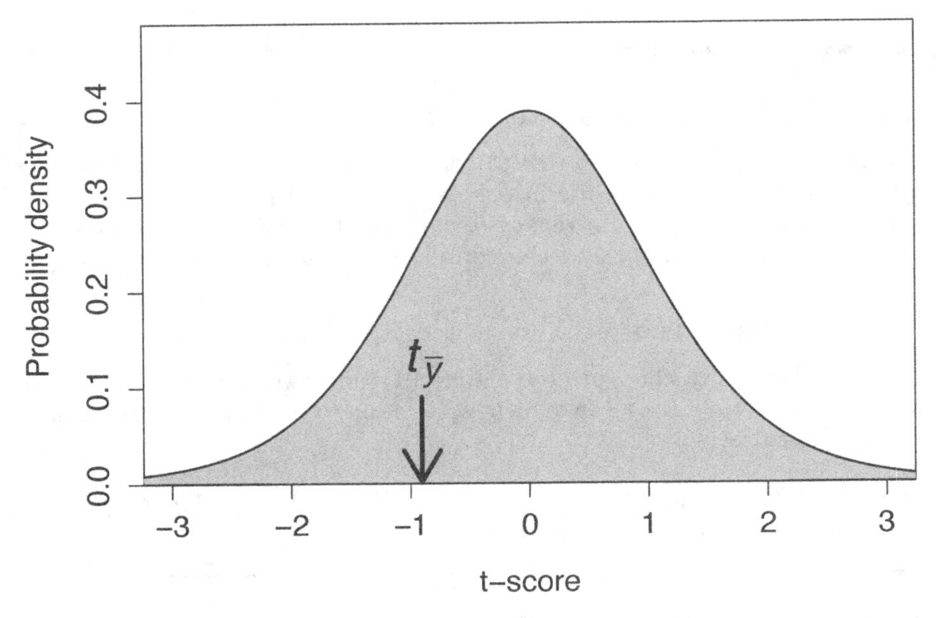

FIGURE 22.3 A t-distribution is shown with a calculated t-statistic of −0.9016175 indicated with a downward arrow.

22.4 Assumptions of t-tests

We make some potentially important assumptions when using t-tests. A consequence of violating these assumptions is a misleading Type I error rate. That is, if our data do not fit the assumptions of our statistical test, then we might not actually be rejecting our null hypothesis at the $\alpha = 0.05$ level. We might unknowingly be rejecting H_0 when it is true at a much higher α value, and therefore concluding that we have evidence supporting the alternative hypothesis H_A when we really do not. It is important to recognise the assumptions that we are making when using any statistical test (including t-tests). If our assumptions are violated, we might need to use a different test, or perhaps apply some kind of transformation on the data. Assumptions that we make when conducting a t-test are as follows:

- Data are continuous (i.e., not count or categorical data)
- Sample observations are a random sample from the population
- Sample means are normally distributed around the true mean

Note that if we are running a Student's independent samples t-test that pools sample variances (rather than a Welch's t-test), then we are also assuming that sample variances are the same (i.e., homogeneity of variance). The last bullet point concerning normally distributed sample means is frequently misunderstood to mean that the sample data themselves need to be normally distributed. This is not the case (Johnson, 1995; Lumley et al., 2002). Instead, what we are really concerned with is the distribution of sample means ($\bar{y}$) around the true mean (μ). And given a sufficiently large sample size, a normal distribution is assured due to the central limit theorem (see Chapter 16).

Moreover, while a normally distributed variable is not *necessary* for satisfying the assumptions of a t-test (or many other tests introduced in this book), it is *sufficient*. In other words, if the variable being measured is normally distributed, then the sample means will also be normally distributed around the true mean (even for a low sample size). So when is a sample size large enough, or close enough to being normally distributed, for the assumption of normality to be satisfied? There really is not a definitive answer to this question, and the truth is that most statisticians will prefer to use a histogram (or some other visualisation approach) and their best judgement to decide if the assumption of normality is likely to be violated.

This book will take the traditional approach of running a statistical test called the Shapiro-Wilk test to test the null hypothesis that data are normally distributed. If we reject the null hypothesis (when $P < 0.05$), then we will conclude that the assumption of normality is violated, and the t-test is not appropriate. The details of how the Shapiro-Wilk test works are not important for now, but the test can be easily run using jamovi (The jamovi project, 2024). If we reject the null hypothesis that the data are normally distributed, then we can use one of two methods to run our statistical test.

1. Transform the data in some way (e.g., take the log of all values) to improve normality.
2. Use a non-parametric alternative test.

The word 'non-parametric' in this context just means that there are no assumptions (or very few) about the shape of the distribution (Dytham, 2011). We will consider the non-parametric equivalents of the one-sample t-test (the Wilcoxon test) and independent samples t-test (Mann-Whitney U test) in the next section. But first, we can show how transformations can be used to improve the fit of the data to satisfy model assumptions.

Often data will have a skewed distribution (see Chapter 13). For example, in Figure 22.4A, we have a dataset (sample size $N = 200$) with a large positive skew (i.e., it is right-skewed). Most values are in the same general area, but with some values being especially high.

Using a t-test on the variable X shown in Figure 22.4A is probably not a good idea. But taking the natural log of all the values of X makes the dataset more normally distributed (Figure 22.4B), thereby more convincingly satisfying the normality assumption required by the t-test. This might seem a bit suspicious

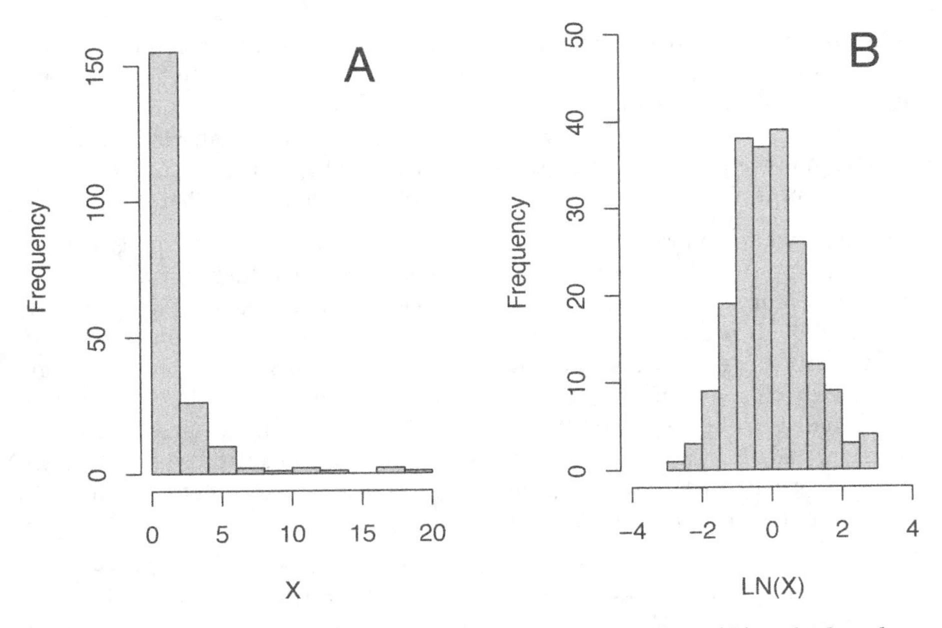

FIGURE 22.4 Set of values with a high positive skew (A), which, when log-transformed (i.e., when we take the natural log of all values), have a normal distribution (B).

at first. Is it really okay to just take the logarithm of all the data instead of the actual values that were measured? Actually, there is no real scientific or statistical reason that we need to use the original scale (Sokal & Rohlf, 1995). Using the log or the square-root of a set of numbers is perfectly fine if it helps satisfy the assumptions of a statistical test.

22.5 Non-parametric alternatives

If we find that the assumption of normality is not satisfied, and a transformation of the data cannot help, then we can consider using non-parametric alternatives to a t-test. These alternatives include the Wilcoxon test and the Mann-Whitney U test.

22.5.1 Wilcoxon test

The Wilcoxon test (also called the 'Wilcoxon signed-rank test') is the non-parametric alternative to a one sample t-test (or a paired t-test). Instead of using the actual data, the Wilcoxon test ranks all of the values in the dataset, then sums up their signs (either positive or negative). The general idea is that we can compare the sum of the *ranks* of the actual data with what would be predicted by the null hypothesis. It tests the null hypothesis that the *median* (M) is significantly different from some given value[5]. An example will make it easier to see how it works. We can use the same hypothetical dataset on student test scores.

```
49.3, 62.9, 73.7, 65.5, 69.6, 70.7, 61.5, 73.4, 61.1, 78.1
```

The first step is to subtract the null hypothesis value ($M = 60$, if we again set H_0 to be that the average student test score equals 60) from each value ($49.3 - 60 = -10.7$, $62.9 - 60 = 2.9$, etc.).

```
-10.7, 2.9, 13.7, 5.5, 9.6, 10.7, 1.5, 13.4, 1.1, 18.1
```

We need to note the sign of each value as negative ($-$) or positive ($+$).

```
-, +, +, +, +, +, +, +, +, +
```

Next, we need to compute the absolute values of the numbers (i.e., $|-10.7| = 10.7$, $|2.9| = 2.9$, $|13.7| = 13.7$, etc.).

```
10.7, 2.9, 13.7, 5.5, 9.6, 10.7, 1.5, 13.4, 1.1, 18.1
```

We then rank these values from lowest to highest and record the sign of each value.

```
6.5, 3.0, 9.0, 4.0, 5.0, 6.5, 2.0, 8.0, 1.0, 10.0
```

Note that both the first and sixth position had the same value (10.7), so instead of ranking them as 6 and 7, we split the difference and rank both as 6.5. Now, we can calculate the sum of the negative ranks (W^-), and the positive ranks W^+. In this case, the negative ranks are easy; there is only one value (the first one), so the sum is just 6.5,

$$W^- = 6.5$$

The positive ranks are in positions 2–10, and the rank values in these positions are 3, 9, 4, 5, 6.5, 2, 8, 1, and 10. The sum of our positive ranks is therefore,

$$W^+ = 3.0 + 9.0 + 4.0 + 5.0 + 6.5 + 2.0 + 8.0 + 1.0 + 10.0 = 48.5$$

[5]Actually, the Wilcoxon signed-rank test technically tests the null hypothesis that two distributions are identical, not just that the medians are identical (Johnson, 1995; Lumley et al., 2002). Note that if one distribution has a higher variance than another, or is differently skewed, it will also affect the ranks. In other words, if we want to compare medians, then we need to assume that the distributions have the same shape (Lumley et al., 2002). Because of this, and because the central limit theorem ensures that means will be normally distributed given a sufficiently high sample size (in which case, t-test assumptions are more likely to be satisfied), some researchers suggest that non-parametric statistics such as the Wilcoxon signed rank test should be used with caution (Johnson, 1995).

Note that W^- plus W^+ (i.e., $6.5 + 48.5 = 55$ in the example here) will always be the same for a given sample size N (in this case, $N = 10$),

$$W = \frac{N(N+1)}{2}.$$

What the Wilcoxon test is doing is calculating the probability of getting a value of W^+ as or more extreme than would be the case if the null hypothesis is true. Note that if, e.g., there are an equal number of values above and below the median, then both W^- and W^+ will be relatively low and about the same value. This is because the ranks of the values below 0 (which we multiply by -1) and above 0 (which we multiply by 1) will be about the same. But if there are a lot more values above the median than expected (as with the example above), then W^+ will be relatively high. And if there are a lot more values below the median than expected, then W^+ will be relatively low.

To find the probability of W^+ being as low or high as it is given that the null hypothesis is true (i.e., the p-value), we need to compare the test statistic W^+ to its distribution under the null hypothesis (note that we can also conduct a one-tailed hypothesis, in which case we are testing if W^+ is either higher or lower than expected given H_0). The old way of doing this is to compare the calculated W^+ to threshold values from a Wilcoxon Signed-Ranks Table. This critical value table is no longer necessary, and statistical software such as jamovi will calculate a p-value for us. For the example above, the p-value associated with $W^+ = 48.5$ and $N = 10$ is $P = 0.037$. Since this p-value is less than our threshold of $\alpha = 0.05$, we can reject the null hypothesis and conclude that the median of our dataset is significantly different from 60.

Note that we can also use a Wilcoxon signed rank test as a non-parametric equivalent to a paired t-test. In this case, instead of subtracting out the null hypothesis of our median value (e.g., $H_0 : M = 60$ in the example above), we just need to subtract the paired values. Consider again the example of the two different tests introduced for the paired samples t-test above (Table 22.2).

TABLE 22.2 Test scores from 10 students (S1–S10) for two different tests in a hypothetical biology education example.

	S1	S2	S3	S4	S5	S6	S7	S8	S9	S10
Test 1	49.3	62.9	73.7	65.5	69.6	70.7	61.5	73.4	61.1	78.1
Test 2	46.6	62.7	73.8	58.3	66.8	69.7	64.5	71.3	64.5	78.8
Change	−2.7	−0.2	0.1	−7.2	−2.8	−1	3	−2.1	3.4	0.7

If the 'Change' values in Table 22.2 were not normally distributed, then we could apply a Wilcoxon test to test the null hypothesis that the median value of $Test\ 2 - Test\ 1 = 0$ (note that a Shapiro-Wilk normality test does not reject the null hypothesis that the difference between test scores is normally distributed, so the paired t-test would be preferred in this case). To do this, we would first note the sign of each value as negative or positive.

```
-, -, +, -, -, -, +, -, +, +
```

Next, we would rank the absolute values of the changes.

```
6, 2, 1, 10, 7, 4, 8, 5, 9, 3
```

If we then sum the ranks, we get $W^- = 6 + 2 + 10 + 7 + 4 + 5 = 34$ and $W^+ = 1 + 8 + 9 + 3 = 21$. The p-value associated with $W^+ = 21$ and $N = 10$ in a two-tailed test is $P = 0.557$, so we do not reject the null hypothesis that Test 1 and Test 2 have the same median.

22.5.2 Mann-Whitney U test

A non-parametric alternative to the independent samples t-test is the Mann-Whitney U test. That is, a Mann-Whitney U test can be used if we want to know whether the median of two independent groups is significantly different. Like the Wilcoxon test, the Mann-Whitney U test uses the ranks of values rather than the values themselves. In the Mann-Whitney U test, the general idea is to rank all of the data across both groups, then see if the sum of the ranks is significantly different (Fryer, 1966; Sokal & Rohlf, 1995). To demonstrate this, we can again consider the same hypothetical dataset used when demonstrating the independent samples t-test above. Test scores from the current year (y_1) are below.

```
49.3, 62.9, 73.7, 65.5, 69.6, 70.7, 61.5, 73.4, 61.1, 78.1
```

We want to know if the median of the above scores is significantly different from the median scores in the previous year (y_2) shown below.

```
57.4, 52.4, 70.5, 71.6, 46.1, 60.4, 70.0, 64.5, 58.8
```

There are 19 values in total, 10 values for y_1 and 9 values for y_2. We therefore rank *all* of the above values from 1 to 19. For y_1, the ranks are below.

```
2, 9, 18, 11, 12, 15, 8, 17, 7, 19
```

For y_2, the ranks are below.

```
4, 3, 14, 16, 1, 6, 13, 10, 5
```

This might be easier to see if we present it as a table showing the test Year (y_1 versus y_2), test Score, and test Rank (Table 22.3).

TABLE 22.3 Test scores from different students across 2 years, and the overall rank of each test score, in a hypothetical biology education example.

Year	Score	Rank
1	49.3	2
1	62.9	9
1	73.7	18
1	65.5	11
1	69.6	12
1	70.7	15
1	61.5	8
1	73.4	17
1	61.1	7
1	78.1	19
2	57.4	4
2	52.4	3
2	70.5	14
2	71.6	16
2	46.1	1
2	60.4	6
2	70	13
2	64.5	10
2	58.8	5

What we need to do now is sum the ranks for y_1 and y_2. If we add up the y_1 ranks, then we get a value of $R_1 = 118$. If we add up the y_2 ranks, then we get a value of $R_2 = 72$. We can then calculate a value U_1 from R_1 and the sample size of y_1 (N_1),

$$U_1 = R_1 - \frac{N_1 (N_1 + 1)}{2}.$$

In the case of our example, $R_1 = 118$ and $N_1 = 10$. We therefore calculate U_1,

$$U_1 = 118 - \frac{10 (10 + 1)}{2} = 63.$$

We then calculate U_2 using the same general formula,

$$U_2 = R_2 - \frac{N_2 (N_2 + 1)}{2}.$$

From the above example of test scores,

$$U_2 = 72 - \frac{9 (9 + 1)}{2} = 27.$$

Whichever of U_1 or U_2 is lower[6] is set to U, so for our example, $U = 27$. Note that if y_1 and y_2 have similar medians, then U will be relatively high. But if y_1 and y_2 have much different medians, then U will be relatively low. One way to think about this is that if y_1 and y_2 are very different, then one of the two samples should have very low ranks, and the other should have very high ranks (Sokal & Rohlf, 1995). But if y_1 and y_2 are very similar, then their summed ranks should be nearly the same. Hence, if the deviation of U is greater than what is predicted given the null hypothesis in which the distributions (and therefore the medians) of y_1 and y_2 are identical, then we can reject H_0.

As with the Wilcoxon test, we could compare our U test statistic to a critical value from a table, but jamovi will also just give us the test statistic U and associated p-value (The jamovi project, 2024). In our example above, $U = 27$, $N_1 = 10$, and $N_2 = 9$ has a p-value of $P = 0.156$, so we do not reject the null hypothesis. We therefore conclude that there is no evidence that there is a difference in the median test scores between the two years.

22.6 Summary

The main focus of this chapter was to provide a conceptual explanation of different statistical tests. In practice, running these tests is relatively straightforward in jamovi (this is good, as long as the tests are understood and interpreted correctly). In general, the first step in approaching these tests is to determine if the data call for a one sample test, an independent samples test, or a paired samples test. In a one sample test, the objective is to test the null hypothesis that the mean (or median) equals a specific value. In an independent samples test, there are two different groups, and the objective is to test the null hypothesis that the groups have the same mean (or median). In a paired samples test, there are two groups, but individual values are naturally paired between each group, and the objective is to test the null hypothesis that the difference between paired values is zero[7]. If assumptions of the t-test are violated, then it might be necessary to use a data transformation or a non-parametric test such as the Wilcoxon test (in place of a one sample or paired samples t-test) or a Mann-Whitney U test (in place of an independent samples t-test).

[6] Note that you could also take the higher of the two values (Sokal & Rohlf, 1995), which is what the statistical programming language R reports (R Core Team, 2022); jamovi reports the lower of the two (The jamovi project, 2024). The null distribution from which the (identical) p-value would be calculated would just be different in this case.

[7] Note, you could also test the null hypothesis that the difference between paired values is some value other than zero, but this is quite rare in practice.

23

Practical. *Hypothesis testing and t-tests*

This chapter focuses on applying the concepts about hypothesis testing from Chapter 21 and the statistical tests from Chapter 22 in jamovi (The jamovi project, 2024). There will be five exercises in total, which will focus on using t-tests and their non-parametric equivalents. The data for this chapter focus on an example of biology and environmental sciences education. Similar to the example in Chapter 22, it uses datasets of student test scores (note that these data are entirely fictional; I have not used scores from real students). We will use 2022 test data[1] and 2023 test data[2]. Variables in these datasets include scores from three different tests.

23.1 One sample t-test

First, we will consider test data from 2022. Open this dataset in jamovi. Before doing any analysis, compute a new column of data called 'Overall' that is each student's mean test score. To do this, double-click on the fifth column of data, then when a new panel opens up, choose 'NEW COMPUTED VARIABLE' from the list of available options. Your input should look like Figure 23.1.

After calculating the overall grade for each student, find the sample size (N), overall sample mean ($\bar{x}$), and sample standard deviation (s). Report these below.

$N =$ ______________________________

$\bar{x} =$ ______________________________

$s =$ ______________________________

Suppose that you have been told that the national average of overall student test scores in classes like yours is $\mu_0 = 60.1$. A concerned colleague approaches you and expresses concern that your test scores are lower than the national average. You decide to test whether or not your students' scores really are lower than the national average, or if your colleague's concerns are unfounded.

[1] https://bradduthie.github.io/stats/data/student_scores_year_1.csv

[2] https://bradduthie.github.io/stats/data/student_scores_year_2.csv

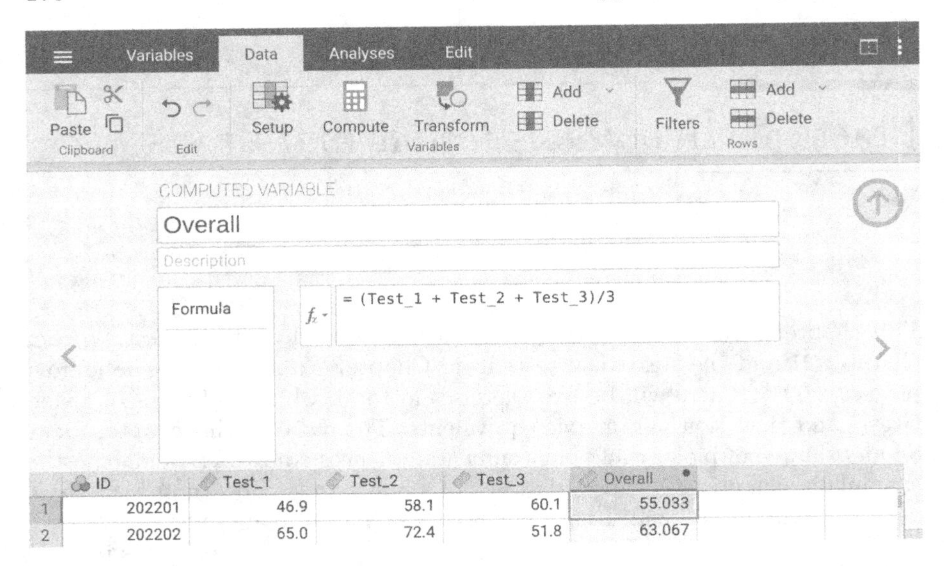

FIGURE 23.1 Jamovi interface for adding a new computed variable called 'Overall' and calculating the mean of the three tests.

What kind(s) of statistical test would be most appropriate to use in this case, and what is the null hypothesis (H_0) of the test?

Test to use: ______________________________________

H_0: ________________________________

What is the alternative hypothesis (H_A), and should you use a one- or two-tailed test?

H_A: ________________________________

One- or two-tailed? ______________________________________

To test if the mean of overall student scores in your class is significantly lower than $\mu_0 = 60.1$ (technically, that your student's scores were sampled from a population with a mean lower than 60.1), we might use a one-tailed, one sample t-test. Recall from Section 22.4 the assumptions underlying a t-test:

- Data are continuous (i.e., not count or categorical data)
- Sample observations are a random sample from the population
- Sample means are normally distributed around the true mean

We will first focus on the first and third assumption. First, the sample data are indeed continuous. But are the sample means normally distributed around the true mean? A sufficient condition to fulfil this assumption is that the sample data are normally distributed (Johnson, 1995; Lumley et al., 2002). We can test whether or not the overall mean student scores are normally distributed. In the jamovi toolbar, find the button that says 'T-Tests' (second from the left), then choose 'One Sample T-test' from the pull-down menu. A new window will open up. In a one sample t-test, there is only one dependent variable, which in this case is 'Overall'. Move the variable 'Overall' to the Dependent Variables box with the right arrow. Below the boxes, find the checkbox options under '**Assumption Checks**'. Check both 'Normality test' (this is a Shapiro-Wilk test of normality) and 'Q-Q Plot'.

After checking the box for 'Normality test' and 'Q-Q Plot', new output will appear in the panel on the right-hand side. The first output is a table showing the results of a Shapiro-Wilk test. The Shapiro-Wilk test tests the null hypothesis that the dependent variable is normally distributed (note, this is *not* our t-test yet; we use the Shapiro-Wilk test to see if the assumptions of the t-test are violated or not). If we reject the null hypothesis, then we will conclude that the data are **not** normally distributed. If we do not reject the null hypothesis, then we will assume that the data **are** normally distributed. From the Normality Test table, what is the p-value of the Shapiro-Wilk test?

$P =$ ______________________________

Based on this p-value, should we reject the null hypothesis? __________

Now have a look at the Q-Q Plot. This plot can also be used to assess whether or not the data are normally distributed. If the data are normally distributed, then the points in the Q-Q Plot should fall roughly along the diagonal black line. These plots take some practice to read and interpret, but for now, just have a look at the Q-Q Plot and think about how it relates to the results of the Shapiro-Wilk test of normality. It might be useful to plot a histogram of the overall scores too in order to judge whether or not they are normally distributed (most researchers will use both a visual inspection of the data and a normality test to evaluate whether or not the assumption of normality is satisfied).

Assuming that we have not rejected our null hypothesis, we can proceed with our one sample t-test. To do this, make sure the variable 'Overall' is still in the Dependent Variables box, then make sure the check box 'Student's' is checked below **Tests**. Underneath **Hypothesis**, change the test value to 60.1 (because $\mu_0 = 60.1$) and put the radio button to '< Test value' to test the null hypothesis that the mean overall score of students is the same as μ_0 against the alternative hypothesis that it is less than μ_0. On the right panel of jamovi, you will see a table with the t-statistic, degrees of freedom, and p-value of the one sample t-test. Write these values down below.

t-statistic: ______________________

degrees of freedom: ______________________

p-value: ______________________

Remember how the t-statistic, degrees of freedom, and p-value are all related. Based on the p-value, should you reject the null hypothesis that your students' mean overall grade is the same as the national average? Why or why not?

Based on this test, how would you respond to your colleague who is concerned that your students are performing below the national average?

Consider again the three assumptions of the t-test listed above. Is there an assumption that might be particularly suspect when comparing the scores of students in a single classroom with a national average? Why or why not?

We will now test whether students in this dataset improved their test scores from Test 1 to Test 2.

23.2 Paired t-test

Suppose we want to test whether or not students in this dataset improved their grades from Test 1 to Test 2. In this case, we are interested in the variables 'Test_1' and 'Test_2', but we are not just interested in whether or not the means of the two test scores are the same. Instead, we want to know if the scores of individual students increased or not. The data are therefore naturally paired. Each test score belongs to a unique student, and each student has a score for Test 1 and Test 2. To test whether or not there has been an increase in test scores, go to the 'T-Tests' button in the jamovi toolbar and select 'Paired Samples T-Test' from the pull-down menu. Place 'Test_1' in the Paired

Variables box first, followed by 'Test_2'. Jamovi will interpret the first variable placed in the box as 'Measure 1', and the second variable placed in the box as 'Measure 2'. Before looking at any test results, first check to see if the difference between Test 1 and Test 2 scores is normally distributed using the same **Assumption Checks** as in the previous exercise. Is there any reason to believe that the data are not normally distributed?

We want to know if student grades have improved. What is the null hypothesis (H_0) and alternative hypothesis (H_A) in this case?

H_0: ______________________________

H_A: ______________________________

To test the null hypothesis against the appropriate alternative hypothesis, select the radio button 'Measure 1 < Measure 2'. As with the one sample t-test, on the right panel of jamovi, you will see a table with the t-statistic, degrees of freedom, and p-value of the one sample t-test. Write these values down below.

t-statistic: ______________________________

degrees of freedom: ______________________________

p-value: ______________________________

Based on this p-value, should you reject or fail to reject your null hypothesis? What can you then conclude about student test scores?

Note that this paired t-test is, mathematically, the exact same as a one sample t-test. If you want to convince yourself of this, you can create a new computed variable that is $Test\,1 - Test\,2$, then run a one sample t-test. You will see that the t-statistic, degrees of freedom, and p-value are all the exact same. Next, we will consider a case in which the assumption of normality is violated.

23.3 Wilcoxon test

Suppose that we want to test the null hypothesis that the scores from Test 3 of the dataset used in Exercises 23.1 and 23.2 were sampled from a population with a mean of $\mu_0 = 62$. We are not interested in whether the scores are higher or lower than 62, just that they are different. Consequently, what should our alternative hypothesis (H_A) be?

H_A: ______________________________

Use the same procedure that you did in Exercise 23.1 to test this new hypothesis concerning Test 3 scores. First, check the assumption that the Test 3 scores are normally distributed. What is the p-value of the Shapiro-Wilk test this time?

$P =$ ______________________________

What inference can you make from the Q-Q plot? Do the points fall along the diagonal line?

Based on the Shapiro-Wilk test and Q-Q plot, is it safe to assume that the Test 3 scores are normally distributed?

Since the Test 3 scores are not normally distributed (and an assumption of the t-test is therefore violated), we can use the non-parametric Wilcoxon signed-rank test instead, as was introduced in Section 22.5.1. To apply the Wilcoxon test, check the box 'Wilcoxon rank' under the **Tests** option. Next, make sure to set the 'Test value' to 62 under the **Hypothesis** option. What are the null and alternative hypotheses of this test?

H_0: ______________________________

H_A: ______________________________

The results of the Wilcoxon signed-rank test will appear in the 'One Sample T-Test' table in the right panel of jamovi in a row called 'Wilcoxon W'. What is the test statistic (not the p-value) for the Wilcoxon test?

Test statistic: ______________________________

Based on what you learnt in Section 22.5.1, what does this test statistic actually mean?

Now look at the p-value for the Wilcoxon test. What is the p-value, and what should you conclude from it?

$P =$ ______________________________

Conclusion: ______________________________

Next, we will introduce a new dataset to test hypotheses concerning the means of two different groups.

23.4 Independent samples t-test

In the first three exercises, we tested hypotheses using data from 2022. In this exercise, we will test the differences in mean test scores between students from 2022 and 2023. As might be expected, the 2022 and 2023 datasets are stored in separate files. What we need to do first is combine the data into a single tidy dataset. To do this, open the Year 2 dataset in a spreadsheet and copy all of the data (but not the column names). You can then paste these data directly into jamovi in the next available row (row 22). Next, add in a new column for 'Year', so that you can differentiate 2022 students from 2023 students in the same dataset. To do this, you can right-click on the 'ID' column and choose 'Add Variable', then 'Insert'. A new column will appear to the left, which you can name 'Year'. Input '2022' for rows 1–21, then '2023' for the remaining rows that you just pasted into jamovi (see Figure 23.2).

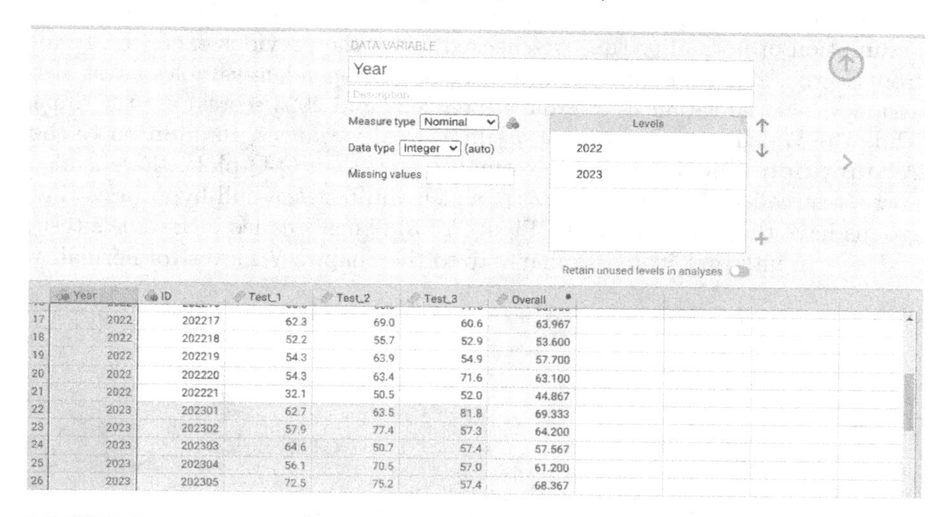

FIGURE 23.2 Jamovi interface for inserting a new variable called 'Year', and the value 2022 pasted into the first 21 rows. The value 2023 has been pasted into the remaining rows of the 'Year' column.

Note that 'Year' does not necessarily need to be in the first column. You could have added it in as the second column, or as the last (i.e., the location of the column will not affect any analyses).

With the new data now included in jamovi, we can compare student test scores between years. Suppose that we first want to test the null hypothesis that the overall student scores have the same mean, $\mu_{2022} = \mu_{2023}$, against the alternative hypothesis that $\mu_{2022} \neq \mu_{2023}$. Is this a one- or a two-tailed hypothesis?

One- or two-tailed? ______________________

It is generally a good idea to plot and summarise your data before running any statistical tests. If you have time, have a look at histograms of the Overall student scores from each year, and look at some summary statistics from the Descriptives output. This will give you a sense of what to expect when running your test diagnostics (e.g., tests of normality) and might alert you to any problems before actually running the analysis (e.g., major outliers that do not make sense, such as a student having an overall score of over 1000 due to a data input error).

After you have had a look at the histograms and summary statistics, go to the jamovi toolbar and navigate to 'T-Tests', then the 'Independent Samples T-Test' from the pull-down menu. Remember that our objective here is to test whether the means of two groups (2022 versus 2023) are the same. We can therefore place the 'Overall' variable in the Dependent Variables box, then 'Year' in the Grouping Variable box. Before running the independent samples t-test, we again need to check our assumptions. In addition to the assumption of normality that we checked for in the previous exercises, recall from Section 22.2 that the standard Student's independent samples t-test also assumes that the variances of groups (i.e., 2022 and 2023 scores) are the same, while the Welch's t-test does not assume equal variances. In addition to the **Assumption Checks** options 'Normality test' and 'Q-Q plot', there is also now a test called 'Homogeneity test', which will test the null hypothesis that groups have the same variances (Figure 23.3). This is called a 'Levene's test', and we can interpret it in a similar way to the Shapiro-Wilk test of normality.

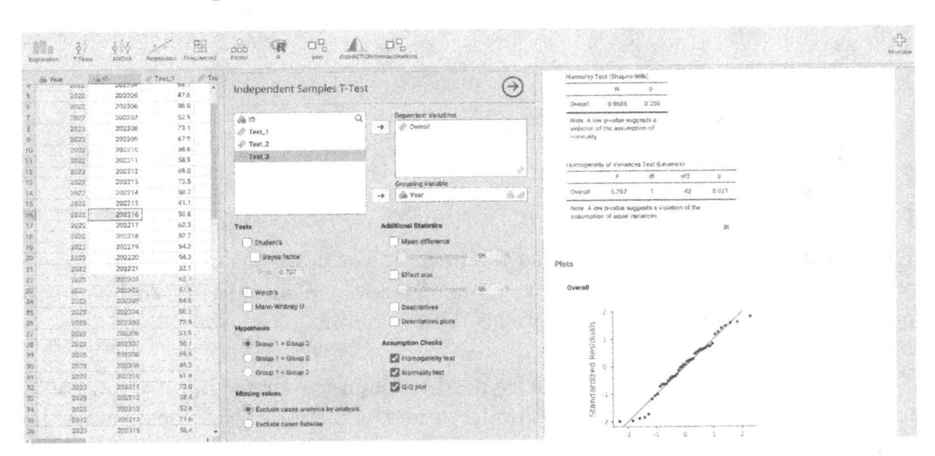

FIGURE 23.3 Jamovi interface for running the assumptions of an Independent Samples T-Test. Here, we are testing if the Overall grades of students differ by Year (2022 versus 2023). Assumption Checks include a test for homogeneity of variances (Homogeneity test) and normal distribution of Overall grades (Normality test and Q-Q plot).

If our p-value is sufficiently low ($P < 0.05$), then we reject the null hypothesis that the groups have the same variances. Based on the **Assumption Checks** in jamovi (and Figure 23.3), what can you conclude about the t-test assumptions?

What is the p-value for the Levene's test?

$P =$ ______________________

It appears from our test of assumptions that we do not reject the null hypothesis that the data are normally distributed, but we should reject the null hypothesis that the groups have equal variances. Based on what you learnt in Section 22.2, what is the appropriate test to run in this case?

Test: ______________________

Note that the appropriate test should be available as a check box underneath the **Tests** options. Check the box for the correct test, then report the test statistic and p-value from the table that appears in the right panel.

Test statistic: ______________________

$P =$ ______________________

What can you conclude from this t-test?

Next, we will compare mean Test 3 scores between years 2022 and 2023.

23.5 Mann-Whitney U Test

Suppose that we now want to use the data from the previous exercise to test whether or not Test 3 scores differ between years. Consequently, in this exercise, Test 3 will be our Dependent Variable and Year will again be the Grouping Variable (also called the 'Independent variable'). Below, summarise the hypotheses for this new test.

H_0 : ______________________

H_A : ______________________

Is this a one- or two-tailed test? ______________________

Next, begin a new Independent Samples T-Test in jamovi and check the assumptions underlying the t-test. Do the variances appear to be the same for Test 3 scores in 2022 versus 2023? How can you make this conclusion?

Next, check to see if the data are normally distributed. What is the p-value of the Shapiro-Wilk test?

$P =$ ______________________

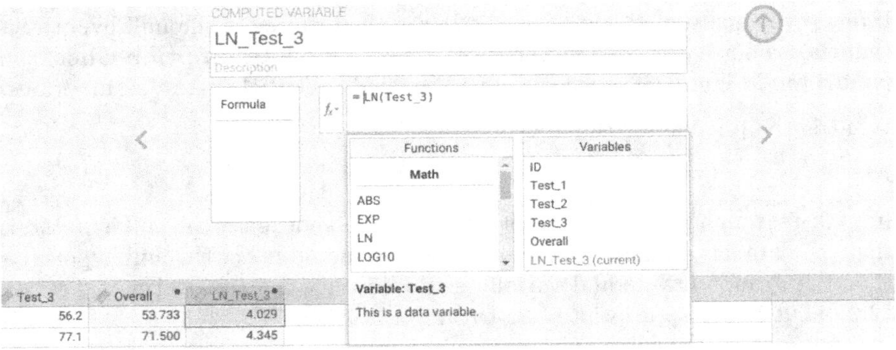

FIGURE 23.4 Jamovi interface for creating a new computed variable that is the natural logarithm of Test 3 data scores in a fictional dataset of student grades.

Now, have a look at the Q-Q plot. What can you infer from this plot about the normality of the data, and why?

Based on what you found from testing the model assumptions above, and the material in Chapter 22, what test is the most appropriate one to use?

Test: ____________________________

Run the above test in jamovi, then report the test statistic and p-value below.

Test statistic: ____________________

$P =$ ____________________

Based on what you learnt in Section 22.5.2, what does this test statistic actually mean?

Finally, what conclusions can you make about Test 3 scores in 2022 versus 2023?

I have not introduced an example in which a transformation (such as a log transformation) is applied to normalise a dataset, as explained in Section 22.4. To do such a transformation, you could create a new computed variable in jamovi and calculate it as the natural log (LN) of a variable (e.g., Test 3). Figure 23.4 illustrates what this should look like.

Lastly, suppose that you wanted to test whether students from 2023 improved their scores from Test 1 to Test 2 more than students from 2022 did. Is there a way to do this with just the tools presented here and in Chapter 22? Think about how the paired t-test works, and how you could apply that logic to test for the difference in the change between two independent samples (2022 versus 2023). What could you do to test the null hypothesis that the change in scores from Test 1 to Test 2 is the same between years?

24

Analysis of variance

An ANalysis Of VAriance (ANOVA) is, as the name implies, a method for analysing variances in a dataset. This is confusing, at first, because the most common application of an ANOVA is to test for differences among group *means*. That is, an ANOVA can be used to test the same null hypothesis as the independent samples Student's t-test introduced in Section 22.2; are two groups sampled from a population that has the same mean? The t-test works fine when we have only two groups, but it does not work when there are three or more groups, and we want to know if the groups all have the same mean. An ANOVA can be used to test the null hypothesis that *all* groups in a dataset are sampled from a population with the same mean. For example, we might want to know if mean wing length is the same for five species of fig wasps sampled from the same area (Duthie et al., 2015). What follows is an explanation of why this can be done by looking at the variance within and between groups (note, 'groups' are also sometimes called 'factors' or 'levels'). Groups are categorical data (see Chapter 5). In the case of the fig wasp example, the groups are the different species (Table 24.1).

TABLE 24.1 Wing lengths (mm) measured for five unnamed species of non-pollinating fig wasps collected from fig trees in 2010 near La Paz in Baja, Mexico.

Het1	Het2	LO1	SO1	SO2
2.122	1.810	1.869	1.557	1.635
1.938	1.821	1.957	1.493	1.700
1.765	1.653	1.589	1.470	1.407
1.700	1.547	1.430	1.541	1.378

Why is any of this necessary? If we want to know if the five species of fig wasps in Table 24.1 have the same mean wing length, can we not just use t-tests to compare the means between each species? There are a couple of problems with this approach. First, there are a lot of group combinations to compare (Het1 vs Het2, Het1 vs LO1, Het1 vs SO1, etc.). For the five fig wasp species in Table 24.1, there are 10 pair-wise combinations that would need to be tested.

And the number of combinations grows exponentially[1] with each new group added to the dataset (Table 24.2).

TABLE 24.2 Number of individual t-tests that would need to be run to compare the means given different numbers of groups (e.g., if a dataset had measurements from 2–10 species).

Groups	2	3	4	5	6	7	8	9	10
Tests	1	3	6	10	15	21	28	36	45

Aside from the tedium of testing every possible combination of group means, there is a more serious problem having to do with the Type I error. Recall from Section 21.3 that a Type I error occurs when we reject the null hypothesis (H_0) and erroneously conclude that H_0 is false when it is actually true (i.e., a false positive). If we reject H_0 at a threshold level of $\alpha = 0.05$ (i.e., reject H_0 when $P < 0.05$, as usual), then we will erroneously reject the null hypothesis about 5% of the time that we run a statistical test and H_0 is true. But if we run 10 separate t-tests to see if the fig wasp species in Table 24.1 have different mean wing lengths, then the probability of making an error increases considerably. The probability of erroneously rejecting **at least 1** of the 10 null hypotheses increases from 0.05 to about 0.40. In other words, about 40% of the time, we would conclude that at least two species differ in their mean wing lengths[2] even when all species *really do* have the same wing length. This is not a mistake that we want to make, which is why we should first test if all of the means are equal:

- H_0 : Mean species wing lengths are all the same
- H_A : Mean species wing lengths are not all the same

We can use an ANOVA to test the null hypothesis above against the alternative hypothesis. If we reject H_0, then we can start comparing pairs of group means (more on this in Chapter 26).

How do we test the above H_0 by looking at *variances* instead of *means*? Before

[1]Technically polynomially, but the distinction really is not important for understanding the concept. In general, the number of possible comparisons between groups is described by a binomial coefficient,

$$\binom{g}{2} = \frac{g!}{2\,(g-2)!}.$$

The number of combinations therefore increases with increasing group number (g).

[2]To get the 0.4, we can first calculate the probability that we (correctly) do not reject H_0 for all 10 pair-wise species combinations $(1 - 0.05)^{10} \approx 0.60$, then subtract from 1, $P(Reject\ H_0) = 1 - (1 - 0.05)^{10} \approx 0.4$. That is, we find the probability of there not being a Type I error in the first test $(1 - 0.05)$, **and** the second test $(1 - 0.05)$, and so forth, thereby multiplying $(1 - 0.05)$ by itself 10 times. This gives the probability of not committing any Type I error across all 10 tests, so the probability that we commit at least one Type I error is 1 minus this probability.

getting into the details of how an ANOVA works, we will first look at the F-distribution. This is relevant because the test statistic calculated in an ANOVA is called an F-statistic, which is compared to an F-distribution in the same way that a t-statistic is compared to a t-distribution for a t-test (see Chapter 22).

24.1 F-distribution

If we want to test whether or not two variances are the same, then we need to know what the null distribution should be if two different samples came from a population with the same variance. The general idea is the same as it was for the distributions introduced in Section 15.4. For example, if we wanted to test whether or not a coin is fair, then we could flip it 10 times and compare the number of times it comes up heads to probabilities predicted by the binomial distribution when $P(Heads) = 0.5$ and $N = 10$ (see Section 15.4.1 and Figure 15.5). To test variances, we will calculate the ratio of variances (F), then compare it to the F probability density function[3]. For example, the ratio of the variances for two samples, 1 and 2, is (Sokal & Rohlf, 1995),

$$F = \frac{\text{Variance 1}}{\text{Variance 2}}.$$

Note that if the variances of samples 1 and 2 are the exact same, then $F = 1$. If the variances are very different, then F is either very low (if Variance 1 $<$ Variance 2) or very high (if Variance 1 $>$ Variance 2). To test the null hypothesis that samples 1 and 2 have the same variance, we therefore need to map the calculated F to the probability density of the F-distribution. Again, the general idea is the same as comparing a t-score to the t-distribution in Section 22.1. Recall that the shape of the t-distribution is slightly different for different degrees of freedom (df). As df increases, the t-distribution starts to resemble the normal distribution. For the F-distribution, there are actually two degrees of freedom to consider. One degree of freedom is needed for Variance 1, and a second degree of freedom is needed for Variance 2. Together, these two degrees of freedom will determine the shape of the F-distribution (Figure 24.1).

[3]The F-distribution was originally discovered in the context of the ratio of random variables with Chi-square distributions, with each variable being divided by its own degree of freedom (Miller & Miller, 2004). We will look at the Chi-square distribution in Chapter 29.

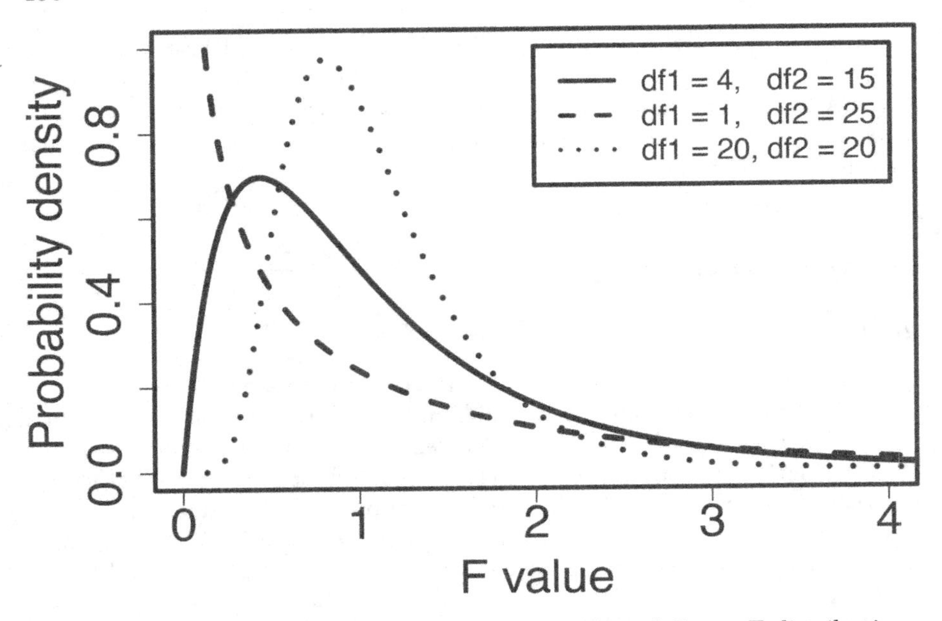

FIGURE 24.1 Probability density functions for three different F-distributions, each of which has different degrees of freedom for the variances in the numerator (df1) and denominator (df2).

Figure 24.1 shows an F-distribution for three different combinations of degrees of freedom. The F-distribution changes its shape considerably given different df values (visualising this is easier with an interactive application[4]).

It is not necessary to memorise how the F-distribution changes with different degrees of freedom. The point is that the probability distribution changes given different degrees of freedom, and that the relationship between probability and the value on the x-axis (F) works like other distributions such as the normal or t-distribution. The entire area under the curve must sum to 1, and we can calculate the area above and below any F value (rather, we can get jamovi to do this for us). Consequently, we can use the F-distribution as the null distribution for the ratio of two variances. If the null hypothesis that the two variances are the same is true (i.e., $F = 1$), then the F-distribution gives us the probability of the ratio of two variances being as or more extreme (i.e., further from 1) than a specific value.

[4]https://bradduthie.github.io/stats/app/f_distribution/

24.2 One-way ANOVA

We can use the F-distribution to test the null hypothesis mentioned at the beginning of the chapter (that fig wasp species have the same mean wing length). The general idea is to compare the mean variance among groups to the mean variance within groups, so our F value (i.e., 'F-statistic') is calculated,

$$F = \frac{\text{Mean variance among groups}}{\text{Mean variance within groups}}.$$

The rest of this section works through the details of how to calculate this F-statistic. It is easy to get lost in these details, but the calculations that follow do not need to be done by hand. As usual, jamovi will do all of this work for us (The jamovi project, 2024). The reason for going through the ANOVA step-by-step process is to show how the total variation in the dataset is being partitioned into the variance among versus within groups, and to provide some conceptual understanding of what the numbers in an ANOVA output actually mean.

24.2.1 ANOVA mean variance among groups

To get the mean variance among groups (i.e., mean squares; MS_{among}), we need to use the sum of squares (*SS*). The *SS* was introduced to show how the variance is calculated in Section 12.3,

$$SS = \sum_{i=1}^{N} \left(x_i - \bar{x}\right)^2.$$

Instead of dividing *SS* by $N - 1$ (i.e., the total *df*), as we would do to get a sample variance, we will need to divide it by the *df among groups* (df_{among}) and *df* within groups (df_{within}). We can then use these $SS_{\text{among}}/df_{\text{among}}$ and $SS_{\text{within}}/df_{\text{within}}$ values to calculate our F[5].

This all sounds a bit abstract at first, so an example will be helpful. We can again consider the wing length measurements from the five species of fig wasps shown in Table 24.1. First, note that the **grand mean** (i.e., the mean across all species) is $\bar{\bar{x}} = 1.6691$. We can also get the sample mean values of each group, individually. For example, for Het1,

[5]Note that the *SS* divided by the degrees of freedom ($N - 1$) is a variance. For technical reasons (Sokal & Rohlf, 1995), we cannot simply calculate the mean variance of groups (i.e., the mean of s^2_{Het1}, s^2_{Het2}, etc.). We need to sum up all the squared deviations from group means *before* dividing by the relevant degrees of freedom (i.e., dfs for the among and within group variation).

$$\bar{x}_{\text{Het1}} = \frac{2.122 + 1.938 + 1.765 + 1.7}{4} = 1.88125.$$

We can calculate the means for all five fig wasps (Table 24.3).

TABLE 24.3 Mean wing lengths (mm) from five unnamed species of non-pollinating fig wasps collected from fig trees in 2010 near La Paz in Baja, Mexico. Each species mean was calculated from four wasps ($N = 4$).

Het1	Het2	LO1	SO1	SO2
1.88125	1.70775	1.71125	1.51525	1.53000

To get the mean variance among groups, we need to calculate the sum of the squared deviations of each species wing length ($\bar{x}_{\text{Het1}} = 1.88125$, $\bar{x}_{\text{Het2}} = 1.70775$, etc.) from the grand mean ($\bar{\bar{x}} = 1.6691$). We also need to weigh the squared deviation of each species by the number of samples for each species[6]. For example, for Het1, the squared deviation would be $4(1.88125 - 1.6691)^2$ because there are four fig wasps, so we multiply the squared deviation from the mean by 4. We can then calculate the sum of squared deviations of the species means from the grand mean,

$$SS_{\text{among}} = 4(1.88125 - 1.6691)^2 + 4(1.70775 - 1.6691)^2 + ... + 4(1.53 - 1.6691)^2.$$

Calculating the above across the five species of wasps gives a value of $SS_{\text{among}} = 0.3651868$. To get our mean variance among groups, we now just need to divide by the appropriate degrees of freedom (df_{among}). Because there are five total species ($N_{\text{species}} = 5$), $df_{\text{among}} = 5 - 1 = 4$. The mean variance among groups is therefore $MS_{\text{among}} = 0.3651868/4 = 0.0912967$.

24.2.2 ANOVA mean variance within groups

To get the mean variance within groups (MS_{within}), we need to calculate the sum of squared deviations of wing lengths from *species means*. That is, we need to take the wing length of each wasp, subtract the mean species wing length, then square it. For example, for Het1, we calculate,

$$SS_{\text{Het1}} = (2.122 - 1.88125)^2 + (1.765 - 1.88125)^2 + ... + (1.7 - 1.88125)^2.$$

If we subtract the mean and square each term of the above,

[6] In this case, weighing by sample size is not so important because each species has the same number of samples. But when different groups have different numbers of samples, we need to multiply by sample number so that each group contributes proportionally to the SS.

$$SS_{\text{Het1}} = 0.0579606 + 0.0032206 + 0.0135141 + 0.0328516 = 0.1075467.$$

Table 24.4 shows what happens after taking the wing lengths from Table 24.1, subtracting the means, then squaring.

TABLE 24.4 Squared deviations from species means for each wing length presented in Table 24.1.

Het1	Het2	LO1	SO1	SO2
0.057960563	0.010455063	0.02488506	0.0017430625	0.01102500
0.003220563	0.012825563	0.06039306	0.0004950625	0.02890000
0.013514062	0.002997562	0.01494506	0.0020475625	0.01512900
0.032851562	0.025840562	0.07910156	0.0006630625	0.02310400

If we sum each column (i.e., do what we did for SS_{Het1} for each species), then we get the SS for each species (Table 24.5).

TABLE 24.5 Sum of squared deviations from species means for each wing length presented in Table 24.1.

Het1	Het2	LO1	SO1	SO2
0.10754675	0.05211875	0.17932475	0.00494875	0.07815800

If we sum the squared deviations in Table 24.5, we get $SS_{\text{within}} = 0.422097$. Note that each species included four wing lengths. We lose a degree of freedom for each of the five species (because we had to calculate the species mean), so our total df is 3 for each species, and $5 \times 3 = 15$ degrees of freedom within groups (df_{within}). To get the mean variance within groups (denominator of F), we calculate $MS_{\text{within}} = SS_{\text{within}}/df_{\text{within}} = 0.0281398$.

24.2.3 ANOVA F-statistic calculation

From Section 24.2.1, we have the mean variance among groups,

$$MS_{\text{among}} = 0.0912967.$$

From Section 24.2.2, we have the mean variance within groups,

$$MS_{\text{within}} = 0.0281398.$$

To calculate F, we just need to divide MS_{among} by MS_{within},

$$F = \frac{0.0912967}{0.0281398} = 3.2443976.$$

Remember that if the mean variance among groups is the same as the mean variance within groups (i.e., $MS_{\text{among}} = MS_{\text{within}}$), then $F = 1$. We can test the null hypothesis that $MS_{\text{among}} = MS_{\text{within}}$ against the alternative hypothesis that there is more variation among groups than within groups ($H_A : MS_{\text{among}} > MS_{\text{within}}$) using the F-distribution (note that this is a one-tailed test). In the example of five fig wasp species, $df_{\text{among}} = 4$ and $df_{\text{within}} = 15$, so we need an F-distribution with four degrees of freedom in the numerator and 15 degrees of freedom in the denominator[7]. We can use an interactive app (see Section 24.1) to get the F-distribution and p-value (Figure 24.2).

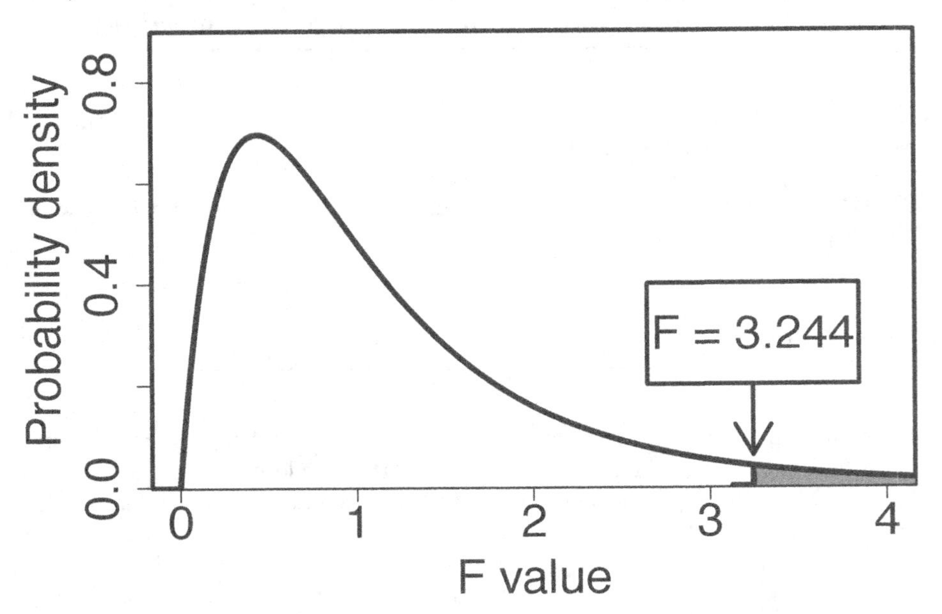

FIGURE 24.2 F-distribution with $df = 4$ for the numerator and $df = 15$ for the denominator. The arrow indicates an F value calculated from fig wasp species wing length measurements for five different species and four measurements per species. Fig wasp wing lengths were collected from a site near La Paz in Baja, Mexico 2010.

[7]Note that $df_{\text{among}} = 4$ and $df_{\text{within}} = 15$ sum to 19, which is the total df of the entire dataset ($N - 1 = 20 - 1 = 19$). This is always the case for the ANOVA; the overall df constrains the degrees of freedom among and within groups.

The area shaded in grey in Figure 24.2, where $F > 3.2443976$, is approximately $P = 0.041762$. This is our p-value. Since $P < 0.05$, we can reject the null hypothesis that all mean species wing lengths are the same, because the variance among species wing lengths is significantly higher than the variance within species wing lengths. Note that the critical value of F (i.e., for which $P = 0.05$) is 3.0555683, so for any F value above this (for $df1 = 5$ and $df2 = 19$), we would reject H_0.

When running an ANOVA in a statistical program, output includes (at least) the calculated F-statistic, degrees of freedom, and the p-value. Figure 24.3 shows the one-way ANOVA output of the test of fig wasp wing lengths in jamovi.

One-Way ANOVA (Fisher's)

	F	df1	df2	p
wing_length	3.24440	4	15	0.04176

FIGURE 24.3 Jamovi output for a one-way ANOVA of wing length measurements in five species of fig wasps collected in 2010 near La Paz in Baja, Mexico.

Jamovi is quite minimalist for a one-way ANOVA (The jamovi project, 2024), but these four statistics (F, $df1$, $df2$, and p) are all that is really needed. Most statistical programs will show ANOVA output that includes the SS and mean squares among (MS_{among}) and within (MS_{within}) groups.

```
Analysis of Variance Table

Response: wing_length
          Df  Sum Sq  Mean Sq F value  Pr(>F)
Species    4 0.36519 0.091297  3.2444 0.04176 *
Residuals 15 0.42210 0.028140
---
Signif. codes:  0 '***' 0.001 '**' 0.01 '*' 0.05 '.' 0.1 ' ' 1
```

The above output, taken from R (R Core Team, 2022), includes the same information as jamovi (F, $df1$, $df2$, and p), but also includes SS and mean variances. We can also get this information from jamovi if we want it (see Chapter 27).

24.3 Assumptions of ANOVA

As with the t-test (see Section 22.4), there are some important assumptions that we make when using an ANOVA. Violating these assumptions will mean that our Type I error rate (α) is, again, potentially misleading. Assumptions of ANOVA include the following (Box et al., 1978; Sokal & Rohlf, 1995):

1. Observations are sampled randomly
2. Observations are independent of one another
3. Groups have the same variance
4. Errors are normally distributed

Assumption 1 just means that observations are not biased in any particular way. For example, if the fig wasps introduced at the start of this chapter were used because they were the largest wasps that were collected for each species, then this would violate the assumption that the wing lengths were sampled randomly from the population.

Assumption 2 means that observations are not related to one another in some confounding way. For example, if all of the Het1 wasps came from one fig tree, and all of the Het2 wasps came from a different fig tree, then wing length measurements are not really independent within species. In this case, we could not attribute differences in mean wing length to species. The differences could instead be attributable to wasps being sampled from different trees (more on this in Chapter 27).

Assumption 3 is fairly self-explanatory. The ANOVA assumes that all of the groups in the dataset (e.g., species in the case of the fig wasp wing measurements) have the same variance, that is, we assume homogeneity of variances (as opposed to heterogeneity of variances). In general, ANOVA is reasonably robust to deviations from homogeneity, especially if groups have similar sample sizes (Blanca et al., 2018). This means that the Type I error rate is about what we want it to be (usually $\alpha = 0.05$), even when the assumption of homogeneity of variances is violated. In other words, we are not rejecting the null hypothesis when it is true more frequently than we intend! We can test the assumption that group variances are the same using a Levene's test in the same way that we did for the independent samples t-test in Chapter 23. If we reject the null hypothesis that groups have the same variance, then we should potentially consider a non-parametric alternative test such as the Kruskal-Wallis H test (see Chapter 26).

Assumption 4 is the equivalent of the t-test assumption from Section 22.4 that sample means are normally distributed around the true mean. What the assumption means is that if we were to repeatedly resample data from a population, the sample means that we calculate would be normally distributed. For the fig wasp wing measurements, this means that if we were to go back out and repeatedly collect four fig wasps from each of the five species, then sample means of species wing length and overall wing length would be normally distributed around the true means. Due to the central limit theorem (see Chapter 16), this assumption becomes less problematic with increasing sample size. We can test if the sample data are normally distributed using a Q-Q plot or Shapiro-Wilk test (the same procedure used for the t-test). Fortunately, the ANOVA is quite robust to deviations from normality (Schmider et al., 2010), but if data are not normally distributed, we should again consider a non-parametric alternative test such as the Kruskal-Wallis H test (see Chapter 26).

25

Multiple comparisons

In the Section 24.2 ANOVA example, we rejected the null hypothesis that all fig wasp species have the same mean wing length. We can therefore conclude that at least one species has a different mean wing length than the rest; can we determine which one(s)? We can try to find this out using a post hoc comparison (*post hoc* is Latin for 'after the event'). That is, after we have rejected the null hypothesis in the one-way ANOVA, we can start comparing individual groups (Het1 vs Het2, Het1 vs LO1, etc.). Nevertheless, we need some way to correct for the Type I error problem explained at the beginning of Chapter 24. That is, if we run a large enough number of t-tests, then we are almost guaranteed that we will find a significant difference between means ($P < 0.05$) where none really exists. A way to avoid this inflated Type I error rate is to set our significance threshold to be lower than the usual $\alpha = 0.05$. We can, for example, divide our α value by the total number of pair-wise t-tests that we run. This is called a Bonferroni correction (Dytham, 2011), and it is an especially cautious approach to post hoc comparisons between groups (Narum, 2006). For the fig wasp wing lengths, recall that there are 10 possible pairwise comparisons between the five species. This means that if we were to apply a Bonferroni correction and run 10 separate t-tests, then we would only conclude that species mean wing lengths were different when $P < 0.005$ instead of $P < 0.05$.

Another approach to correcting for multiple comparisons is a Tukey's honestly significant difference test (Tukey's HSD, or just a 'Tukey test'). The general idea of a Tukey test is the same as the Bonferroni. Multiple t-tests are run in a way that controls the Type I error rate so that the probability of making a Type I error across the whole set of comparisons is fixed (e.g., at $\alpha = 0.05$). The Tukey test does this by using a modified t-test, with a t-distribution called the 'studentised range distribution' that applies the range of mean group values (i.e., $\max(\bar{x}) - \min(\bar{x})$) and uses the sample variance across the groups with the highest and lowest sample means (Box et al., 1978; Tukey, 1949).

Multiple comparisons tests can be run automatically in jamovi. Figure 25.1 shows a post hoc comparisons table for all pair-wise combinations of fig wasp species wing length means.

The column 'p' in Figure 25.1 is the uncorrected p-value, i.e., the p-value that a t-test would produce without any correction for multiple comparisons. The

Post Hoc Comparisons - Species

Comparison									
Species		Species	Mean Difference	SE	df	t	p	p_{tukey}	$p_{bonferroni}$
Het1	-	Het2	0.17350	0.11862	15.00000	1.46270	0.164	0.600	1.000
	-	LO1	0.17000	0.11862	15.00000	1.43319	0.172	0.617	1.000
	-	SO1	0.36600	0.11862	15.00000	3.08557	0.008	0.050	0.075
	-	SO2	0.35125	0.11862	15.00000	2.96122	0.010	0.063	0.097
Het2	-	LO1	-0.00350	0.11862	15.00000	-0.02951	0.977	1.000	1.000
	-	SO1	0.19250	0.11862	15.00000	1.62288	0.125	0.506	1.000
	-	SO2	0.17775	0.11862	15.00000	1.49853	0.155	0.579	1.000
LO1	-	SO1	0.19600	0.11862	15.00000	1.65238	0.119	0.489	1.000
	-	SO2	0.18125	0.11862	15.00000	1.52803	0.147	0.561	1.000
SO1	-	SO2	-0.01475	0.11862	15.00000	-0.12435	0.903	1.000	1.000

Note. Comparisons are based on estimated marginal means

FIGURE 25.1 Jamovi output showing a table of 10 post hoc comparisons between species mean wing lengths for five different species of fig wasps. The last three columns show the uncorrected p-value (p), a p-value obtained from a Tukey test, and a p-value obtained from a Bonferroni correction. Species wing length measurements were collected in 2010 near La Paz in Baja, Mexico.

columns p_{tukey} and $p_{\text{bonferroni}}$ show corrected p-values for the Tukey test and Bonferroni corrected t-test, respectively. We can interpret these p-values as usual, concluding that two species have different means if $P < 0.05$ (i.e., jamovi does the correction for us; we do not need to divide $\alpha = 0.05$ to figure out what the significance threshold should be, given the Bonferroni correction).

Note that from Figure 25.1, it appears that both the Tukey test and the Bonferroni correction fail to find that any pair of species have significantly different means. This does not mean that we have done the test incorrectly. The multiple comparisons tests are asking a slightly different question than the one-way ANOVA. The multiple comparisons tests are testing the null hypothesis that two *particular* species have the same mean wing lengths. The one-way ANOVA tested the null hypothesis that all species have the same mean, and our result for the ANOVA was barely below the $\alpha = 0.05$ threshold ($P = 0.042$). The ANOVA also has more statistical power because it makes use of all 20 measurements in the dataset, not just a subset of measurements between two of the five species. It is therefore not particularly surprising or concerning that we rejected H_0 for the ANOVA, but the multiple comparisons tests failed to find any significant difference between group means.

26

Kruskal-Wallis H test

If the assumptions of the one-way ANOVA are violated, then we can consider using a Kruskal-Wallis test. The Kruskal-Wallis test is essentially an extension of the Mann-Whitney U test from Section 22.5.2 for samples with more than two groups (Kruskal & Wallis, 1952). Like the Mann-Whitney U test, it uses the *ranks* of values instead of the actual values in the dataset. The idea is to rank all values in the dataset, ignoring group, then calculate a test statistic (H) that increases as the difference among group ranks increases, relative to the difference within group ranks. We can again use the example of the fig wasp wing lengths introduced in Chapter 24. For convenience, Table 24.1 is reproduced here as Table 26.1.

TABLE 26.1 Reproduction of Table 24.1. Wing lengths (mm) measured for five unnamed species of non-pollinating fig wasps collected from fig trees in 2010 near La Paz in Baja, Mexico. Note, for readability, this table is not presented in a tidy format.

Het1	Het2	LO1	SO1	SO2
2.122	1.810	1.869	1.557	1.635
1.938	1.821	1.957	1.493	1.700
1.765	1.653	1.589	1.470	1.407
1.700	1.547	1.430	1.541	1.378

Recall that in the one-way ANOVA from Section 24.2, we rejected the null hypothesis that all species had the same mean wing length ($P = 0.042$). But we had not actually tested the assumptions of the one-way ANOVA before running the test! If we had tested the ANOVA assumptions, we would not reject the null hypothesis that wing length is normally distributed (Shapiro-Wilk test $P = 0.698$). But a Levene's test of homogeneity of variances convincingly rejects the null hypothesis that the group variances are equal ($P = 0.008$). Consequently, we should probably have considered the non-parametric Kruskal-Wallis H test instead. To do this, we first need to rank all of the values in Table 26.1. There are 20 total values, so we rank them from 1 to 20 (Table 26.2).

TABLE 26.2 Ranks of wing lengths (mm) measured for five unnamed species of non-pollinating fig wasps collected from fig trees in 2010 near La Paz in Baja, Mexico.

Het1	Het2	LO1	SO1	SO2
20	15	17	8	10
18	16	19	5	12.5
14	11	9	4	2
12.5	7	3	6	1

From these ranks, we calculate a test statistic H from the overall sample size (N), the sample size (N_i) of each of the groups (g), the mean of group ranks ($\bar{R}_i$), and the overall mean rank ($\bar{\bar{R}}$). Of course, we do not need to do this by hand. But the formula shows how a statistical program will do the calculation (Kruskal, 1952),

$$H = (N-1)\frac{\sum_{i=1}^{g} N_i \left(\bar{R}_i - \bar{\bar{R}}\right)^2}{\sum_{i=1}^{g} \sum_{j=1}^{N_i} \left(R_{ij} - \bar{\bar{R}}\right)^2}.$$

For our purposes, the mathematical details are not important. The equation is included here only to show the similarity between the calculation of H versus F from Section 24.2. In the numerator of the equation for H, we are calculating the squared deviation of mean group ranks from the overall mean rank $(\bar{R}_i - \bar{\bar{R}})^2$, as weighted by the group sample size N_i. We are then dividing by the sum of squared deviations of all ranks (R_{ij}) from the overall mean rank $\bar{\bar{R}}$. All of this gets multiplied by $N - 1$ to give the test statistic H. We can then compare H to a suitable null distribution, which might be calculated precisely using a specific algorithm (e.g., Choi et al., 2003). But most statistical programs such as jamovi compare H to a Chi-square distribution (see Chapter 29), which is an effective approximation (Miller & Miller, 2004).

The output of a Kruskal-Wallis H test is quite minimal and easy to read. Jamovi reports a Chi-square (χ^2) test statistic, degrees of freedom, and p-value (The jamovi project, 2024). Figure 26.1 shows the output of a Kruskal-Wallis H test on the fig wasp wing lengths data.

Kruskal-Wallis

	χ^2	df	p
wing_length	8.75301	4	0.06758

FIGURE 26.1 Jamovi output table for a non-parametric Kruskal-Wallis H test, which tests the null hypothesis that species mean wing lengths are the same for five different species of fig wasps. Species wing length measurements were collected in 2010 near La Paz in Baja, Mexico.

Given the Kruskal-Wallis H test output (Figure 26.1), we should not reject the null hypothesis that species have different wing lengths because $P > 0.05$. This is in contrast to our one-way ANOVA result, for which we did reject the same null hypothesis. The Kruskal-Wallis H test does not assume that group variances are the same, unlike the one-way ANOVA. Since we know that the homogeneity of variances assumption is violated for the fig wasp data, it is probably best to be cautious and conclude that there is no evidence that mean wing lengths differ among species.

27

Two-way ANOVA

The one-way ANOVA tested the null hypothesis that all group means are the same. But, we might also have a dataset in which there is more than one type of group. For example, suppose we know that the fig wasps used in Chapters 24–26 actually came from two different trees (Tree A and Tree B). This would mean that there could be a confounding variable affecting wing length, violating assumption 2 from Section 24.3. If, for whatever reason, fig wasps on different trees have different wing lengths, then we should include tree as an explanatory variable in the ANOVA. Because we would then have two group types (species and tree), we would need a two-way ANOVA[1]. Here, to illustrate the key concepts as clearly as possible, we will run an example of a two-way ANOVA with just two of the five species (Het1 and SO2). Table 27.1 shows the fig wasp wing length dataset for the two species in a tidy format, which includes columns for two group types: species and tree.

TABLE 27.1 Wing lengths (mm) measured for two unnamed species of non-pollinating fig wasps collected from two fig trees in 2010 near La Paz in Baja, Mexico.

Species	Tree	Wing Length
Het1	A	2.122
Het1	A	1.938
Het1	B	1.765
Het1	B	1.700
SO2	A	1.635
SO2	A	1.700
SO2	B	1.407
SO2	B	1.378

We could run a one-way ANOVA (or t-test) to see if wing lengths differ between species, then run another one-way ANOVA to see if wing lengths differ between trees. But by including both group types in the same model (species and tree),

[1]Yes, it is possible to have a three-way or a four-way ANOVA, but having so many group types can get very messy very quickly, especially if we want to look at interactions between group types.

we can test how one group affects wing length in the context of the other group (e.g., how tree affects wing length, while also accounting for any effects of species on wing length). We can also see if there is any synergistic effect between groups, which is called an **interaction effect**. For example, if Het1 fig wasps had longer wing lengths than SO2 on Tree A, but shorter wing lengths than SO2 on Tree B, then we would say that there is an interaction between species and tree. Given this kind of interaction effect, it would not make sense to say that Het1 fig wasps have longer or shorter wings than SO2 because this would depend on the tree!

This chapter will not delve into the mathematics of the two-way ANOVA[2]. Working out a two-way ANOVA by hand requires similar, albeit more laborious, calculations of sum of squares and degrees of freedom than is needed for the one-way ANOVA in Section 24.2. But the general concept is the same. The idea is to calculate the amount of the total variation attributable to different groups, or the interaction among group types. Note that the assumptions for the two-way ANOVA are the same as for the one-way ANOVA.

Unlike previous statistical tests in this book, we can actually test three separate null hypotheses with a two-way ANOVA. The first test is the same as the one-way ANOVA in Chapter 24, which focuses on mean species wing lengths:

- H_0 : Mean wing lengths are the same for both species
- H_A : Mean wing lengths are not the same for both species

The second test focuses on the other group type, Tree:

- H_0 : Mean wing lengths are the same in both trees
- H_A : Mean wing lengths are not the same in both trees

These two hypotheses address the **main effects** of the independent variables (species and tree) on the dependent variable (wing length). In other words, the mean effect of one group type (either species or tree) by itself, holding the other constant. The third hypothesis addresses the interaction effect:

- H_0 : There is no interaction between species and tree
- H_A : There is an interaction between species and tree

Interaction effects are difficult to understand at first, so we will look at two concrete examples of two-way ANOVAs. The first will use the Table 27.1 data in which no interaction effect exists. The second will use a different species pairing in which an interaction does exist.

In a two-way ANOVA, the jamovi output (Figure 27.1) appears similar to that of a one-way ANOVA (see Section 24.2.3), but there are four rows of output. Rows 1 and 2 test the null hypotheses associated with the main effects of species and tree, respectively. Row 3 tests the interaction effect. And row

[2]For those interested, chapter 11 of Sokal & Rohlf (1995) works all of this out step by step.

4 shows us the variation in the data that cannot be attributed to either the main effects or the interaction (i.e., residual variation). This is equivalent to the variation within groups in the one-way ANOVA.

ANOVA - Wing length (mm)

	Sum of Squares	df	Mean Square	F	p
Species	0.24675	1	0.24675	45.75115	0.00249
Tree	0.16388	1	0.16388	30.38508	0.00529
Species * Tree	0.00025	1	0.00025	0.04693	0.83909
Residuals	0.02157	4	0.00539		

FIGURE 27.1 Jamovi output table for a two-way ANOVA, which tests for the effects of species (Het1 and SO2), tree, and their interaction on fig wasp wing lengths. Species wing length measurements were collected in 2010 near La Paz in Baja, Mexico.

This is a lot more information than the one-way ANOVA, but it helps to think of each row separately as a test of a different null hypothesis. From the first row, given the high $F = 45.75$ and the low $P = 0.002$, we can reject the null hypothesis that species mean wing lengths are the same. Similarly, from the second row, we can reject the null hypothesis that wing lengths are the same in both trees ($F = 30.39$; $P = 0.005$). In contrast, from the third row, we should not reject the null hypothesis that there is no interaction between species and tree ($F = 0.05$; $P = 0.839$). Figure 27.2 shows these results visually (The jamovi project, 2024).

In Figure 27.2, there is no interaction between species and tree. How can we infer this from Figure 27.2? Wing length is always consistently higher in Tree A than it is in Tree B, regardless of the species of wasp. Similarly, wing length is consistently higher for Het1 than it is for SO2, regardless of tree. Consequently, while wasp species is important for predicting wing length, as is the tree from which the wasp was collected, we do not need to consider one variable to know the effect of the other. This is reflected in the lines having a similar slope, or, more technically, slopes that are not significantly different from one another. If the slopes of the two lines were significantly different, then this would be evidence for an interaction between species and tree.

What would Figure 27.2 look like if there was an interaction effect between species and tree? We can run another two-way ANOVA, this time with a different pair of species: SO1 and SO2. Table 27.2 shows the data.

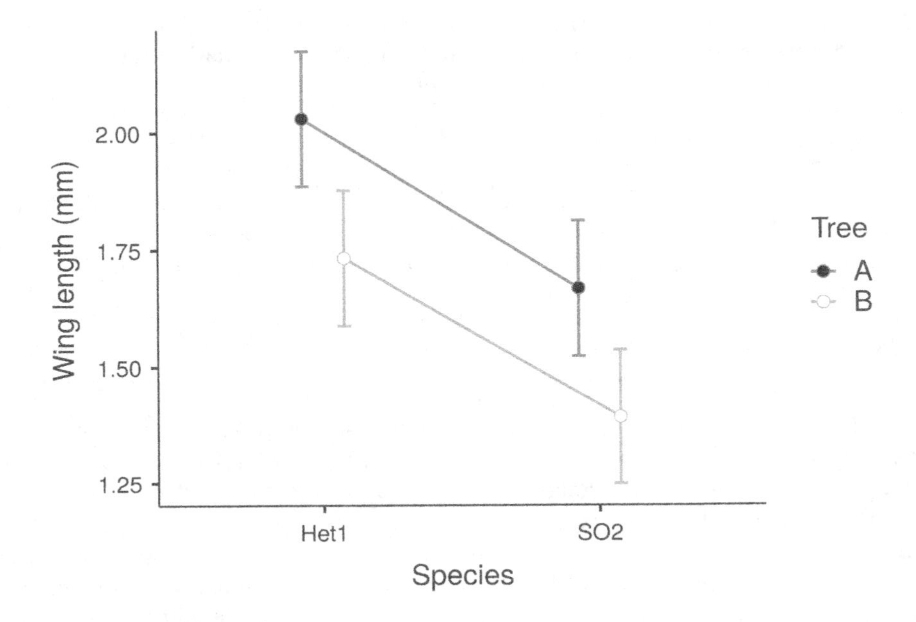

FIGURE 27.2 Interaction plot for a two-way ANOVA of fig wasp wing measurements as affected by species and tree. Points show mean values for the four species and tree combinations (e.g., Het1 from Tree A in the upper left). Error bars show standard errors around the means. Species wing length measurements were collected in 2010 near La Paz in Baja, Mexico.

TABLE 27.2 Wing lengths (mm) measured for two unnamed species of non-pollinating fig wasps collected from two fig trees in 2010 near La Paz in Baja, Mexico.

Species	Tree	Wing Length
SO1	A	1.557
SO1	A	1.493
SO1	B	1.470
SO1	B	1.541
SO2	A	1.635
SO2	A	1.700
SO2	B	1.407
SO2	B	1.378

If we run the two-way ANOVA in jamovi, then we can see the ANOVA output table (Figure 27.3).

ANOVA - Wing length (mm)

	Sum of Squares	df	Mean Square	F	p
Species	0.00044	1	0.00044	0.24509	0.64652
Tree	0.04337	1	0.04337	24.42590	0.00780
Species * Tree	0.03264	1	0.03264	18.38492	0.01277
Residuals	0.00710	4	0.00178		

FIGURE 27.3 Jamovi output table for a two-way ANOVA, which tests for the effects of species (SO1 and SO2), tree, and their interaction on fig wasp wing lengths. Species wing length measurements were collected in 2010 near La Paz in Baja, Mexico.

Figure 27.3 shows that the effect of species is not significant ($P > 0.05$), but the effect of tree is significant ($P < 0.05$). Consequently, in terms of the main effects, it appears that species does not affect wing length, but tree does affect wing length. Unlike the example test between Het1 and SO2, the interaction between species and tree is significant ($P < 0.05$) in Figure 27.3. This means that we cannot interpret the main effects in isolation because the effect of each on wing length will change in the presence of the other. In other words, the effect of tree on wing length will depend on the species. Figure 27.4 shows this interaction visually.

Note that for SO1, wing length is basically the same for Trees A and B (the two left points in Figure 27.4). In contrast, for SO2, tree has a very noticeable effect on wing length! Fig wasps in species SO2 clearly have higher wing lengths on Tree A than they do on Tree B. *Overall*, however, the mean wing length for SO2 appears to be about the same as it is for SO1 (i.e., the middle of the two points for SO2 is at about the same wing length as it is for SO1, around 1.5 mm). Species therefore does not really have an effect on wing length, at least not by itself. But species *does* have a clear effect if you also take tree into account. In other words, to predict how tree will affect wing length (not at all for SO1, or a lot for SO2), it is necessary to know what species of fig wasp is being considered. This is how interactions work in ANOVA, and more generally in statistics.

There is one last point that is relevant to make for the two-way ANOVA. Recall that we tested three null hypotheses simultaneously. Should we not then apply some kind of correction for multiple comparisons, as we did in Chapter 25? This is actually not necessary. The reason is subtle. With the multiple t-tests, we wanted to know if *any* of the pair-wise differences between groups were significant. Each t-test was not really a separate question in this case. We just tried all possible combinations in search of a pair of species with significantly different wing lengths. With the two-way ANOVA, we are asking

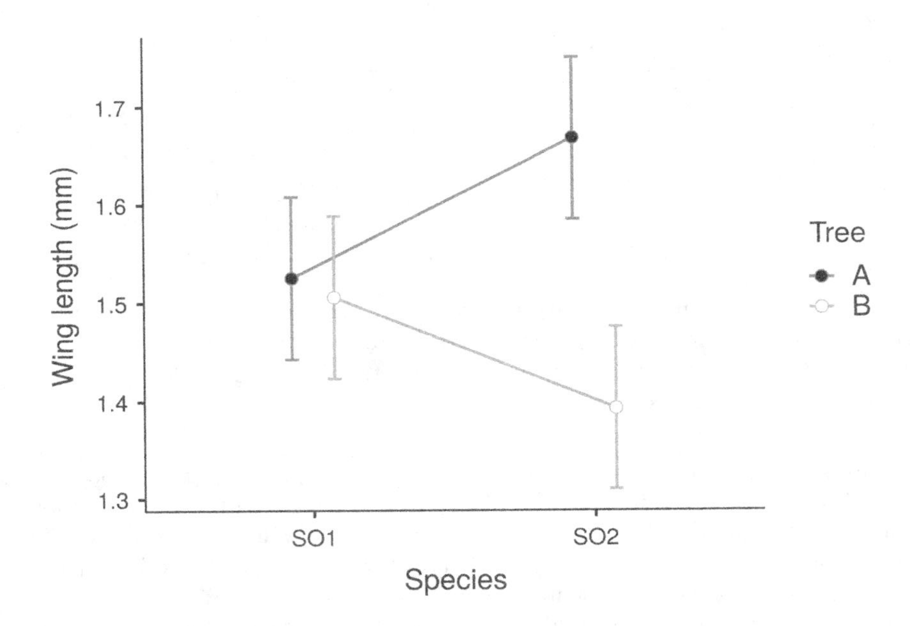

FIGURE 27.4 An interaction plot for a two-way ANOVA of fig wasp wing measurements as affected by species and tree. Points show mean values for the four species and tree combinations (e.g., SO2 from Tree A in the upper right). Error bars show standard errors around the means. Species wing length measurements were collected in 2010 near La Paz in Baja, Mexico.

three separate questions, and accepting a Type I error rate of 0.05 for each of them, individually.

28

Practical. *ANOVA and associated tests*

This chapter focuses on applying the concepts from Chapters 24-27 in jamovi (The jamovi project, 2024). The focus will be on ANOVA and associated tests, with five exercises in total. The data for this chapter are inspired by the work of Dr Lidia de Sousa Teixeira at the University of Stirling (de Sousa Teixeira, 2022). This doctoral work included a nutrient analysis of agricultural soil in different regions of Angola. Measuring soil nutrient concentrations is essential for assessing soil quality, and these data include measurements of Nitrogen (N), Phosphorus (P), and Potassium (K) concentrations. Here we will focus on testing whether or not the concentrations of N, P, and K differ among two different sites and three soil profiles (upper, middle, and lower). This chapter uses the Angola soils dataset[1]. All concentrations of Nitrogen, Phosphorus, and Potassium are given in parts per million (ppm).

28.1 One-way ANOVA (site)

Suppose that we first want to test whether or not mean Nitrogen concentration is the same in different sites. Notice that there are only two sites in the dataset: Funda and Bailundo. We could therefore also use an Independent samples t-test. We will do this first, then compare the t-test output to the ANOVA output. What are the null (H_0) and alternative (H_A) hypotheses for the t-test?

H_0 : ________________________________

H_A : ________________________________

Before running a t-test, remember that we need to check the assumptions of a t-test to see if any of them are violated (see Section 22.4). Use the **Assumption Checks** in jamovi as we did in Section 23.4 to test for normality and homogeneity of variances in Nitrogen concentration. What can you conclude from these two tests?

Normality conclusion: ____

[1] https://bradduthie.github.io/stats/data/Angola_soils.csv

Homogeneity of variances conclusion: ____________________

Given the conclusions from the checks of normality and homogeneity of variances above, what kind of test should you use to see if the mean Nitrogen concentration is significantly different in Funda versus Bailundo?

Test: ____________________

Run the test above in jamovi. What is the p-value of the test, and what conclusion do you make about Nitrogen concentration at the two sites?

$P =$ ____________________

Conclusion: ____________________

Now we will use an ANOVA to test if the mean Nitrogen concentration differs between sites. Remember from Chapter 24 that the ANOVA compares the variance among groups to the variance within groups, calculating an F-statistic and finding where it is in the null F-distribution. To run an ANOVA, navigate to the 'Analyses' tab in jamovi, then select the 'ANOVA' button in the toolbar. From the ANOVA pull-down, select 'One-Way ANOVA'. After selecting the one-way ANOVA, a familiar interface will open up. Place 'Nitrogen' in the Dependent Variables box and 'Site' in the Grouping Variable box. Although we have already checked the assumptions of normality and homogeneity of variances when we ran the t-test, check these boxes under **Assumption Checks** too (Figure 28.1).

Confirm that the Shapiro-Wilk test of normality and Levene's test of homogeneity of variances are consistent with what you concluded when testing the assumptions of the t-test above. Since there is no reason to reject the null hypothesis that group variances are equal, we can use Fisher's One-Way ANOVA by checking 'Assume equal (Fisher's)' under **Variances**. A table called 'One-Way ANOVA' will appear in the panel on the right. Write down the test statistic (F), degrees of freedom, and p-value from this table below.

$F =$ ____________________

$df1 =$ ____________________

$df2 =$ ____________________

$P =$ ____________________

Remember from Chapter 24 that the ANOVA calculates an F-statistic (mean variance among groups divided by mean variance withing groups). This F-statistic is then compared to the null F-distribution with the correct degrees of freedom to calculate the p-value. You can use the interactive app to visualise this from the above jamovi output[2]. To do this, move the 'Variance 1' slider in the app until it is approximately equal to F, then change *df1* and *df2* to the

[2] https://bradduthie.github.io/stats/app/f_distribution/

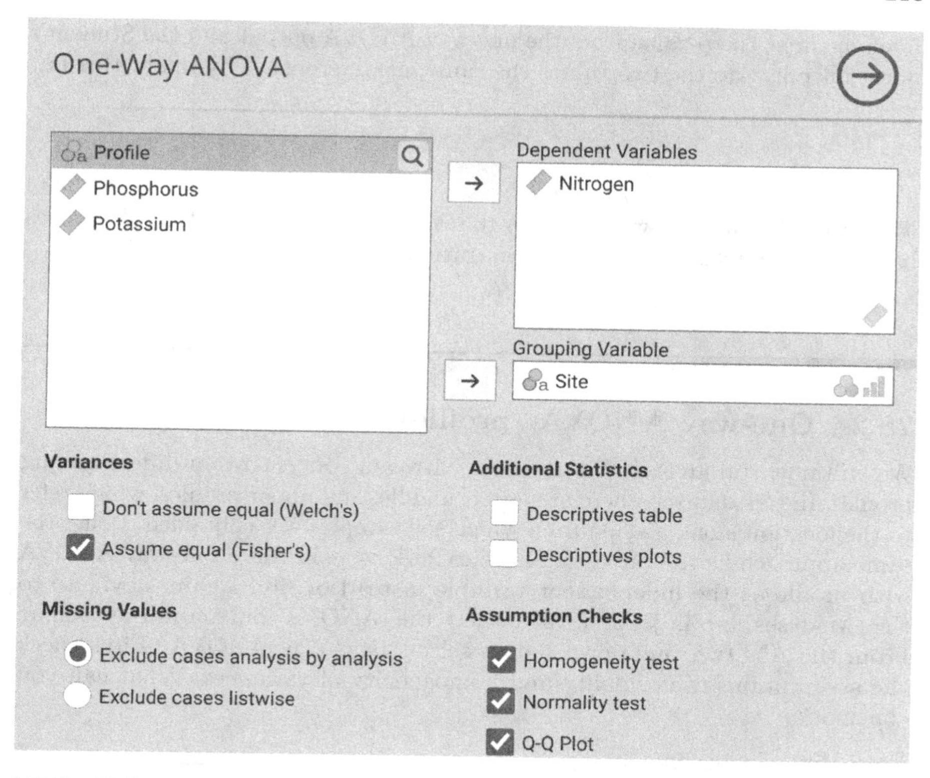

FIGURE 28.1 Jamovi interface for running a one-way ANOVA to test if Nitrogen concentration (ppm) differs among sites in the soils of Angola. Data for this test were inspired by the doctoral thesis of Dr Lidia de Sousa Teixeira.

values above. From the interactive app, what is the approximate area under the curve (i.e., orange area) where the F value on the x-axis is greater than your calculated F?

$P =$ ______________________

Slide the 'Variance 1' to the left now until you find the F value where the probability density in the tail (orange) is about $P = 0.05$. Approximately, what is this threshold F value above which we will reject the null hypothesis?

Approximate threshold F: ______________________

What should you conclude regarding the null hypothesis that sites have the same mean?

Conclusion: ______________________

Look again at the p-value from the one-way ANOVA output and the Student's t-test output. Are the two values the same, or different? Why might this be?

Next, we will run a one-way ANOVA to test the null hypothesis that all profiles have the same mean Nitrogen concentration.

28.2 One-way ANOVA (profile)

We will now run an ANOVA to see if Nitrogen concentration differs among profiles. In this dataset, there are lower, middle, and upper profiles, which refer to the location along a slope from which soil samples were obtained. Using the same approach as the previous Exercise 28.1, we will run a one-way ANOVA with profile as the independent variable instead of site. Again, navigate to the 'Analyses' tab in jamovi, then select the 'ANOVA' button in the toolbar. From the ANOVA pull-down menu, select 'One-Way ANOVA'. First check the assumptions of normality and homogeneity of variances. What can you conclude?

Normality conclusion: ____________________________________

Homogeneity of variances conclusion: ____________________

It appears that the assumptions of normality and homogeneity of variances are met. We can therefore proceed with the one-way ANOVA. Run the one-way ANOVA with the assumption of equal variances (i.e., Fisher's test). What are the output statistics in the One-Way ANOVA table?

$F =$ ____________________

$df1 =$ ____________________

$df2 =$ ____________________

$P =$ ____________________

From these statistics, what do you conclude about the difference in Nitrogen concentration among profiles?

Conclusion: ____________________________

In the previous Exercise 28.1, we used an interactive app[3] to visualise the relationship between the F-statistic and the p-value. We can do the same thing with the distrACTION module in jamovi. To do this, go to the distrACTION option in the jamovi toolbar and select 'F-distribution' from the pull-down menu. Place the *df1* and *df2* from the One-Way ANOVA table into the df1 and df2 boxes under **Parameters** (ignore λ). Under **Function**, select 'Compute probability', then place the F value from the One-Way ANOVA table in the box for x1. We want the upper tail of the F probability distribution, so choose $P(X \geq x1)$ from the radio buttons below. Write down the 'Probability' value from the Results table in the panel to the right.

Probability: ______________________

Note that this is the same value (perhaps with a rounding error) as the p-value from the One-Way ANOVA table above. We can also find the threshold value of F, above which we will reject the null hypothesis. To do this, check the 'Compute quantile' box and set p = 0.95 in the box below. From the Results table, what is the critical F value ('Quantile'), above which we would reject the null hypothesis that all groups have the same mean?

Critical F value: ______________________

Note that the objective of working this out in the distrACTION module (and with the interactive app) is to help explain what these different values in the One-Way ANOVA table actually mean. To actually test the null hypothesis, the One-Way ANOVA output table is all that we really need.

Finally, note that in the ANOVA pull-down menu from the jamovi toolbar, the option 'ANOVA' is just below the 'One-way ANOVA' that we used in this exercise and Exercise 28.1. This is just a more general tool for running an ANOVA, which includes the two-way ANOVA that we will use in Exercise 28.5 below. For now, give this a try by selecting 'ANOVA' from the pull-down menu. In the ANOVA interface, place 'Nitrogen' into the 'Dependent Variable' box and 'Profile' in the 'Fixed Factors' box (Figure 28.2).

The output in the right panel shows an ANOVA table. It includes the sum of squares of the among-group (Profile) and within-group (Residuals) sum of squares and mean square. This is often how ANOVA results are presented in the literature. Fill in the table below (Table 28.1) with the information for degrees of freedom, F, and p.

[3]https://bradduthie.github.io/stats/app/f_distribution/

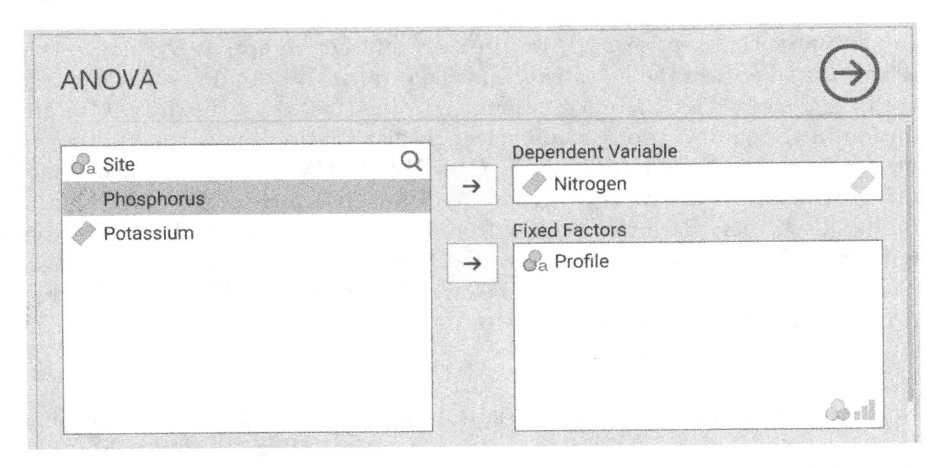

FIGURE 28.2 Jamovi interface for running an ANOVA to test if Nitrogen concentration (ppm) differs among soil profiles in Angola. Data for this test were inspired by the doctoral thesis of Dr Lidia de Sousa Teixeira.

TABLE 28.1 ANOVA output testing the null hypothesis that mean Nitrogen concentration is the same across three different soil profiles in Angola. Data for this test were inspired by the doctoral thesis of Dr Lidia de Sousa Teixeira.

	Sum of Squares	df	Mean Square	F	p
Profile	16888.18606		8444.09303		
Residuals	118092.02927		2460.25061		

Now that we have established from the one-way ANOVA that mean Nitrogen concentration is not the same across all soil profiles, we can use a test of multiple comparisons to test which profile(s) are significantly different from one another.

28.3 Multiple comparisons

In this exercise, we will pick up where we left off in the ANOVA of Exercise 28.2. We have established that not all soil profiles have the same mean. Next, we will run a post hoc multiple comparisons test to evaluate which, if any, soil profiles have different means. In the ANOVA input panel, scroll down to the pull-down option called 'Post Hoc Tests'. Move 'Profile' to the box to the right, then select the 'Tukey' checkbox under **Correction**. Doing this will run Tukey's honestly significant difference (HSD) test introduced in Chapter 25. The output will appear in the panel on the right in a table called 'Post Hoc Tests'. Note that these post hoc tests use the t-distribution to test for significance. Find the p-values associated with Tukey's HSD (P_{tukey}) for each profile pairing. Report these below.

Tukey's HSD Lower - Middle: $P =$ ____________________

Tukey's HSD Lower - Upper: $P =$ ____________________

Tukey's HSD Middle - Upper: $P =$ ____________________

From this output, what can we conclude about the difference among soil profiles?

Next, instead of running Tukey's HSD test, we will use a series of t-tests with a Bonferroni correction. Check the box for 'Bonferroni' in the ANOVA Post Hoc Tests input panel, then find the p-values for the Bonferroni correction ($p_{\text{bonferroni}}$). Note that we do not need to change the α threshold ourselves (i.e., we do not need to see if P is less than $\alpha = 0.05/3 = 0.016667$ instead of $\alpha = 0.05$). Jamovi modifies the p-values appropriately for the Bonferroni correction (we can see the difference by clicking the checkbox for 'No correction' in the Post Hoc Tests input panel). Report the p-values for the Bonferroni correction below.

Bonferroni Lower - Middle: $P =$ ____________________

Bonferroni Lower - Upper: $P =$ ____________________

Bonferroni Middle - Upper: $P =$ ____________________

In general, how are the p-values different between Tukey's HSD and the Bonferroni correction? Are they about the same, higher, or lower?

What does this difference mean in terms of making a Type I error? In other words, based on this output, are we more likely to make a Type I error with Tukey's HSD test or the Bonferroni test?

Note that we ran Tukey's HSD test and the Bonferroni test separately. This is because, when doing a post hoc test, we should choose which test to use in advance. This will avoid biasing our results to get the conclusion that we *want* rather than the conclusion that is *accurate*. If, for example, we first decided to use a Bonferroni correction, but then found that none of our p-values were below 0.05, it would not be okay to try a Tukey's HSD test instead in hopes of changing this result. This kind of practice is colloquially called 'p-hacking' (or 'data dredging'), and it causes an elevated risk of Type I error and a potential for bias in scientific results. Put more simply, trying to game the system to get results in which $P < 0.05$ can lead to mistakes in science (Head et al., 2015). Specifically, p-hacking can lead us to believe that there are patterns in nature where none really exist, which is definitely something that we want to avoid!

28.4 Kruskal-Wallis H test

In this exercise, we will apply the non-parametric equivalent of the one-way ANOVA: the Kruskal-Wallis H test. Suppose that we now want to know if Potassium concentration differs among soil profiles. We therefore want to test the null hypothesis that the mean Potassium concentration is the same for all soil profiles. Before opening the ANOVA input panel, have a look at a histogram of Potassium concentration. How would you describe the distribution? Do the data appear to be normally distributed?

We can test the assumption of normality using a Shapiro-Wilk test. This can be done in the Descriptives panel of jamovi, or we can do it in the One-Way ANOVA panel. To do it in the one-way ANOVA panel, first select 'ANOVA' from the pull-down menu as we did at the end of Exercise 28.2. In the ANOVA interface, place 'Potassium' into the 'Dependent Variable' box and 'Profile' in the 'Fixed Factors' box. Next, scroll down to the 'Assumption Checks' pull-down menu and select all three options. From Levene's test, the Shapiro-Wilk test, and the Q-Q plot, what assumptions of ANOVA might be violated?

Given the violation of ANOVA assumptions, we should consider a non-parametric option. As introduced in Chapter 26, the Kruskal-Wallis H test is a non-parametric alternative to a one-way ANOVA. Like other non-parametric tests introduced in this book, the Kruskal-Wallis H test uses the ranks of a dataset instead of the actual values. To run a Kruskal-Wallis H test, select the Analyses tab, then the 'ANOVA' button from the jamovi toolbar. In the pull-down ANOVA menu, choose 'One-Way ANOVA: Kruskal-Wallis'.

The Kruskal-Wallis input is basically the same as the one-way ANOVA input. We just need to put 'Potassium' in the dependent variable list and 'Profile' in the Grouping Variable box. The output table includes the test statistic (jamovi uses a χ^2 value as a test statistic, which I will introduce in Chapter 29), degrees of freedom, and p-value. Report these values below.

$\chi^2 =$ ____________________

df = ____________________

P = ____________________

From the above output, should we reject or not reject our null hypothesis?

H_0 : ________________________________

Note that the Kruskal-Wallis test in jamovi also includes a type of multiple comparisons test (DSCF pairwise comparisons checkbox). We will not use the Dwass-Steel-Critchlow-Fligner pairwise comparisons, but the general idea is the same as Tukey's HSD test for post hoc multiple comparisons in the ANOVA.

28.5 Two-way ANOVA

Since we have two types of categorical variables (site and profile), we might want to know if either has a significant effect on the concentration of an element, and if there is any interaction between site and profile. The two-way ANOVA was introduced in Chapter 27 with an example of fig wasp wing lengths. Here we will test the effects of site, profile, and their interaction on Nitrogen concentration. Recall from Chapter 27 that a two-way ANOVA actually tests three separate null hypotheses. Write these null hypotheses down below (the order does not matter).

First H_0: ____

Second H_0: ____

Third H_0: ____

To test these null hypotheses again, select 'ANOVA' from the pull-down menu as we did at the end of Exercise 28.2. In the ANOVA interface, place 'Nitrogen' into the 'Dependent Variable' box and both 'Site' and 'Profile' in the 'Fixed Factors' box. Next, scroll down to the 'Assumption Checks' pull-down menu and select all three options. From the assumption checks output tables, is there any reason to be concerned about using the two-way ANOVA?

In the two-way ANOVA output, we see the same ANOVA table as in Exercise 28.2 (Table 28.1). This time, however, there are four rows in total. The first two rows correspond with tests of the main effects of Site and Profile, and the third row tests the interaction between these two variables. Fill in Table 28.2 with the relevant information from the two-way ANOVA output.

TABLE 28.2 Two-way ANOVA output testing the effects of two sites and three different soil profiles on soil Nitrogen concentration in Angola. Data for this test were inspired by the doctoral thesis of Dr Lidia de Sousa Teixeira.

	Sum of Squares	df	Mean Square	F	p
Site	21522.18384		21522.18384		
Profile	22811.1368		11405.5684		
Site * Profile	16209.13035		8104.56517		
Residuals	80497.68348		1788.83741		

From this output table, should you reject or not reject your null hypotheses?

First H_0: ____________________________________

Second H_0: ____________________________________

Third H_0: ____________________________________

In non-technical language, what should you conclude from this two-way ANOVA?

Lastly, we can look at the interaction effect between Site and Profile visually. To do this, scroll down to the 'Estimated Marginal Means' pull-down option. Move 'Site' and 'Profile' from the box on the left to the 'Marginal Means' box on the right (Figure 28.3).

In the panel on the right-hand side, a plot will appear under the heading 'Estimated Marginal Means'. Based on what you learnt in Chapter 27 about interaction effects, what can you say about the interaction between Site and Profile? Does one Profile, in particular, appear to be causing the interaction to be significant? How can you infer this from the Estimated Marginal Means plot?

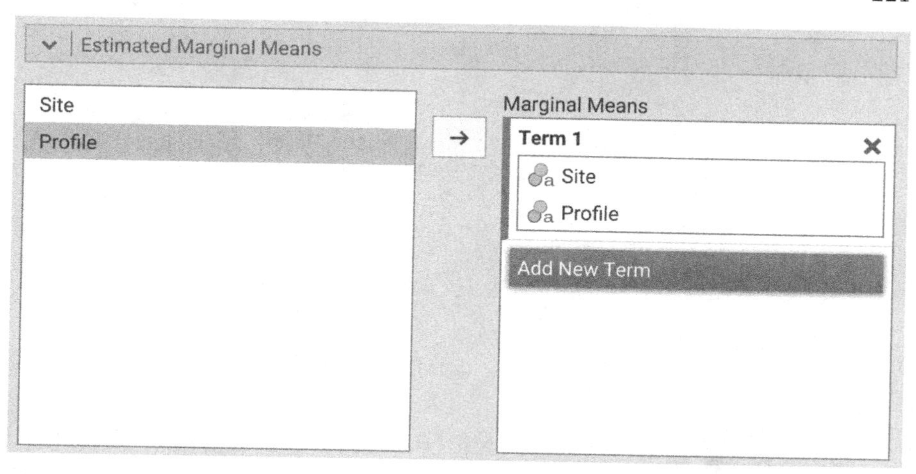

FIGURE 28.3 Jamovi two-way ANOVA test with the pull-down menu for Estimated Marginal Means, which will produce a plot showing the interaction effect of the two-way ANOVA.

Try running a two-way ANOVA to test the effects of Site and Profile on Phosphorus concentration. Based on the ANOVA output, what can you conclude?

29

Frequency and count data

In this book, we have introduced hypothesis testing as a tool to determine if variables were sampled from a population with a specific mean (one sample t-test in Section 22.1), or if different groups of variables were sampled from a population with the same mean (the hypothesis testing in Section 22.2 and ANOVA in Chapter 24). In these tests, the variables for which we calculated the means were always continuous (e.g., fig wasp wing lengths, nitrogen concentration in parts per million). That is, the dependent variables of the t-test and ANOVA could always, at least in theory, take any real value (i.e., any decimal). And the comparison was always between the means of categorical groups (e.g., fig wasp species or study sites). But not every variable that we measure will be continuous. For example, in Chapter 5, we also introduced discrete variables, which can only take discrete counts (1, 2, 3, 4, and so forth). Examples of such **count data** might include the number of species of birds in a forest or the number of days in the year for which an extreme temperature is recorded. Chapter 15 included some examples of count data when introducing probability distributions (e.g., counts of heads or tails in coin flips, or the number of people testing positive for COVID-19). Count data are discrete because they can only take integer values. For example, there cannot be 14.24 bird species in a forest; it needs to be a whole number.

In the biological and environmental sciences, we often want to test whether or not observed counts are significantly different from some expectation. For example, we might hypothesise that the probability of flowers being red versus blue in a particular field is the same. In other words, $P(flower = red) = 0.5$ and $P(flower = Blue) = 0.5$. By this logic, if we were to collect 100 flowers at random from the field, we would expect 50 to be red and 50 to be blue. If we actually went out and collected 100 flowers at random, but found 46 to be red and 54 to be blue, would this be sufficiently different from our expectation to reject the null hypothesis that the probability of sampling a red versus blue flower is the same? We could test this null hypothesis using a Chi-square goodness of fit test (Section 29.1). Similarly, we might want to test if two different count variables (e.g., flower colour and flower species) are associated with one another (e.g., if blue flowers are more common in one species than another species). We could test this kind of hypothesis using a Chi-square test of association.

Before introducing the Chi-square goodness of fit test or the Chi-square test of association, it makes sense to first introduce the Chi-square (χ^2) distribution. The general motivation for introducing the Chi-square distribution is the same as it was for the t-distribution (Chapter 19) or F-distribution (Section 24.1). We need some probability density distribution that is our null distribution, which is what we predict if our null hypothesis is true. We then compare this null distribution to our test statistic to find the probability of sampling a test statistic as or more extreme if the null hypothesis is really true (i.e., a p-value).

29.1 Chi-square distribution

The Chi-square (χ^2) distribution is a continuous distribution in which values of χ^2 can be any real number greater than or equal to 0. We can generate a χ^2 distribution by adding up squared values that are sampled from a standard normal distribution (Sokal & Rohlf, 1995), hence the 'square' in 'Chi-square'. There is a lot to unpack in the previous sentence, so we can go through it step by step. First, we can take another look at the standard normal distribution from Section 15.4.4 (Figure 15.9). Suppose that we randomly sampled four values from the standard normal distribution.

- $x_1 = -1.244$
- $x_2 = 0.162$
- $x_3 = -2.214$
- $x_4 = 2.071$

We can square all of these values, then add up the squares,

$$\chi^2 = (-1.244)^2 + (0.162)^2 + (-2.214)^2 + (2.071)^2.$$

Note that χ^2 cannot be negative because when we square a number that is either positive or negative, we always end up with a positive value (e.g., $-2^2 = 4$, see Section 1.1). The final value is $\chi^2 = 10.76462$. Of course, this χ^2 value would have been different if our x_i values (x_1, x_2, x_3, and x_4) had been different. And if we are sampling randomly from the normal distribution, we should not expect to get the same χ^2 value from four new random standard normal deviates. We can therefore ask: if we were to keep sampling four standard normal deviates and calculating new χ^2 values, what would be the distribution of these χ^2 values? The answer is shown in Figure 29.1.

Looking at the shape of Figure 29.1, we can see that most of the time, the sum of deviations from the mean of $\mu = 0$ will be about 2. But sometimes we will get a much lower or higher value of χ^2 by chance, if we sample particularly low or high values of x_i.

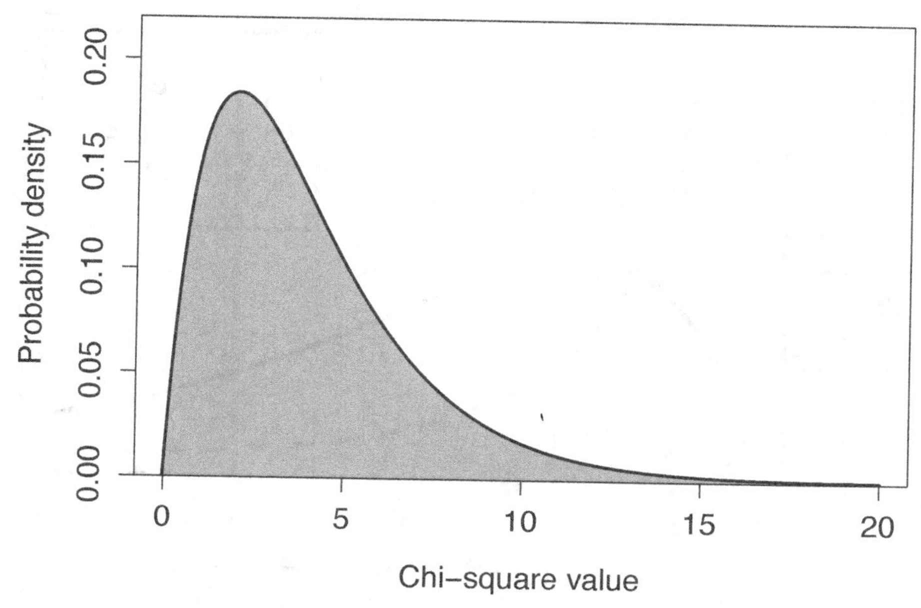

FIGURE 29.1 A Chi-square distribution that is the expected sum of four squared standard normal deviates, i.e., the sum of four values sampled from a standard normal distribution and squared.

If we summed a different number of squared x_i values, then we would expect the distribution of χ^2 to change. Had we sampled fewer than four x_i values, the expected χ^2 would be lower just because we are adding up fewer numbers. Similarly, had we sampled more than four x_i values, the expected χ^2 would be higher just because we are adding up more numbers. The shape of the χ^2 distribution[1] is therefore determined by the number of values sampled (n), or more specifically the degrees of freedom (*df*, or sometimes v), which in a sample is $df = n - 1$. This is the same idea as the t-distribution from Chapter 19, which also changed shape depending on the degrees of freedom. Figure 29.2 shows the different χ^2 probability density distributions for different degrees of freedom.

As with the F-distribution from Section 24.1, visualising the χ^2 distribution is easier using an interactive application[2]. And as with the F-distribution, it

[1]A random variable X has a χ^2 distribution if and only if its probability density function is defined by (Miller & Miller, 2004),

$$f(x) = \begin{cases} \frac{1}{2^{\frac{2}{v}}\Gamma(\frac{v}{2})} x^{\frac{v-2}{2}} e^{-\frac{x}{2}} & for\ x > 0 \\ 0 & elsewhere \end{cases}$$

In this equation, v is the degrees of freedom of the distribution.

[2]`https://bradduthie.github.io/stats/app/chi-square/`

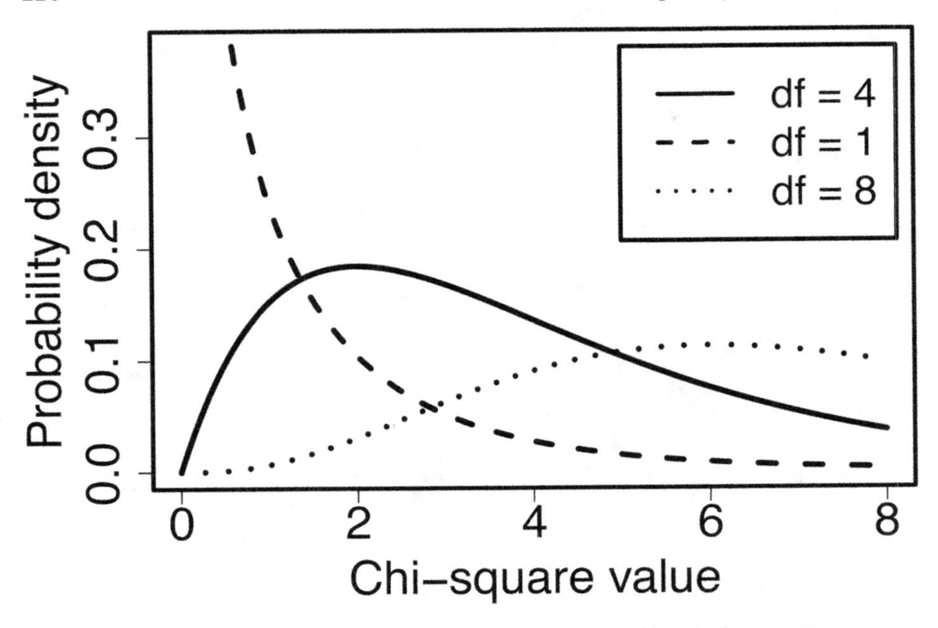

FIGURE 29.2 Probability density functions for three different Chi-square distributions, each of which has different degrees of freedom (*df*).

is not necessary to memorise how the χ^2 distribution changes with different degrees of freedom. The important point is that the distribution changes with different degrees of freedom, and we can map probabilities to the χ^2 value on the x-axis in the same way as any other distribution.

What does any of this have to do with count data? It actually is a bit messy. The χ^2 distribution is not a perfect tool for comparing observed and expected counts (Sokal & Rohlf, 1995). After all, counts are integer values, and the χ^2 distribution is clearly continuous (unlike, e.g., the binomial or Poisson distributions from Section 15.4). The χ^2 distribution is in fact a useful approximation for testing counts, and one that becomes less accurate when sample size (Slakter, 1968) or expected count size (Tate & Hyer, 1973) is small. Nevertheless, we can use the χ^2 distribution as a tool for testing whether observed counts are significantly different from expected counts. The first test that we will look at is the goodness of fit test.

29.2 Chi-square goodness of fit

The first kind of test that we will consider for count data is the goodness of fit test. In this test, we have some number of counts that we expect to observe (e.g., expected counts of red versus blue flowers), then compare this expectation to the counts that we actually observe. If the expected and observed counts differ by a lot, then we will get a large test statistic and reject the null hypothesis. A simple concrete example will make this a bit more clear.

Recall the practical in Chapter 17, in which players of the mobile app game 'Power Up!' chose a small, medium, or large dam at the start of the game. Suppose that we are interested in the size of dam that policy-makers choose to build when playing the game, so we find 60 such people in Scotland and ask them to play the game. Perhaps we do not think that the policy-makers will have any preference for a particular dam size (and therefore just pick one of the three dam sizes at random). We would therefore expect an equal number of small, medium, and large dams to be selected among the 60 players. That is, for our expected counts of each dam size (E_{size}), we expect 20 small ($E_{\text{small}} = 20$), 20 medium ($E_{\text{medium}} = 20$), and 20 large ($E_{\text{large}} = 20$) dams in total (because $60/3 = 20$).

Of course, even if our players have no preference for a particular dam size, the number of small, medium, and large dams will not always be *exactly* the same. The expected counts might still be a bit different from the observed counts of each dam size (O_{size}). Suppose, for example, we find that out of our total 60 policy-makers, we observe 18 small ($O_{\text{small}} = 18$), 24 medium ($O_{\text{medium}} = 24$), and 18 large ($O_{\text{large}} = 18$), dams were actually chosen by game players. What we want to test is the null hypothesis that there is no significant difference between expected and observed counts.

- H_0: There is no significant difference between expected and observed counts.
- H_A: There is a significant difference between expected and observed counts.

To get our test statistic[3], we now just need to take each observed count, subtract the expected count, square this difference, divide by the expected count, then add everything up,

$$\chi^2 = \frac{(18-20)^2}{20} + \frac{(24-20)^2}{20} + \frac{(18-20)^2}{20}.$$

[3] A lot of statisticians will use X^2 to represent the test statistic here instead of χ^2 (Sokal & Rohlf, 1995). The difference is the upper case 'X' versus the Greek letter Chi, 'χ'. The X is used since the test statistic we calculate here is not *technically* from the χ^2 distribution, just an approximation. We will not worry about the distinction here, and to avoid confusion, we will just go with χ^2.

We can calculate the values in the numerator. Note that all of these numbers must be positive (e.g., $18 - 20 = -2$, but $-2^2 = 4$),

$$\chi^2 = \frac{4}{20} + \frac{16}{20} + \frac{4}{20}.$$

When we sum the three terms, we get a value of $\chi^2 = 1.2$. Note that if all of our observed values had been the same as the expected values (i.e., 20 small, medium, and large dams actually chosen), then we would get a χ^2 value of 0. The more the observed values differ from the expectation of 20, the higher the χ^2 will be. We can now check to see if the test statistic $\chi^2 = 1.2$ is sufficiently large to reject the null hypothesis that our policy-makers have no preference for small, medium, or large dams. There are $n = 3$ categories of counts (small, medium, and large), meaning that there are $df = 3 - 1 = 2$ degrees of freedom. As it turns out, if the null hypothesis is true, then the probability of observing a value of $\chi^2 = 1.2$ or higher (i.e., the p-value) is $P = 0.5488$. Figure 29.3 shows the appropriate χ^2 distribution plotted, with the area above the test statistic $\chi^2 = 1.2$ shaded in grey.

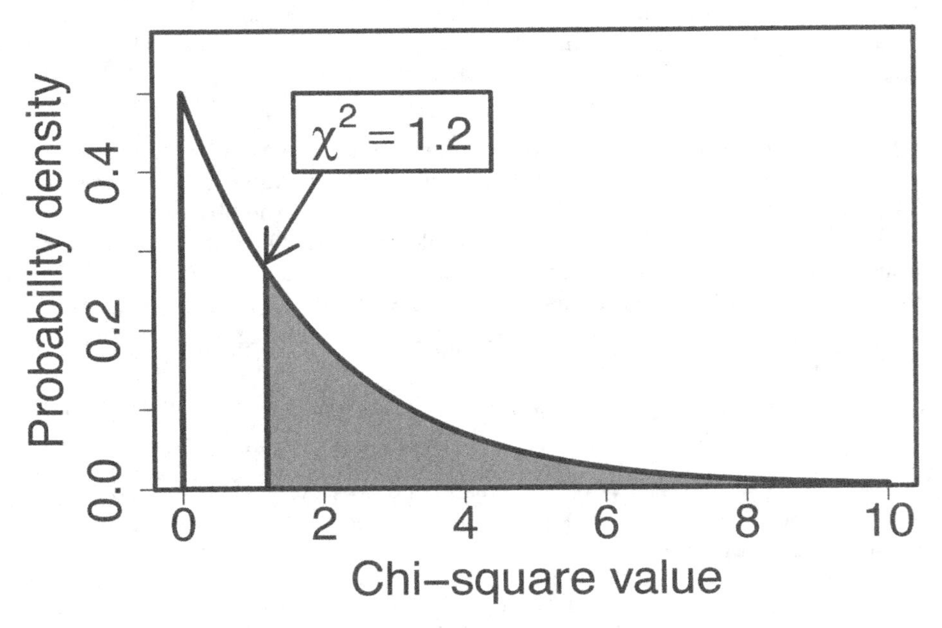

FIGURE 29.3 A Chi-square distribution with 2 degrees of freedom, used in a goodness of fit test to determine if all players of the mobile game 'Power Up!' have the same preference for initial dam size. An arrow indicates a χ^2 value of 1.2 on the x-axis, and grey shading indicates the probability of a more extreme value under the null hypothesis.

Because $P > 0.05$, we do not reject the null hypothesis that there is no significant difference between expected and observed counts of chosen dam sizes.

Note that this was a simple example. For a goodness of fit test, we can have any number of different count categories (at least, any number greater than 1). The expectations also do not need to be integers. For example, if we only managed to find 59 policy makers instead of 60, then our expected counts would have been $59/3 = 19.33$ instead of $60/3 = 20$. The expectations also do not *need* to be the same. For example, we could have tested the null hypothesis that twice as many policy-makers would choose large dams (i.e., $E_{\mathrm{large}} = 40$, $E_{\mathrm{medium}} = 10$, and $E_{\mathrm{small}} = 10$). For n categories, the more general equation for the χ^2 statistic is,

$$\chi^2 = \sum_{i=1}^{n} \frac{(O_i - E_i)^2}{E_i}.$$

We can therefore use this general equation to calculate a χ^2 for any number of categories (n). Next, we will look at testing associations between counts in different types of categories.

29.3 Chi-square test of association

The second kind of test that we will consider for count data is the Chi-square test of association. While the goodness of fit test focused on a single categorical variable (dam sizes in the example above), the Chi-square test of association focuses on two different categorical variables. What we want to know is whether or not the two categorical variables are independent of one another (Box et al., 1978). In other words, does knowing something about one variable tell us anything about the other variable? A concrete example will make it easier to explain. We can again make use of the Chapter 17 game 'Power Up!'. As mentioned in the previous section, game players choose a small, medium, or large dam at the start of the game. Players can play the game on either an Android or macOS mobile device. We therefore have two categorical variables: dam size and OS type. We might want to know, do Android users choose the same dam sizes as macOS users? In other words, are dam size and OS type associated? We can state this as a null and alternative hypothesis.

- H_0: There is no association between OS and dam size choice.
- H_A: There is an association between OS and dam size choice.

Consider the data in Table 29.1, which show counts of Android versus macOS users and their dam choices.

TABLE 29.1 Counts ($N = 60$) from a mobile game called 'Power Up!', in which players are confronted with trade-offs between energy output, energy justice, and biodiversity. Players can use one of two categories of Operating System (Android or macOS) and build one of three types of dam in the game (Small, Medium, or Large).

	Small	Medium	Large
Android	8	16	6
macOS	10	8	12

Just looking at the counts in Table 29.1, it appears that there might be an association between the two variables. For example, Android users appear to be more likely to choose a medium dam than macOS users. Medium dams are the most popular choice for Android users, but they are the least popular choice for macOS users. Nevertheless, could this just be due to chance? If it were due to chance, then how unlikely are the counts in Table 29.1? In other words, if Android and macOS users in the whole population really do choose dam sizes at the same frequencies, then what is the probability of getting a sample of 60 players in which the choices are as or more unbalanced as this? This is what we want to answer with our Chi-square test of association.

The general idea is the same as with the Chi-square goodness of fit test. We have our observed values (Table 29.1). We now need to find the expected values to calculate a χ^2 value. But the expected values are now a bit more complicated. With the goodness of fit test in Section 29.2, we just assumed that all categories were equally likely (i.e., the probability of choosing each size dam was the same). There were 60 players and three dam sizes, so the expected frequency of each dam choice was $60/3 = 20$. Now it is different. We are not testing if dam sizes or OS choices are the same. We want to know if they are *associated* with one another. That is, regardless of the popularity of Android versus macOS, or the frequency with which small, medium and large dams are selected, do Android users choose different dam sizes than macOS users? If dam size is not associated with OS, then we would predict that the relative frequency of small, medium, and large dams would be the same for both Android and macOS.

To find the expected counts of each variable combination (e.g., Android and Small, or macOS and Large), we need to get the probability that each category is selected independently. For example, what is the probability of a player selecting a large dam, regardless of the OS that they are using? Table 29.2 shows these probabilities as additional rows and columns added onto Table 29.1

TABLE 29.2 Counts ($N = 60$) from a mobile game called 'Power Up!', in which players are confronted with trade-offs between energy output, energy justice, and biodiversity. Players can use one of two categories of Operating System (Android or macOS) and build one of three categories of dam in the game (Small, Medium, or Large). Outer rows and columns show the probabilities of categories being selected.

	Small	Medium	Large	Probability
Android	8	16	6	0.5
macOS	10	8	12	0.5
Probability	0.3	0.4	0.3	–

Since there are 30 total Android users ($8 + 16 + 6 = 30$) and 30 total macOS users ($10 + 8 + 12 = 30$), the probability of a player having an Android OS is $30/60 = 0.5$, and the probability of a player having a macOS is also $30/60 = 0.5$. Similarly, there are 18 small, 24 medium, and 18 large dam choices in total. Hence, the probability of a player choosing a small dam is $18/60 = 0.3$, medium is $24/60 = 0.4$, and large is $18/60 = 0.3$. If these probabilities combine independently[4], then we can multiply them to find the probability of a particular combination of categories. For example, the probability of a player using Android is 0.5 and choosing a small dam is 0.3, so the probability of a player having both Android **and** a small dam is $0.5 \times 0.3 = 0.15$ (see Chapter 16 for an introduction to probability models). The probability of a player using Android **and** choosing a medium dam is $0.5 \times 0.4 = 0.2$. We can fill in all of these joint probabilities in a new Table 29.3.

TABLE 29.3 Probabilities for each combination of categorical variables from a dataset in which players on either an Android or macOS choose a dam size in the mobile app game 'Power Up!', assuming that variables are independent of one another.

	Small	Medium	Large	Probability
Android	0.15	0.2	0.15	0.5
macOS	0.15	0.2	0.15	0.5
Probability	0.3	0.4	0.3	–

From Table 29.3, we now have the probability of each combination of variables. Note that all of these probabilities sum to 1.

$$0.15 + 0.2 + 0.15 + 0.15 + 0.2 + 0.15 = 1.$$

[4]We can call these the 'marginal probabilities'.

To get the expected count of each combination, we just need to multiply the probability by the sample size, i.e., the total number of players ($N = 60$). For example, the expected count of players who use Android and choose a small dam will be $0.15 \times 60 = 9$. Table 29.4 fills in all of the expected counts. Note that the sum of all the counts equals our sample size of 60.

TABLE 29.4 Expected counts for each combination of categorical variables from a dataset in which players on either an Android or macOS choose a dam size in the mobile app game 'Power Up!', assuming that variables are independent of one another.

	Small	Medium	Large	Sum
Android	9	12	9	30
macOS	9	12	9	30
Sum	18	24	18	–

We now have both the observed (Table 29.2) and expected (Table 29.4) counts (remember that the expected counts do not *need* to be integers). To get our χ^2 test statistic, we use the same formula as in Section 29.2,

$$\chi^2 = \sum_{i=1}^{n} \frac{(O_i - E_i)^2}{E_i}.$$

There are six total combinations of OS and dam size, so there are $n = 6$ values to sum up,

$$\chi^2 = \frac{(8-9)^2}{9} + \frac{(16-12)^2}{12} + ... + \frac{(16-12)^2}{12} + \frac{(8-9)^2}{9}.$$

If we sum all of the six terms, we get a value of $\chi^2 = 4.889$. We can compare this to the null χ^2 distribution as we did in the Section 29.2 goodness of fit test, but we need to know the correct degrees of freedom. The correct degrees of freedom[5] is the number of categories in variable 1 (n_1) minus 1, times the number of categories in variable 2 (n_2) minus 1,

[5]This formula works due to a bit of a mathematical trick (Sokal & Rohlf, 1995). The actual logic of the degrees of freedom is a bit more involved. From our total of $k = 6$ different combinations, we actually need to subtract one degree of freedom for the total sample size ($N = 60$), then a degree of freedom for each variable probability estimated (i.e., subtract $n_1 - 1$ and $n_2 - 1$ because we need this many degrees of freedom to get the n_1 and n_2 probabilities, respectively; if we have all but one probability, then we know the last probability because the probabilities must sum to 1). Since we lose $n_1 - 1$ and $n_2 - 1$ degrees of freedom, and 1 for the sample size, this results in $df = k - (n_1 - 1) - (n_2 - 1) - 1$. In the case of the 'Power Up!' example, we get $df = 6 - (3 - 1) - (2 - 1) - 1 = 2$. The $df = (n_1 - 1) \times (n_2 - 1)$ formulation is possible because $k = n_1 \times n_2$ (Sokal & Rohlf, 1995).

$$df = (n_1 - 1) \times (n_2 - 1).$$

In our example, the degrees of freedom equals the number of dam types minus 1 ($n_{\text{dam}} = 3 - 1$) times the number of operating systems minus 1 ($n_{\text{OS}} = 2 - 1$). The correct degrees of freedom is therefore $df = 2 \times 1 = 2$. We now just need to find the p-value for a Chi-square distribution with two degrees of freedom and a test statistic of $\chi^2 = 4.889$. We get a value of about $P = 0.0868$. In other words, if H_0 is true, then the probability of getting a χ^2 of 4.889 or higher is $P = 0.0868$. Consequently, because $P > 0.05$, we would not reject the null hypothesis. We should therefore conclude that there is no evidence for an association between OS and dam size choice.

Jamovi will calculate the χ^2 value and get the p-value for the appropriate degrees of freedom (The jamovi project, 2024). To do this, it is necessary to input the categorical data (e.g., Android, macOS) in a tidy format, which will be a focus of Chapter 31.

There is one final point regarding expected and observed values of the Chi-square test of association. There is another way of getting these expected values that is a bit faster (and more widely taught) but does not demonstrate the logic of expected counts as clearly. If we wanted to, we could sum the rows and columns of our original observations. Table 29.5 shows the original observations with the sum of each row and column.

TABLE 29.5 Observed counts for each combination of categorical variables from a dataset in which players on either an Android or macOS choose a dam size in the mobile app game 'Power Up!'. The last row and column shows the sum of observed dam sizes and OS users, respectively.

	Small	Medium	Large	Sum
Android	8	16	6	30
macOS	10	8	12	30
Sum	18	24	18	–

We can get the expected counts from Table 29.5 if we multiply each row sum by each column sum, then divide by the total sample size ($N = 60$). For example, to get the expected counts of Android users who choose a small dam, we can multiply $(18 \times 30)/60 = 9$. To get the expected counts of macOS users who choose a medium dam, we can multiply $(30 \times 24)/60 = 12$. This works for all of combinations of rows and columns, so we could do it to find all of the expected counts from Table 29.4.

30

Correlation

This chapter focuses on the association between types of variables that are quantitative (i.e., represented by numbers). It is similar to the Chi-square test of association from Chapter 29 in the sense that it is about how variables are associated. The focus of the Chi-square test of association was on the association when data were categorical (e.g., 'Android' or 'macOS' operating system). Here we focus instead on the association when data are numeric. But the concept is generally the same; are variables independent, or does knowing something about one variable tell us something about the other variable? For example, does knowing something about the latitude of a location tell us something about its average yearly temperature?

30.1 Scatterplots

The easiest way to visualise the concept of a correlation is by using a scatterplot. Scatterplots are useful for visualising the association between two quantitative variables. In a scatterplot, the values of one variable are plotted on the x-axis, and the values of a second variable are plotted on the y-axis. Consider two fig wasp species of the genus *Heterandrium* (Figure 30.1).

Both fig wasp species in Figure 30.1 are unnamed. We can call the species in Figure 30.1A 'Het1' and the species in Figure 30.1B 'Het2'. We might want to collect morphological measurements of fig wasp head, thorax, and abdomen lengths in these two species (Duthie et al., 2015). Table 30.1 shows these measurements for 11 wasps.

TABLE 30.1 Body segment length measurements (mm) from fig wasps of two species. Data were collected from Baja, Mexico.

Species	Head	Thorax	Abdomen
Het1	0.566	0.767	1.288
Het1	0.505	0.784	1.059
Het1	0.511	0.769	1.107
Het1	0.479	0.766	1.242

Species	Head	Thorax	Abdomen
Het1	0.545	0.828	1.367
Het1	0.525	0.852	1.408
Het2	0.497	0.781	1.248
Het2	0.450	0.696	1.092
Het2	0.557	0.792	1.240
Het2	0.519	0.814	1.221
Het2	0.430	0.621	1.034

Intuitively, we might expect most of these measurements to be associated with one another. For example, if a fig wasp has a relatively long thorax, then it probably also has a relatively long abdomen (i.e., it could just be a big wasp). We can check this visually by plotting one variable on the x-axis and the other on the y-axis. Figure 30.2 does this for wasp thorax length (x-axis) and abdomen length (y-axis).

In Figure 30.2, each point is a different wasp from Table 30.1. For example, in the last row of Table 30.1, there is a wasp with a particularly low thorax length (0.621 mm) and abdomen length (1.034 mm). In the scatterplot, we can see this wasp represented by the point that is lowest and furthest to the left (Figure 30.2).

There is a clear association between thorax length and abdomen length in Figure 30.2. Fig wasps that have low thorax lengths also tend to have low abdomen lengths, and wasps that have high thorax lengths also tend to have high abdomen lengths. In this sense, the two variables are associated. More specifically, they are positively correlated. As thorax length increases, so does abdomen length.

30.2 Correlation coefficient

The **correlation coefficient** formalises the association described in the previous section. It gives us a single number that defines how two variables are correlated. We represent this number with the letter 'r', which can range from values of -1 to 1.[1] Positive values indicate that two variables are positively correlated, such that a higher value of one variable is associated with higher values of the other variable (as was the case with fig wasp thorax and abdomen measurements). Negative values indicate that two variables are negatively

[1]Note that r is the sample correlation coefficient, which is an estimate of the population correlation coefficient. The population correlation coefficient is represented by the Greek letter 'ρ' ('rho'), and sometimes the sample correlation coefficient is represented as $\hat{\rho}$.

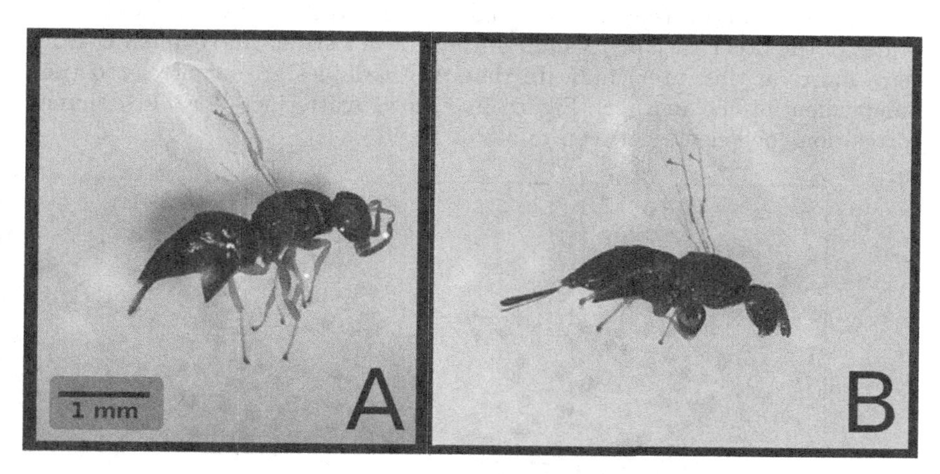

FIGURE 30.1 Fig wasps from two different species, (A) Het1 and (B) Het2. Wasps were collected from Baja, Mexico. Image modified from Duthie, Abbott, and Nason (2015).

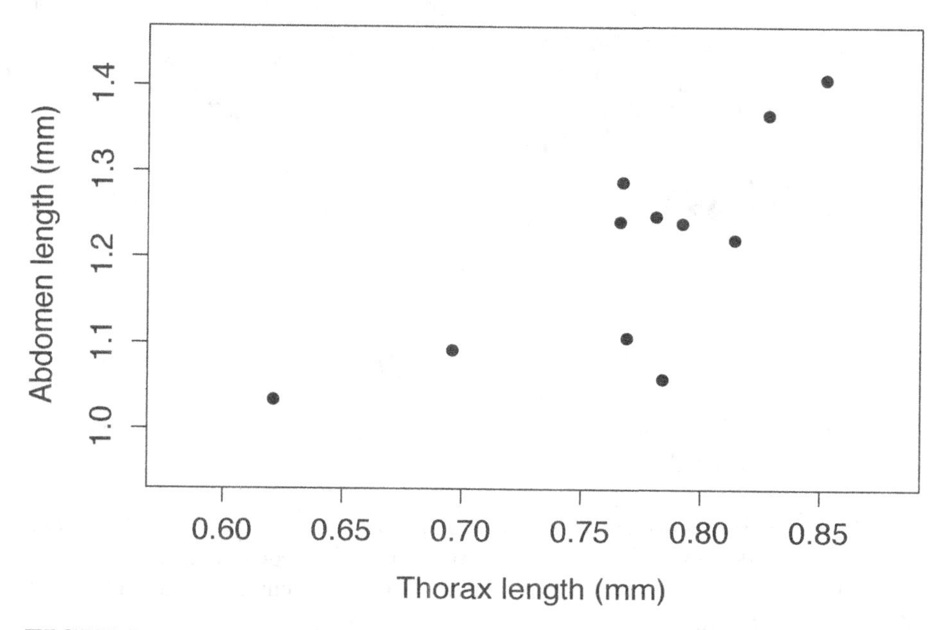

FIGURE 30.2 Example of a scatterplot in which fig wasp thorax length (x-axis) is plotted against fig wasp abdomen length (y-axis). Each point is a different fig wasp. Wasps were collected in 2010 in Baja, Mexico.

correlated, such that a higher values of one variable are associated with lower values of the other variable. Values of zero (or not significantly different from zero, more on this later) indicate that two variables are uncorrelated (i.e., independent of one another). Figure 30.3 shows scatterplots for four different correlation coefficients between values of x and y.

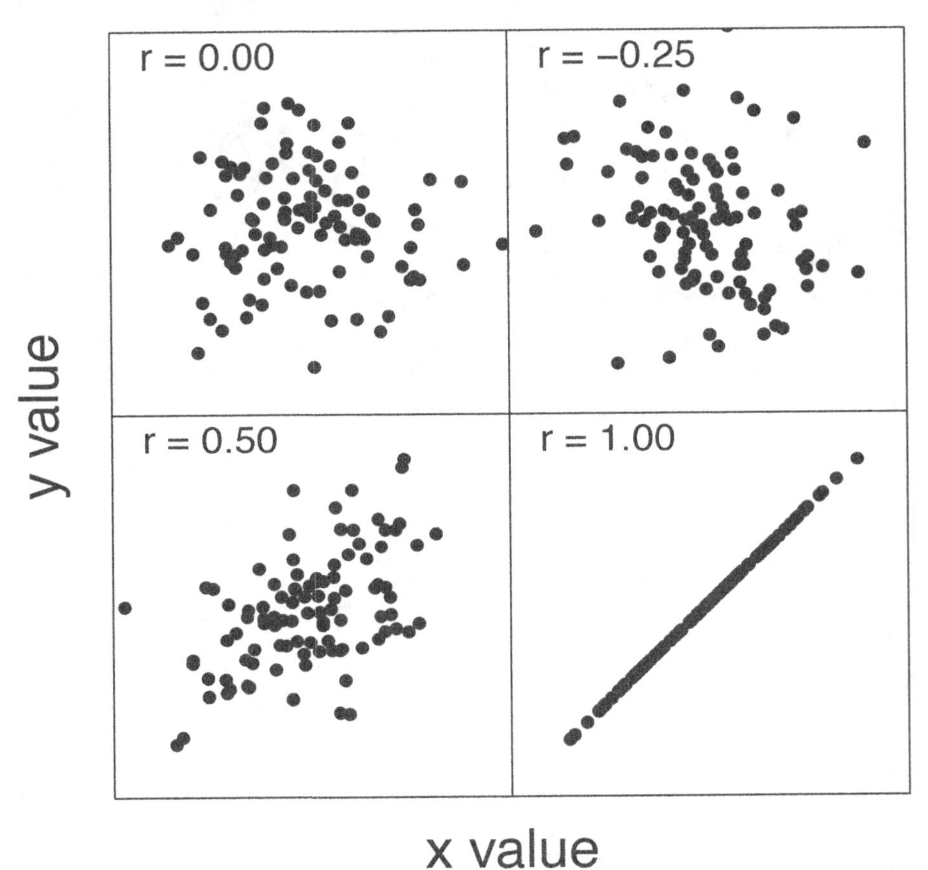

FIGURE 30.3 Examples of scatterplots with different correlation coefficients (r) between two variables (x and y).

We will look at two types of correlation coefficient, the Pearson product moment correlation coefficient and Spearman's rank correlation coefficient. The two are basically the same; Spearman's rank correlation coefficient is just a correlation of the ranks of values instead of the actual values.

30.2.1 Pearson product moment correlation coefficient

To understand the correlation coefficient, we need to first understand covariance. Section 12.3 introduced the variance (s^2) as a measure of spread in some variable x,

$$s^2 = \frac{1}{N-1}\sum_{i=1}^{N}(x_i - \bar{x})^2 .$$

The variance is actually just a special case of a covariance. The variance describes how a variable x covaries with itself. The covariance ($cov_{x,y}$) describes how a variable x covaries with another variable y,

$$cov_{x,y} = \frac{1}{N-1}\sum_{i=1}^{N}(x_i - \bar{x})(y_i - \bar{y}) .$$

The $\bar{x}$ and $\bar{y}$ are the means of x and y, respectively. Note that if $x_i = y_i$, then the equation for $cov_{x,y}$ is identical to the equation for s^2 because $(x_i - \bar{x})(x_i - \bar{x}) = (x_i - \bar{x})^2$.

What the equation for $cov_{x,y}$ is describing is how variation in x relates to variation in y. If a value of x_i is much higher than the mean $\bar{x}$, and a value of y_i is much higher than the mean $\bar{y}$, then the product of $(x_i - \bar{x})$ and $(y_i - \bar{y})$ will be especially high because we will be multiplying two large positive numbers together. If a value of x_i is much higher than the mean $\bar{x}$, but the corresponding y_i is much lower than the mean $\bar{y}$, then the product of $(x_i - \bar{x})$ and $(y_i - \bar{y})$ will be especially low because we will be multiplying a large positive number and a large negative number. Consequently, when x_i and y_i tend to deviate from their means $\bar{x}$ and $\bar{y}$ in a consistent way, we get either high or low values of $cov_{x,y}$. If there is no such relationship between x and y, then we will get $cov_{x,y}$ values closer to zero.

The covariance can, at least in theory, be any real number. How low or high it is will depend on the magnitudes of x and y, just like the variance. To get the Pearson product moment correlation coefficient[2] (r), we need to standardise the covariance so that the minimum possible value of r is -1 and the maximum possible value of r is 1. We can do this by dividing $cov_{x,y}$ by the product of the standard deviation of x (s_x) and the standard deviation of y (s_y),

$$r = \frac{cov_{x,y}}{s_x \times s_y}.$$

This works because $s_x \times s_y$ describes the total variation between the two variables, and the absolute value of $cov_{x,y}$ cannot be larger than this total. We

[2]We can usually just call this the 'correlation coefficient'.

can again think about the special case in which $x = y$. Since the covariance between x and itself is just the variance of x (s_x^2), and $s_x \times s_x = s_x^2$, we end up with the same value on the top and bottom, and $r = 1$ (i.e., x is completely correlated with itself).

We can expand $cov_{x,y}$, s_x, and s_y to see the details of the equation for r,

$$r = \frac{\frac{1}{N-1}\sum_{i=1}^{N}\left(x_i - \bar{x}\right)\left(y_i - \bar{y}\right)}{\sqrt{\frac{1}{N-1}\sum_{i=1}^{N}\left(x_i - \bar{x}\right)^2}\sqrt{\frac{1}{N-1}\sum_{i=1}^{N}\left(y_i - \bar{y}\right)^2}}.$$

This looks like a lot, but we can clean the equation up a bit because the $1/(N-1)$ expressions cancel on the top and bottom of the equation,

$$r = \frac{\sum_{i=1}^{N}\left(x_i - \bar{x}\right)\left(y_i - \bar{y}\right)}{\sqrt{\sum_{i=1}^{N}\left(x_i - \bar{x}\right)^2}\sqrt{\sum_{i=1}^{N}\left(y_i - \bar{y}\right)^2}}.$$

As with other statistics defined in this book, it is almost never necessary to calculate r by hand. Jamovi will make these calculations for us (The jamovi project, 2024). The reason for working through all of these equations is to help make the conceptual link between r and the variance of the variables of interest (Rodgers & Nicewander, 1988). To make this link a bit clearer, we can calculate the correlation coefficient of thorax and abdomen length from Table 30.1. We can set thorax to be the x variable and abdomen to be the y variable. Mean thorax length is $\bar{x} = 0.770$, and mean abdomen length is $\bar{y} = 1.209636$. The standard deviation of thorax length is $s_x = 0.06366161$, and the standard deviation of abdomen length is $s_y = 0.1231806$. This gives us the numbers that we need to calculate the bottom of the fraction for r, which is $s_x \times s_y = 0.007841875$. We now need to calculate the covariance on the top. To get the covariance, we first need to calculate $\left(x_i - \bar{x}\right)\left(y_i - \bar{y}\right)$ for each row (i) in Table 30.1. For example, for row 1, $(0.767 - 0.770)(1.288 - 1.210) = -0.000234$. For row 2, $(0.784 - 0.770)(1.059 - 1.210) = -0.002114$. We continue this for all rows. Table 30.2 shows the thorax length (x_i), abdomen length (y_i), and $\left(x_i - \bar{x}\right)\left(y_i - \bar{y}\right)$ for rows $i = 1$ to $i = 11$.

TABLE 30.2 Measurements of 11 fig wasp thorax and abdomen lengths (mm). The fourth column shows the product of the deviations of each measurement from the mean, where mean thorax length is 0.770 and mean abdomen length is 1.210.

Row (i)	Thorax	Abdomen	Squared Deviation
1	0.767	1.288	−0.000234
2	0.784	1.059	−0.002114
3	0.769	1.107	0.000103
4	0.766	1.242	−0.000128

Row (i)	Thorax	Abdomen	Squared Deviation
5	0.828	1.367	0.009106
6	0.852	1.408	0.016236
7	0.781	1.248	0.000418
8	0.696	1.092	0.008732
9	0.792	1.240	0.00066
10	0.814	1.221	0.000484
11	0.621	1.034	0.026224

If we sum up all of the values in the column 'Squared Deviation' from Table 30.2, we get a value of 0.059487. We can multiply this value by $1/(N-1)$ to get the top of the equation for r (i.e., $cov_{x,y}$), $(1/(11-1)) \times 0.059487 = 0.0059487$. We now have all of the values we need to calculate r between fig wasp thorax and abdomen length,

$$r_{x,y} = \frac{0.0059487}{0.06366161 \times 0.1231806}.$$

Our final value is $r_{x,y} = 0.759$. As suggested by the scatterplot in Figure 30.2, thorax and abdomen lengths are highly correlated. We will test whether or not this value of r is statistically significant in Section 30.3, but first we will introduce the Spearman's rank correlation coefficient.

30.2.2 Spearman's rank correlation coefficient

We have seen how the *ranks* of data can be substituted in place of the actual values. This has been useful when data violate the assumptions of a statistical test, and we need a non-parametric test instead (e.g., Wilcoxon signed rank test, Mann-Whitney U test, or Kruskal-Wallis H test). We can use the same trick for the correlation coefficient. Spearman's rank correlation coefficient is calculated the exact same way as the Pearson product moment correlation coefficient, except on the ranks of values. To calculate Spearman's rank correlation coefficient for the fig wasp example in the previous section, we just need to rank the thorax and abdomen lengths from 1 to 11, then calculate r using the rank values instead of the actual measurements of length. Table 30.3 shows the same 11 fig wasp measurements as Tables 30.1 and 30.2, but with columns added to show the ranks of thorax and abdomen lengths.

TABLE 30.3 Measurements of 11 fig wasp thorax and abdomen lengths (mm) and their ranks.

Wasp (i)	Thorax	Thorax Rank	Abdomen	Abdomen Rank
1	0.767	4	1.288	9

Wasp (i)	Thorax	Thorax Rank	Abdomen	Abdomen Rank
2	0.784	7	1.059	2
3	0.769	5	1.107	4
4	0.766	3	1.242	7
5	0.828	10	1.367	10
6	0.852	11	1.408	11
7	0.781	6	1.248	8
8	0.696	2	1.092	3
9	0.792	8	1.240	6
10	0.814	9	1.221	5
11	0.621	1	1.034	1

Note from Table 30.3 that the lowest value of Thorax is 0.621, so it gets a rank of 1. The highest value of Thorax is 0.852, so it gets a rank of 11. We do the same for abdomen ranks. To get Spearman's rank correlation coefficient, we just calculate r using the ranks. The ranks number from 1 to 11 for both variables, so the mean rank is 6 and the standard deviation is 3.317 for both thorax and abdomen ranks. We can then go through each row and calculate $(x_i - \bar{x}) \times (y_i - \bar{y})$ using the ranks. For the first row, this gives us $(4-6)(9-6) = -6$. If we do this same calculation for each row and sum them up, then multiply by $1/(N\text{-}1)$, we get a value of 6.4. To calculate r,

$$r_{\text{rank}(x),\text{rank}(y)} = \frac{6.4}{3.317 \times 3.317}$$

Our Spearman's rank correlation coefficient is therefore $r = 0.582$, which is a bit lower than our Pearson product moment correlation was. The key point here is that the definition of the correlation coefficient has not changed; we are just using the ranks of our measurements instead of the measurements themselves. The reason why we might want to use Spearman's rank correlation coefficient instead of the Pearson product moment correlation coefficient is explained in the next section.

30.3 Correlation hypothesis testing

We often want to test if two variables are correlated. In other words, is r significantly different from zero? We therefore want to test the null hypothesis that r is not significantly different from zero.

- H_0 : The population correlation coefficient is zero.
- H_A : The correlation coefficient is significantly different from zero.

Note that H_A above is for a two-tailed test, in which we do not care about direction. We could also use a one-tailed test if our H_A is that the correlation coefficient is greater than (or less than) zero.

How do we actually test the null hypothesis? As it turns out, the sample correlation coefficient (r) will be approximately t-distributed around a true correlation (ρ) with a t-score defined by $r - \rho$ divided by its standard error (SE(r))[3],

$$t = \frac{r - \rho}{\text{SE}(r)}.$$

Since our null hypothesis is that variables are uncorrelated, $\rho = 0$. Jamovi will use this equation to test whether or not the correlation coefficient is significantly different from zero (The jamovi project, 2024). The reason for presenting it here is to show the conceptual link to other hypothesis tests in earlier chapters. In Section 22.1, we saw that the one sample t-test defined t as the deviation of the sample mean from the true mean, divided by the standard error. Here we are doing the same for the correlation coefficient. One consequence of this is that, like the one sample t-test, the test of the correlation coefficient assumes that r will be normally distributed around the true correlation ρ. If this is not the case, and the assumption of normality is violated, then the test might have a misleading Type I error rate.

To be cautious, we should check whether or not the variables that we are correlating are normally distributed (especially if the sample size is small). If they are normally distributed, then we can use the Pearson's product moment correlation to test the null hypothesis. If the assumption of normality is violated, then we might consider using the non-parametric Spearman's rank correlation coefficient instead. The fig wasp thorax and abdomen lengths from Table 30.1 are normally distributed, so we can use the Pearson product moment correlation coefficient to test whether or not the correlation between these two variables is significant. In jamovi, the t-score is not even reported as output when using a correlation test. We only see r and the p-value (Figure 30.4).

[3]We can calculate the standard error of r as (Rahman, 1968),

$$\text{SE}(r) = \sqrt{\frac{1 - r^2}{N - 2}}.$$

But this is not necessary to ever do by hand. Note that we lose two degrees of freedom ($N - 2$), one for calculating each variable mean.

Correlation Matrix

		Thorax_Length_mm	Abdomen_Length_mm
Thorax_Length_mm	Pearson's r	—	
	p-value	—	
Abdomen_Length_mm	Pearson's r	0.75858	—
	p-value	0.00680	—

FIGURE 30.4 Jamovi output for a test of the null hypothesis that thorax length and abdomen length are not significantly correlated in a sample of fig wasps collected in 2010 from Baja, Mexico.

From Figure 30.4, we can see that the sample $r = 0.75858$, and the p-value is $P = 0.00680$. Since our p-value is less than 0.05, we can reject the null hypothesis that fig wasp thorax and abdomen lengths are not correlated (an interactive application[4] can make this more intuitive).

[4] `https://bradduthie.github.io/stats/app/corr_click/`

31

Practical. *Analysis of counts and correlations*

This chapter focuses on applying the concepts from Chapter 29 and Chapter 30 in jamovi (The jamovi project, 2024). Exercises in this chapter will use the Chi-square goodness of fit test, the Chi-square test of association, and the correlation coefficient. For all of these examples, this chapter will use a dataset inspired by Burrows et al. (2022). This experimental work tested the effects of radiation on bumblebee nectar consumption, carbon dioxide output, and body mass in different bee colonies.

The chapter will use the bumblebee dataset[1]. This dataset includes variables for the radiation level experienced by the bee (radiation), the colony from which the bee came (colony), whether or not the bee survived to the end of the 30-day experiment (survived), the mass of the bee in grams at the beginning of the experiment (mass), the output of carbon dioxide put out by the bee (CO_2) in micromoles per minute, and the daily volume of nectar consumed by the bee in millilitres (nectar).

31.1 Survival goodness of fit

Suppose that we want to run a simple goodness of fit test to determine whether or not bees are equally likely to survive versus die in the experiment. If this is the case, then we would expect to see the same number of living and dead bees in the dataset. We can use a Chi-square goodness of fit test to answer this question. What are the null and alternative hypotheses for this χ^2 goodness of fit test?

- H_0: ______________________________
- H_A: ______________________________

[1] https://bradduthie.github.io/stats/data/bumblebee.csv

What is the sample size (N) of the dataset?

$N =$ ______________________________

Based on this sample size, what are the expected counts for bees that survived and died?

Survived (E_{surv}): ______________________________

Died (E_{died}): ________________________

Next, we can find the observed counts of bees that survived and died. To do this, we need to use the Frequency tables option in jamovi. We did this once in Chapter 17 for calculating probabilities. As a reminder, to find the counts of bumblebees that survived (Yes) or did not survive (No), we need to go to the Exploration toolbar in jamovi, then choose 'Descriptives'. Place 'Survival' in the Variables box, then check the box for 'Frequency tables' below. A Frequencies table will appear in the panel on the right. Write down the observed counts of bees that survived and died.

Survived (O_{surv}): ______________________________

Died (O_{died}): ________________________

Try to use the formula in Section 29.2 to calculate the χ^2 test statistic. Here is what it should look like for the two counts in this dataset,

$$\chi^2 = \frac{(O_{\text{surv}} - E_{\text{surv}})^2}{E_{\text{surv}}} + \frac{(O_{\text{died}} - E_{\text{died}})^2}{E_{\text{died}}}.$$

What is the χ^2 value?

$\chi^2 =$ ______________________

There are two categories for survival (Yes and No). How many degrees of freedom are there?

$df =$ ____________________________

Now we can try to use jamovi to replicate the analysis above and find a p-value. In the jamovi Analyses tab, select 'Frequencies' from the toolbar, then select 'N Outcomes: χ^2 Goodness of fit' from the pull-down menu. A new window will open up.

After selecting the option 'N Outcomes: χ^2 Goodness of fit', a new window will appear called 'Proportion Test (N Outcomes)'. To run a χ^2 goodness of fit test on bee survival, move the 'survived' variable into the 'Variable' box. Leave the Counts box empty (Figure 31.1).

The χ^2 Goodness of Fit table will appear in the panel to the right. From this table, we can see the χ^2 test statistic, degrees of freedom (df), and the p-value (p). Write these values below, and check to see if the χ^2 and *df* match the values you calculated above by hand.

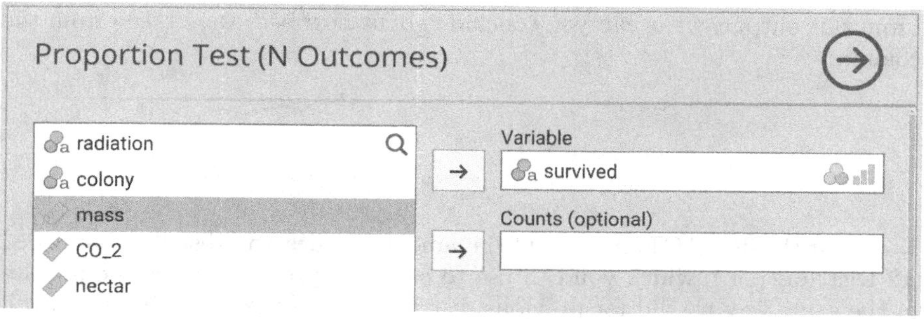

FIGURE 31.1 Jamovi interface for running a Chi-square goodness of fit test on bumblebee survival in a dataset.

$\chi^2 =$ ____________________

$df =$ ____________________

$P =$ ____________________

Next, we will try another goodness of fit test, but this time to test whether or not bees were taken from all colonies with the same probability.

31.2 Colony goodness of fit

Next, suppose that we want to know if bees were sampled from the colonies with the same expected frequencies. What are the null and alternative hypotheses in this scenario?

- H_0: ____________________
- H_A: ____________________

How many colonies are there in this dataset?

Colonies: ____________________

Run the χ^2 goodness of fit test using the same procedure in jamovi that you used in the previous exercise. What is the output from the Goodness of Fit table?

$\chi^2 =$ ____________________

$df =$ ____________________

$P =$ ____________________

From this output, what can you conclude about how bees were taken from the colonies?

Note that the distrACTION module in jamovi includes a χ^2 distribution (called 'x2-Distribution'), which you can use to compute probabilities and quantiles in the same way we did for previous distributions in the book. Next, we will move on to a χ^2 test of association between colony and survival.

31.3 Chi-square test of association

Suppose we want to know if there is an association between bee colony and bee survival. We can use a χ^2 test of association to investigate this question. What are the null and alternative hypotheses for this test of association?

- H_0: ________________________
- H_A: ________________________

To run the χ^2 test of association, choose 'Frequencies' from the jamovi toolbar, but this time select 'Independent Samples: χ^2 test of association' from the pull-down menu. To test for an association between bee colony and survival, place 'colony' in the 'Rows' box and 'survived' in the 'Columns' box. Leave the rest of the boxes blank (Figure 31.2).

There is a pull-down called 'Statistics' below the Contingency Tables input. Make sure that the χ^2 checkbox is selected. Output from the χ^2 test of association will appear in the panel to the right. Report the key statistics in the output table below.

$\chi^2 =$ ____________________

$df =$ ____________________

$P =$ ____________________

From these statistics, should you reject or not reject the null hypothesis?

H_0: ____________________

Note that scrolling down further in the left panel (Contingency Tables) reveals an option for plotting. Have a look at this and create a barplot by checking 'Bar Plot' under **Plots**. Note that there are various options for changing bar

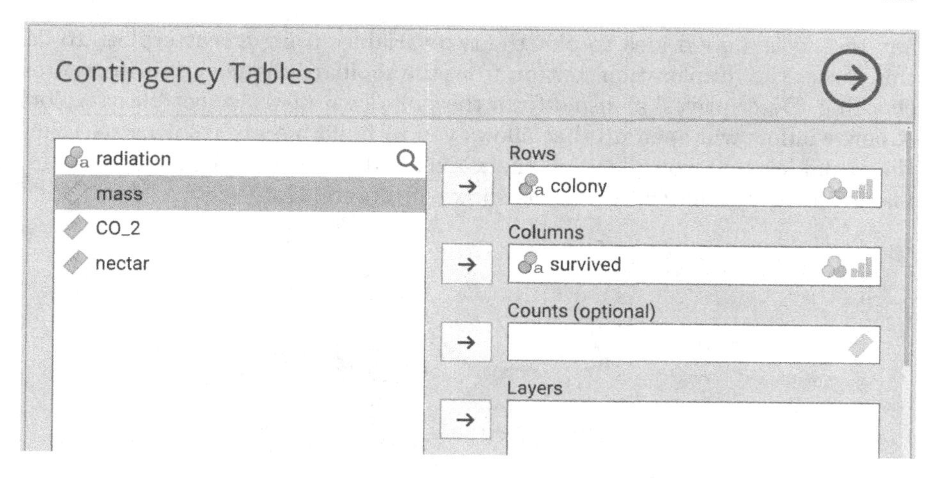

FIGURE 31.2 Jamovi interface for running a Chi-square test of association on bumblebee survival versus colony in a dataset.

types (side by side or stacked), y-axis limits (counts versus percentages), and bar groupings (by rows or columns).

Now try running a χ^2 test of association to see if there is an association between radiation and bee survival (hint, you just need to swap 'colony' for 'radiation' in the Rows box). What can you conclude from this test? Explain your conclusion as if you were reporting the results of the test to someone who was unfamiliar with statistical hypothesis testing.

Lastly, did the order in which you placed the two variables matter? What if you switched Rows and Columns? In other words, put 'survived' in the Rows box and 'radiation' in the Columns box. Does this give you the same answer?

Next, we will look at correlations between variables.

31.4 Pearson product moment correlation test

Suppose that we want to test if bumblebee mass at the start of the experiment (mass) is associated with carbon dioxide output (CO_2). Specifically, we want to know if more massive bees also output less carbon dioxide. Before running any test, it is a good idea to plot the two variables using a scatterplot. To do this, select the 'Exploration' button from the toolbar in jamovi, but instead of choosing 'Descriptives' as usual from the pull-down menu, select 'Scatterplot'. A new window will open up that allows you to build a scatterplot by selecting the variables that you place on the x-axis and the y-axis. Put mass on the x-axis and CO_2 on the y-axis, as shown in Figure 31.3.

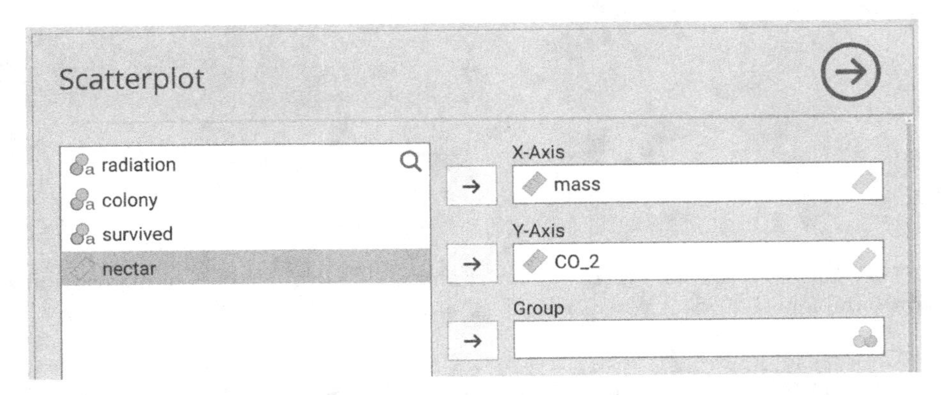

FIGURE 31.3 Jamovi interface for building a scatterplot with bumblebee mass on the x-axis and carbon dioxide output on the y-axis.

Notice that the scatterplot appears in the panel on the right. Each point in the scatterplot is a different bee (i.e., row). Just looking at the scatterplot, does it appear as though bee mass and CO_2 output are correlated? Why or why not?

Note that it is possible to separate points in the scatterplot by group., Try placing 'survived' in the box 'Group'.

Now we can test whether or not bee mass and CO2 output are negatively correlated. What are the null and alternative hypotheses of this test?

- H_0: ______________________
- H_A: ______________________

Before we test whether or not the correlation coefficient (r) is significant, we need to know which correlation coefficient to use. Remember from Section 30.3 that a test of the Pearson product moment correlation assumes that the sample r is normally distributed around the true correlation coefficient. If both of our variables (mass and CO2) are normally distributed, then we can be confident that this assumption will not be violated. But if one or both variables are not normally distributed, then we should consider using Spearman's rank correlation coefficient instead. To test if mass and CO2 are normally distributed, navigate to the Descriptives panel in jamovi (where we usually find the summary statistics of variables). Place mass and CO2 in the 'Variables' box, then scroll down and notice that there is a checkbox under **Normality** for 'Shapiro-Wilk'. Check this box, then find the p-values for the Shapiro-Wilk test of normality in the panel to the right. Write these p-values down below.

Mass: $P =$ ____________________

CO2: $P =$ __________________

Based on these p-values, which type of correlation coefficient should we use to test H_0, and why?

To run the correlation coefficient test, choose the button in the jamovi toolbar called 'Regression', then select the first option 'Correlation Matrix' from the pull-down menu. The Correlation Matrix option will pull up a new window in jamovi (Figure 31.4).

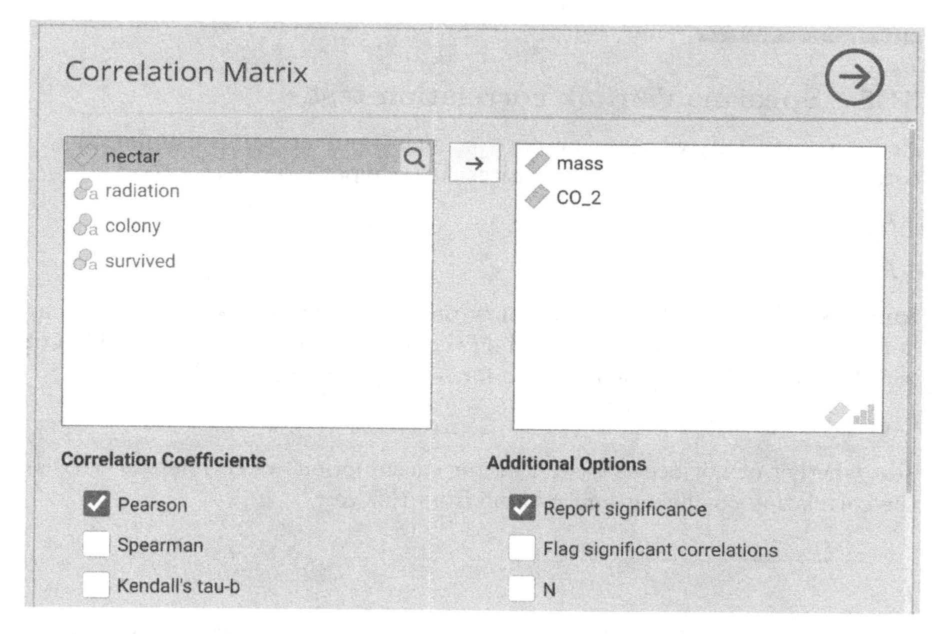

FIGURE 31.4 Jamovi interface for testing correlation coefficients.

Notice that the Pearson product moment correlation is selected in the checkbox of Figure 31.4 ('Pearson'). Immediately below this checkbox is a box called 'Spearman', which would report Spearman's rank correlation coefficient test. Below the **Correlation Coefficients** options, there are options for **Hypothesis**. Remember that we are interested in the alternative hypothesis that mass and CO_2 are negatively correlated, so we should select the radio button 'Correlated negatively'.

The output of the correlation test appears in the panel on the right in the form of a table called 'Correlation Matrix'. This table reports both the correlation coefficient (here called 'Pearson's r') and the p-value. Write these values below.

$r =$ ______________________

$P =$ ______________________

Based on this output, what should we conclude about the association between bumblebee mass and carbon dioxide output?

Next, we will test whether or not bee mass is associated with nectar consumption.

31.5 Spearman's rank correlation test

Next, we will test whether or not bee mass and nectar consumption are correlated. What are the null and alternative hypotheses of this test?

- H_0: ______________________
- H_A: ______________________

Run a Shapiro-Wilk test of normality on each of the two variables, as was done in the previous exercise. Based on the output of these tests, what kind of correlation coefficient should we use for testing the null hypothesis?

Correlation coefficient: ______________________

Test whether or not bee mass and nectar consumption are correlated. What is the correlation coefficient and p-value from this test?

$r =$ ______________________

$P =$ ______________________

Based on these results, should we reject or not reject the null hypothesis?

H_0: ______________________

Suppose that we had used the Pearson product moment correlation coefficient instead of Spearman's rank correlation coefficient. Would we have made the same conclusion about the correlation (or lack thereof) between bee mass and nectar consumption? Why or why not?

31.6 Untidy goodness of fit

In Exercise 31.1, we ran a χ^2 test using data in a tidy format, in which each row corresponded to a single observation, and categorical data were listed over $N = 256$ rows. For the 'survived' variable, this meant 256 rows of 'Yes' or 'No'. But there is a shortcut in jamovi if we do not have a full tidy dataset. If you know that the dataset included 139 'Yes' counts and 117 'No' counts, you could set up the data as a table of counts (Table 31.1).

TABLE 31.1 Counts of bees that did not survive (No) or did survive (Yes) in an experiment involving radiation.

Survived	Count
No	117
Yes	139

Open a new data frame in jamovi, then recreate the small dataset in Table 31.1. Column names should be 'Survived' and 'Count', as shown in Figure 31.5.

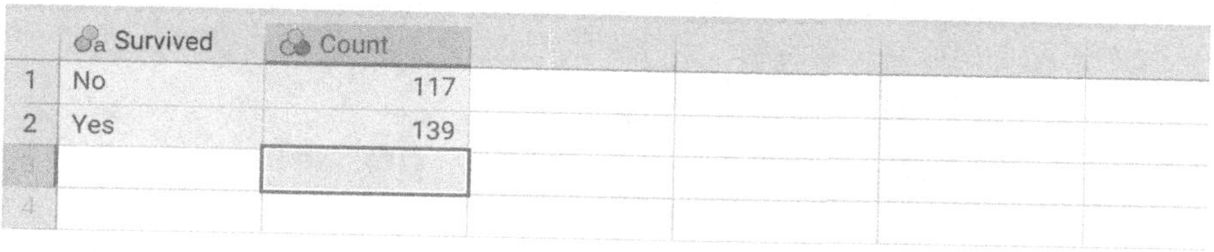

	Survived	Count
1	No	117
2	Yes	139
3		
4		

FIGURE 31.5 Jamovi data frame with a simple organisation of count data.

Next, navigate to the 'Analyses' tab and choose 'N Outcomes' to do a goodness of fit test. Place 'Survived' in the Variable box, then place 'Count' in the Counts (optional) box. Notice that you will get the same χ^2, *df*, and p-values in the output table as you did in Exercise 31.1.

We could do the same for a χ^2 test of association, although it would be a bit more complicated. To test for an association between radiation and survival, as we did at the end of Exercise 31.3, we would need three columns and eight rows of data (Table 31.2).

TABLE 31.2 Counts of bees that did not survive (No) or did survive (Yes) for different levels of radiation.

Survived	Radiation	Count
No	None	12
Yes	Low	52
No	Medium	29
Yes	High	35
No	None	39
Yes	Low	25
No	Medium	37
Yes	High	27

If we put Table 31.2 into jamovi, we can run a χ^2 test of association by navigating to the 'Frequencies' button in the jamovi toolbar and selecting 'Independent Samples: χ^2 test of association' from the pull-down. In the Contingency Tables input panel, we can put 'Survived' in the Rows box, 'Radiation' in the Columns box, then place 'Count' in the Counts (optional) box. The panel on the right will give us the output of the χ^2 test of association.

32

Simple linear regression

Linear regression focuses on the association between two or more quantitative variables. In the case of simple linear regression, which is the focus of this chapter, there are only two variables to consider. At first, this might sound similar to correlation, which was introduced in Chapter 30. Simple linear regression and correlation are indeed similar, both conceptually and mathematically, and the two are frequently confused. Both methods focus on two quantitative variables, but the general aim of regression is different from correlation. The aim of correlation is to describe how the variance of one variable is associated with the variance of another variable. In other words, the correlation measures the intensity of covariance between variables (Sokal & Rohlf, 1995). But there is no attempt to predict what the value of one variable will be based on the other.

Linear regression, in contrast to correlation, focuses on prediction. The aim is to predict the value of one quantitative variable Y given the value of another quantitative variable X. In other words, regression focuses on an association of dependence in which the value of Y depends on the value of X (Rahman, 1968). The Y variable is therefore called the **dependent variable**; it is also sometimes called the response variable or the output variable (Box et al., 1978; Sokal & Rohlf, 1995). The X variable is called the **independent variable**; it is also sometimes called the predictor variable or the regressor (Box et al., 1978; Sokal & Rohlf, 1995). Unlike correlation, the distinction between the two variable types matters because the aim is to understand how a change in the independent variable will affect the dependent variable. For example, if we increase X by 1, how much will Y change?

32.1 Visual interpretation of regression

A visual example using a scatterplot can illustrate one way to think about regression. Suppose that we have sampled fig fruits from various latitudes (Figure 32.1), and we want to use latitude to predict fruit volume (Duthie & Nason, 2016).

A sample of fig fruits from different latitudes is shown in Table 32.1.

FIGURE 32.1 Fruits of the Sonoran Desert Rock Fig in the desert of Baja, Mexico with different fig wasps on the surface (A and B). A full fig tree is shown to the right (C) with the author attempting to collect fig fruits from a branch of the tree.

TABLE 32.1 Volumes (mm^3) of fig fruits collected from different latitudes from trees of the Sonoran Desert Rock Fig in Baja, Mexico.

Latitude	23.7	24.0	27.6	27.2	29.3	28.2	28.3
Volume	2399.0	2941.7	2167.2	2051.3	1686.2	937.3	1328.2

How much does fruit volume change with latitude? To start answering this question, we can plot the relationship between the two variables. We want to predict fruit volume from latitude, meaning that fruit volume *depends on* latitude. Fruit volume is therefore the dependent variable, and we should plot it on the y-axis. Latitude is our independent variable, and we should plot it on the x-axis (Figure 32.2).

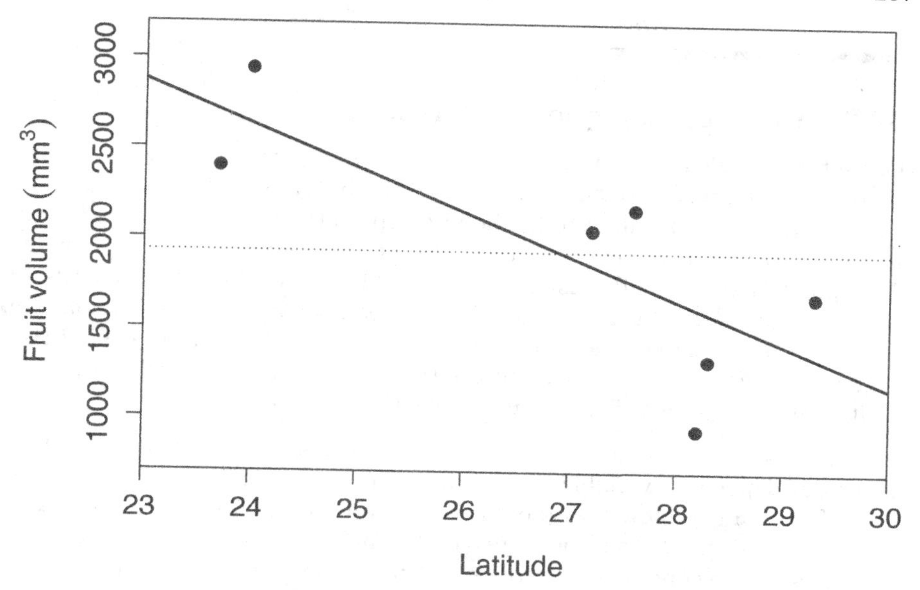

FIGURE 32.2 Relationship between latitude and fruit volume for seven fig fruits collected from Baja, Mexico in 2010. The solid line shows the regression line of best fit, and the thin dotted line shows the mean of fruit volume.

In Figure 32.2, each of the seven points is a different fig fruit. The x-axis shows the latitude from which the fruit was collected, and the y-axis shows the volume of the fruit in mm^3. The thin dotted line shows the mean fruit volume for the seven fruits, $\bar{y} = 1930.1$. The thick black line trending downwards in Figure 32.2 is the regression line, also called the line of best fit. How this line is calculated will be explained later, but for now there are two important concepts to take away from Figure 32.2. First, the regression line gives us the best prediction of what fruit volume will be for any given latitude. For example, if we wanted to predict what fruit volume would be for a fruit collected at 28 degrees north latitude, we could find the value 28 on the x-axis, then find what fruit value this corresponds to on the y-axis using the regression line. At an x-axis value of 28, the regression line has a y-axis value of approximately 1660, so we would predict that a fig fruit collected at 28 degrees north latitude would have a volume of 1660 mm^3.

This leads to the second important concept to take away from Figure 32.2. In the absence of any other information (including latitude), our best guess of what any given fruit's volume will be is just the mean ($\bar{y} = 1930.1$). A key aim of regression is to test if the regression line can do a significantly better job of predicting what fruit volume will be. In other words, is the solid line of Figure 32.2 really doing that much better than the horizontal dotted line? Before answering this question, a few new terms are needed.

32.2 Intercepts, slopes, and residuals

Given the latitude of each fruit (i.e., each point in Figure 32.2), we can predict its volume from three numbers. These three numbers are the intercept (b_0), the slope (b_1), and the residual (ϵ_i). The intercept is the point on the regression line where $x = 0$, i.e., where latitude is 0 in the example of fig fruit volumes. This point is not actually visible in Figure 32.2 because the lowest latitude on the x-axis is 23. At a latitude of 23, we can see that the regression line predicts a fruit volume of approximately 2900 mm^3. If we were to extend this regression line all the way back to a latitude of 0, then we would predict a fruit volume of 8458.3. This is our intercept[1] in Figure 32.2.

The slope is the direction and steepness of the regression line. It describes how much our dependent variable changes if we increase the independent variable by 1. For example, how do we predict fruit volume to change if we increase latitude by 1 degree? From the regression line in Figure 32.2, whenever latitude increases by 1, we predict a decrease in fruit volume of 242.7. Consequently, the slope is -242.7. Since we are predicting using a straight line, this decrease is the same at every latitude. This means that we can use the slope to predict how much our dependent variable will change given any amount of units of change in our independent variable. For example, we can predict how fruit volume will change for any amount of change in degrees latitude. If latitude increases by 2 degrees, then we would predict a $2 \times -242.7 = -485.4$ mm^3 change in fruit volume (i.e., a decrease of 485.4). If latitude decreases by 3 degrees, then we would predict a $-3 \times -242.7 = 728.1$ mm^3 change in fruit volume (i.e., an increase of 728.1).

We can describe the regression line using just the intercept and the slope. For the example in Figure 31.2, this means that we can predict fruit volume for any given latitude with just these two numbers. But prediction almost always comes with some degree of uncertainty. For example, if we could perfectly predict fruit volume from latitude, then all of the points in Figure 32.2 would fall exactly on the regression line. But this is not the case. None of the seven points in Figure 32.2 fall exactly on the line, so there is some unexplained variation (i.e., some error) in predicting fruit volume from latitude. To map each fruit's latitude to its corresponding volume, we therefore need one more number. This number is the **residual**, and it describes how far away a point is from the regression line (Figure 32.3).

The residual of each of the seven points is shown with a dashed line in Figure 32.3. Residual values are positive when they are higher than the value predicted

[1] Biologically, a fruit volume of 8458.3 might be entirely unrealistic, which is why we need to be careful when extrapolating beyond the range of our independent variable (more on this later).

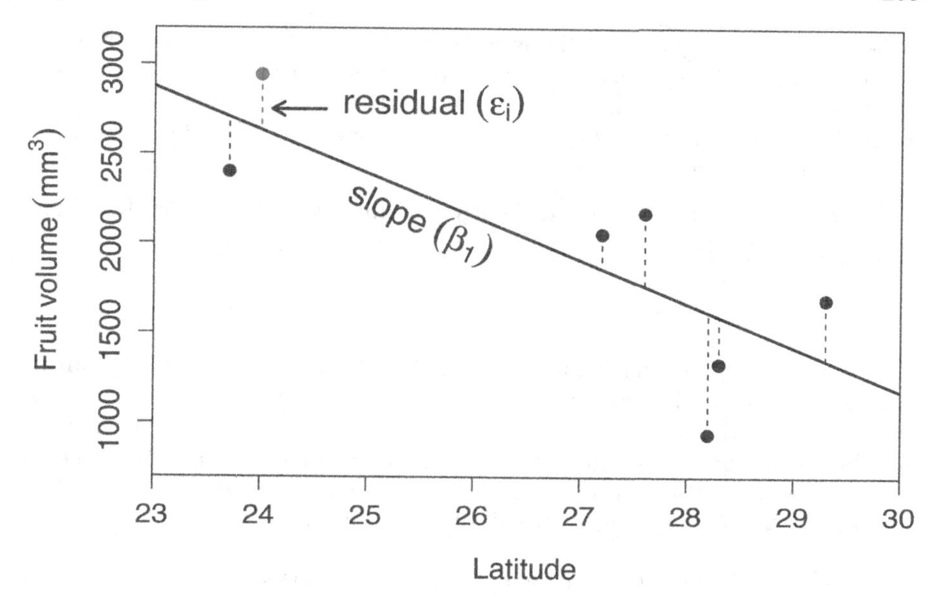

FIGURE 32.3 Relationship between latitude and fruit volume for seven fig fruits collected from Baja, Mexico in 2010. The solid line shows the regression line of best fit, and the vertical dashed lines show the residuals for each point.

by the regression line, and they are negative when they are lower than the value predicted by the regression line. In the example of Figure 32.3, the residual indicated by the arrow, at a latitude of 24, is 307.8 because the volume of the fig fruit collected from this latitude deviates from the predicted volume on the regression line by 307.8. For the point just to the left where the latitude from which the fruit was sampled is 23.7 degrees, the residual is −307.7. For any fig fruit i, we can therefore find its volume using the intercept (b_0), the slope (b_1), and the residual value (ϵ_i). Next, we will see how these different values relate to one another mathematically.

32.3 Regression coefficients

Simple linear regression predicts the dependent variable (y) from the independent variable (x) using the intercept (b_0) and the slope (b_1),

$$y = b_0 + b_1 x.$$

The equation for y mathematically describes the regression line in Figures 32.2 and 32.3. This gives us the expected value of y for any value of x. In other words, the equation tells us what y will be *on average* for any given x. Sometimes different letters are used to represent the same mathematical relationship, such as $y = a + bx$ or $y = mx + b$, but the symbols used are not really important[2]. Here, b_0 and b_1 are used to make the transition to multiple regression in Chapter 33 clearer.

For any specific value of x_i, the corresponding y_i can be described more generally,

$$y_i = b_0 + b_1 x_i + \epsilon_i.$$

For example, for any fig fruit i, we can find its *exact* volume (y_i) from its latitude (x_i) using the intercept (b_0), the slope (b_1), and the residual (ϵ_i). We can do this for the residual indicated by the arrow in Figure 32.3. The latitude at which this fruit was sampled was $x_i = 24$, its volume is $y_i = 2941.7$, and its residual value is 307.8. From the previous section, we know that $b_0 = 8458.3$ and $b_1 = -242.7$. If we substitute all of these values,

$$2941.7 = 8458.3 - 242.68(24) + 307.84.$$

Note that if we remove the residual 307.84, then we get the predicted volume for our fig fruit at 24 degrees latitude,

$$2633.98 = 8458.3 - 242.68(24).$$

Visually, this is where the dotted residual line meets the solid regression line in Figure 32.3.

This explains the relationship between the independent and dependent variables using the intercept, slope, and residuals. But how do we actually define the line of best fit? In other words, what makes the regression line in this example better than some other line that we might use instead? The next section explains how the regression line is actually calculated.

32.4 Regression line calculation

The regression line is defined by its relationship to the residual values. Figure 32.4 shows the same regression as in Figures 32.2 and 32.3, but with the values of the residuals written next to each point.

Some of the values are positive, and some are negative. An intuitive reason

[2]Another common way to represent the above is, $y = \hat{\beta}_0 + \hat{\beta}_1 x$, where $\hat{\beta}_0$ and $\hat{\beta}_1$ are sample estimates of the true parameters β_0 and β_1.

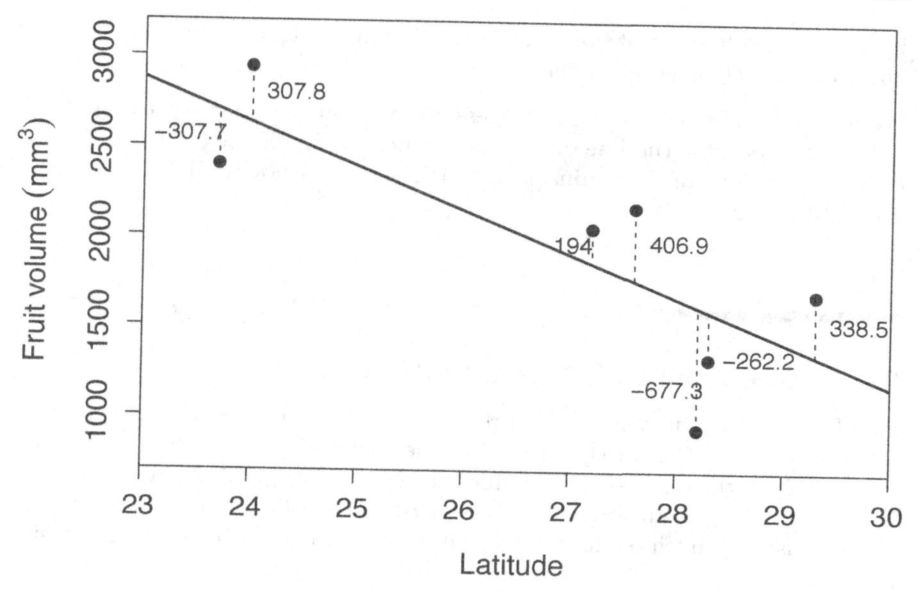

FIGURE 32.4 Relationship between latitude and fruit volume for seven fig fruits collected from Baja, Mexico in 2010. The solid black line shows the regression line of best fit, and the vertical dashed lines show the residuals for each point. Residuals are rounded to one decimal place.

the line in Figure 32.4 is the line of best fit is the positive and negative values exactly balance each other out. In other words, the sum of all the residual values in Figure 32.4 is 0,

$$0 = -307.715 + 307.790 + 193.976 + 406.950 - 677.340 - 262.172 + 338.511.$$

If we were to move the regression line, then the sum of residuals would no longer be 0. There is only one line that fits.

More technically, the line of best fit minimises the sum of squared residuals (SS_{residual}). In other words, when we take all of the residual values, square them, then add up the squares, the sum should be lower than any other line we could draw,

$$SS_{\mathrm{residual}} = (-307.715)^2 + (307.790)^2 + ... + (338.511)^2.$$

For the regression in Figure 32.4, $SS_{\mathrm{residual}} = 1034772$. Any line other than the regression line shown in Figure 32.4 would result in a higher SS_{residual} (to get a better intuition for how this works, we can use an interactive application[3]

[3] `https://bradduthie.github.io/stats/app/regr_click/`

in which a random set of points are placed on a scatterplot and the intercept and slope are changed until the residual sum of squares is minimised).

We have seen how key terms in regression are defined, what regression coefficients are, and how the line of best fit is calculated. The next section focuses on the coefficient of determination, which describes how well data points fit around the regression line.

32.5 Coefficient of determination

We often want to know how well a regression line fits the data. In other words, are most of the data near the regression line (indicating a good fit), or are most far away from the regression line? How closely the data fit to the regression line is described by the **coefficient of determination** (R^2). More formally, the R^2 tells us how much of the total variation in y is explained by the regression line[4],

$$R^2 = 1 - \frac{SS_{\text{residual}}}{SS_{\text{total}}}.$$

Mathematically, the coefficient of determination compares the sum of squared residuals from the linear model (SS_{residual}) to what the sum of squared residuals would be had we just used the mean value of y (SS_{total}). If SS_{residual} is very small compared to SS_{total}, then subtracting $SS_{\text{residual}}/SS_{\text{total}}$ from 1 will give a large R^2 value. This large R^2 means that the model is doing a good job of explaining variation in the data. Figure 32.5 shows some examples of scatterplots with different R^2 values.

We can calculate the R^2 value for our example of fig fruit volumes over a latitudinal gradient. To do this, we need to calculate the sum of the squared residuals (SS_{residual}) and the total sum of squared deviations of y_i from the mean $\bar{y}$ (SS_{total}). From the previous section, we have already found that $SS_{\text{residual}} = 1034772$. Now, to get SS_{total}, we just need to get the sum of squares for fruit volume (see Section 12.3). We can visualise this as the sum of squared deviations from the mean fruit volume of $\bar{y} = 1930.1$ instead of the value predicted by the regression line (Figure 32.6).

The numbers in Figure 32.6 show the deviations of each point from the regression line, just like in Figure 32.4. New numbers have been added to

[4]Note that, mathematically, R^2 is in fact the square of the correlation coefficient. Intuitively this should make some sense; when two variables are more strongly correlated (i.e., r is near -1 or 1), data are also more tightly distributed around the regression line. But it is also important to understand R^2 conceptually in terms of variation explained by the regression model.

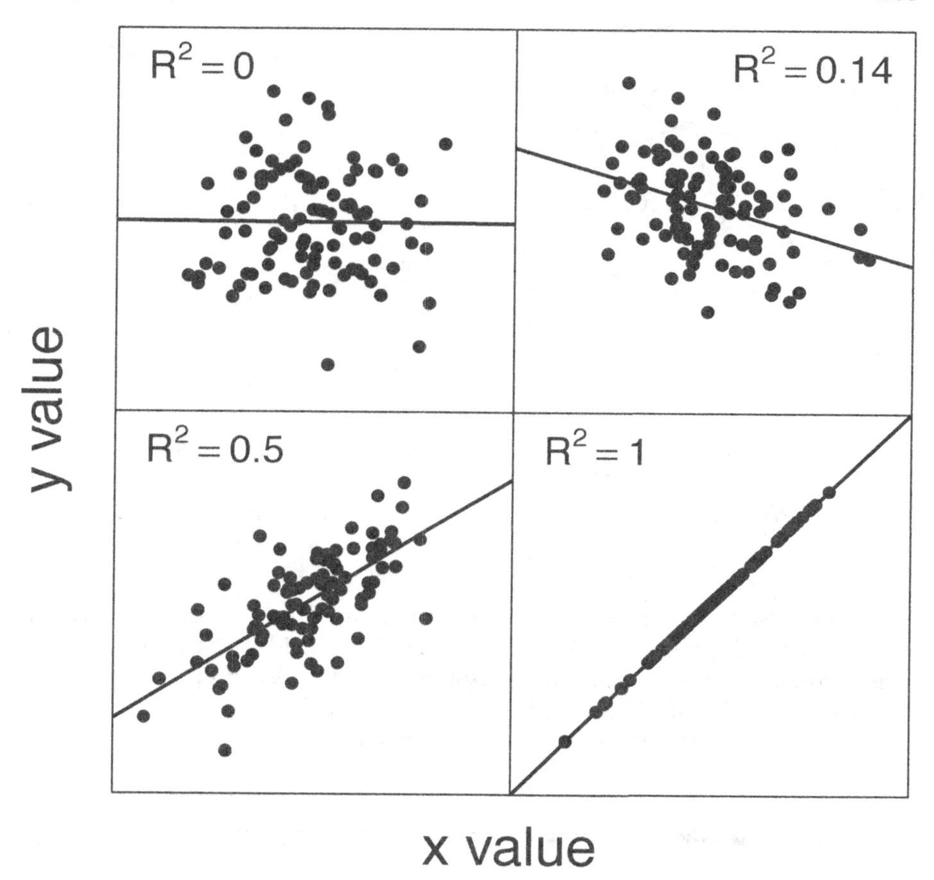

FIGURE 32.5 Examples of scatterplots with different coefficients of determination (R-squared).

Figure 32.6 to show the deviation of each point from the mean fruit volume. Summing the squared values of residuals from the regression line gives a value of 1034772 Summing the squared deviations of values from the mean $\bar{y} = 1930.1$ gives a value of 2721530. To calculate R^2,

$$R^2 = 1 - \frac{1034772}{2721530}.$$

The above gives us a value of $R^2 = 0.619783$. In other words, about 62% of the variation in fruit volume is explained by latitude.

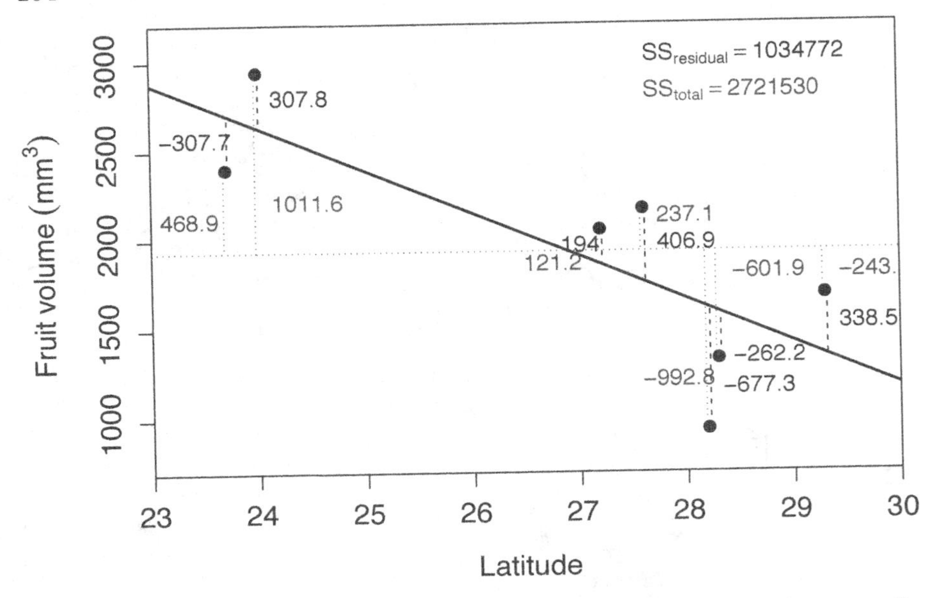

FIGURE 32.6 Relationship between latitude and fruit volume for seven fig fruits collected from Baja, Mexico in 2010. The solid black line shows the regression line of best fit, and the horizontal dotted line shows the mean of fruit volume. Vertical dashed lines show the model residuals (dashed) and deviations from the mean (dotted). Residuals are rounded to one decimal place.

32.6 Regression assumptions

It is important to be aware of the assumptions underlying linear regression. There are four key assumptions underlying the simple linear regression models described in this chapter (Sokal & Rohlf, 1995):

1. Measurement of the independent variable (x) is completely accurate. In other words, there is no measurement error for the independent variable. Of course, this assumption is almost certainly violated to some degree because every measurement has some associated error (see Section 6.1 and Chapter 7).
2. The relationship between the independent and dependent variables is linear. In other words, we assume that the relationship between x and y can be defined by a straight line satisfying the equation $y = b_0 + b_1x$. If this is not the case (e.g., because the relationship between x and y is described by some sort of curved line), then a simple linear regression might not be appropriate.

3. For any value of x_i, y_i values are independent and normally distributed. In other words, the *residual* values (ϵ_i) should be normally distributed around the regression line, and they should not have any kind of pattern (such as, e.g., ϵ_i values being negative for low x but positive for high x). If we were to go out and resample the same values of x_i, the corresponding y_i values should be normally distributed around the predicted y.

4. For all values of x, the variance of residuals is identical. In other words, the variance of y_i values around the predicted y should not change over the range of x. The term for this is 'homoscedasticity', meaning that the variance is constant. This is in contrast to heteroscedasticity, which means that the variance is not constant.

Figure 32.7 shows a classic example of heteroscedasticity. Notice that the variance of y_i values increases with increasing x.

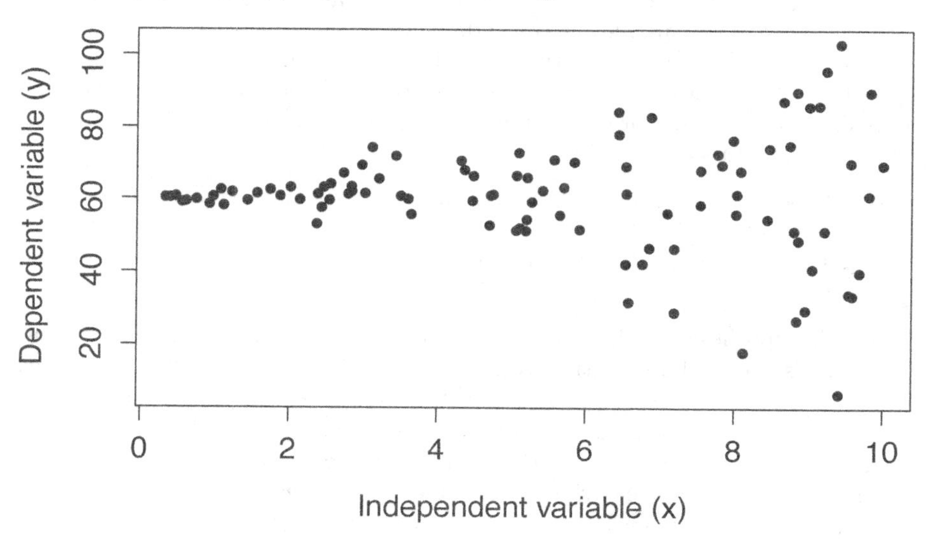

FIGURE 32.7 Hypothetical dataset in which data show heteroscedasticity, thereby violating an assumption of simple linear regression.

Note that even if our assumptions are not perfectly met, this does not completely invalidate the method of linear regression. In reality, linear regression is often robust to minor deviations from the above assumptions (as are other statistical tools), but large violations of one or more of these assumptions might indeed invalidate the use of linear regression.

32.7 Regression hypothesis testing

We typically want to know if our regression model is useful for predicting the dependent variable given the independent variable. There are three specific null hypotheses that we can test, which tell us the significance of (1) the overall model, (2) the intercept, and (3) the slope. We will go through each of these null hypotheses.

32.7.1 Overall model significance

As mentioned in Section 32.1, in the absence of any other information, the best prediction of our dependent variable is the mean. For example, if we did not have any information about latitude in the previous sections, then the best prediction of fruit volume would just be the mean fruit volume, $\bar{y} = 1930.1$ (Figure 32.2). Does including the independent variable latitude result in a significantly better prediction than just using the mean? In other words, does a simple linear regression model with latitude as the independent variable explain significantly more variation in fruit volume than just the mean fruit volume? We can state this more formally as null and alternative hypotheses.

- H_0: A model with no independent variables fits the data as well as the linear model.
- H_A: The linear model fits the data better than the model with no independent variables.

The null hypothesis can be tested using an F-test of overall significance. This test makes use of the F-distribution (see Section 24.1) to calculate a p-value that we can use to reject or not reject H_0. Recall that the F-distribution describes the null distribution for a ratio of variances. In this case, the F-distribution is used to test for the overall significance of a linear regression model by comparing the variation explained by the model to its residual (i.e., unexplained) variation[5]. If the ratio of explained to unexplained variation is sufficiently high, then we will get a low p-value and reject the null hypothesis.

[5]For the fig fruit volume example, the total variation is the sum of squared deviations of fruit volume from the mean is $SS_{\text{deviation}} = 2721530$. The amount of variation explained by the model is $SS_{\text{model}} = 1686758$ with 1 degree of freedom. The remaining residual variation is $SS_{\text{residual}} = 1034772$ with 5 degrees of freedom. To get an F value, we can use the same approach as with the ANOVA in Chapter 24. We calculate the mean squared errors as $MS_{\text{model}} = 1686758/1 = 1686758$ and $MS_{\text{residual}} = 1034772/5 = 206954.4$, then take the ratio to get the value $F = 1686758/206954.4 = 8.150385$.

32.7.2 Significance of the intercept

Just like we test the significance of the overall linear model, we can test the significance of individual model coefficients, b_0 and b_1. Recall that b_0 is the coefficient for the intercept. We can test the null hypothesis that $b_0 = 0$ against the alternative hypothesis that it is different from 0.

- H_0: The intercept equals 0.
- H_A: The intercept does not equal 0.

The estimate of b_0 is t-distributed (see Chapter 19) around the true parameter value β_0. Jamovi will therefore report a t-value for the intercept, along with a p-value that we can use to reject or not reject H_0 (The jamovi project, 2024).

32.7.3 Significance of the slope

Testing the significance of the slope (b_1) works in the same way as testing the significance of the intercept. We can test the null hypothesis that $b_1 = 0$ against the alternative hypothesis that it is different from 0. Visually, this is testing whether the regression line shown in Figures 32.2-32.5 is flat, or if it is trending either upwards or downwards.

- H_0: The slope equals 0.
- H_A: The slope does not equal 0.

Like b_0, the estimate of b_1 is t-distributed (see Chapter 19) around the true parameter value β_1. We can therefore use the t-distribution to calculate a p-value and either reject or not reject H_0. Note that this is often the hypothesis that we are most interested in testing. For example, we often do not care if the intercept of our model is significantly different from 0 (in the case of our fig fruit volumes, this would not even make sense; fig fruits obviously do not have zero volume at the equator). But we often do care if our dependent variable is increasing or decreasing with an increase in the independent variable.

32.7.4 Simple regression output

If we run the simple regression of fig fruit latitude against fruit volume, we can find output statistics $R^2 = 0.6198$, and $P = 0.03562$ for the overall model in jamovi. This means that the model explains about 61.98% of the total variation in fruit volume, and the overall model does a significantly better job of predicting fruit volume than the mean. We therefore reject the null hypothesis and conclude that the model with latitude as an independent variables fits the data significantly better than a model with just the mean of fruit volume (Figure 32.8).

Figure 32.8 reports the R^2 value along with F statistic, degrees of freedom, and the resulting p-value for the overall model in jamovi. We can also see a

Model Fit Measures

			Overall Model Test			
Model	R	R²	F	df1	df2	p
1	0.78726	0.61978	8.15039	1	5	0.03562

FIGURE 32.8 Jamovi output table for a simple linear regression in which latitude is an independent variable and fig fruit volume is a dependent variable.

table of model coefficients, the intercept (b_0) and slope (b_1) associated with latitude (Figure 32.9).

Model Coefficients - Volume

Predictor	Estimate	SE	t	p
Intercept	8458.30961	2293.12354	3.68855	0.01417
Latitude	-242.68331	85.00625	-2.85489	0.03562

FIGURE 32.9 Jamovi output table for a simple linear regression showing model coefficients and their statistical significance.

From the jamovi output shown in Figure 32.9, we can see that the intercept is significant ($P < 0.05$), so we reject the null hypothesis that $b_0 = 0$. Fruit volume decreases with increasing latitude ($b_1 = -242.68$), and this decrease is also significant ($P < 0.05$), so we reject the null hypothesis that $b_1 = 0$. We therefore conclude that fig fruit volume changes with latitude.

32.8 Prediction with linear models

We can use our linear model to predict a given value of y from x. In other words, given a value for the independent variable, we can use the regression equation ($y = b_0 + b_1 x$) to predict the dependent variable. This is possible because our model provides values for the coefficients b_0 and b_1. For the example of predicting fruit volume from latitude, the linear model estimates $b_0 = 8458.3$ and $b_1 = -242.68$. We could therefore write our regression equation,

$$Volume = 8458.3 - 242.68(Latitude).$$

Now, for any given latitude, we can predict fig fruit volume. For example, Figure 32.2 shows that there is a gap in fruit collection between 24 and 27 degrees north latitude. If we wanted to predict how large a fig fruit would be at a volume of 25, then we could set $Latitude = 25$ in our regression equation,

$$Volume = 8458.3 - 242.68(25).$$

Our predicted fig fruit volume at 25 degrees north latitude would be 2391.3 mm^3. Note that this is a point on the regression line in Figure 32.2. To find it visually in Figure 32.2, we just need to find 25 on the x-axis, then scan upwards until we see where this position on the x-axis hits the regression line.

There is an important caveat to consider when making a prediction using regression equations. Predictions might not be valid outside the range of independent variable values on which the regression model was built. In the case of the fig fruit example, the lowest latitude from which a fruit was sampled was 23.7, and the highest latitude was 29.3. We should be very cautious about predicting what volume will be for fruits outside of this latitudinal range because we cannot be confident that the linear relationship between latitude and fruit volume will persist. It is possible that at latitudes greater than 30, fruit volume will no longer decrease. It could even be that fruit volume starts to *increase* with increasing latitudes greater than 30. Since we do not have any data for such latitudes, we cannot know with much confidence what will happen. It is therefore best to avoid **extrapolation**, i.e., predicting outside of the range of values collected for the independent variable. In contrast, **interpolation**, i.e., predicting within the range of values collected for the independent variable, is generally safe.

32.9 Conclusion

There are several new concepts introduced in this chapter with simple linear regression. It is important to understand the intercept, slope, and residuals both visually and in terms of the regression equation. It is also important to be able to interpret the coefficient of determination (R^2), and to understand the hypotheses that simple linear regression can test and the assumptions underlying these tests. In the next chapter, we move on to multiple regression, in which regression models include multiple independent variables.

33

Multiple regression

Multiple regression is an extension of the general idea of simple linear regression, with some important caveats. In multiple regression, there is more than one independent variable (x), and each independent variable is associated with its own regression coefficient (b). For example, if we have two independent variables, x_1 and x_2, then we can predict y using the equation,

$$y = b_0 + b_1 x_1 + b_2 x_2.$$

More generally, for k independent variables,

$$y = b_0 + b_1 x_1 + ... + b_k x_k.$$

Mathematically, this almost seems like a trivial extension of the simple linear regression model. But conceptually, there is an additional consideration necessary to correctly interpret the regression coefficients (i.e., b values). Values of b_i now give us the predicted effects of x_i *if all other independent variables were to be held constant* (Sokal & Rohlf, 1995). In other words, b_i tells us what would happen if we were to increase x_i by a value of 1 in the context of every other independent variable in the regression model. We call these b coefficients **partial regression coefficients**. The word 'partial' is a general mathematical term meaning that we are only looking at the effect of a single independent variable (Borowski & Borwein, 2005). Since multiple regression investigates the effect of each independent variable in the context of all other independent variables, we might sometimes expect regression coefficients to be different from what they would be given a simple linear regression (Morrissey & Ruxton, 2018). It is even possible for the sign of the coefficients to change (e.g., from negative to positive).

To illustrate a multiple regression, consider again the fig fruit volume example from Chapter 32 (Duthie & Nason, 2016). Suppose that in addition to latitude, altitude was also measured in metres for each fruit (Table 33.1).

TABLE 33.1 Volumes (mm^3) of fig fruits collected from different latitudes and altitudes (m) from trees of the Sonoran Desert Rock Fig in Baja, Mexico.

Latitude	23.7	24.0	27.6	27.2	29.3	28.2	28.3
Altitude	218.5	163.5	330.1	542.3	656.0	901.3	709.6
Volume	2399.0	2941.7	2167.2	2051.3	1686.2	937.3	1328.2

We can use a scatterplot to visualise each independent variable on the x-axis against the dependent variable on the y-axis (Figure 33.1).

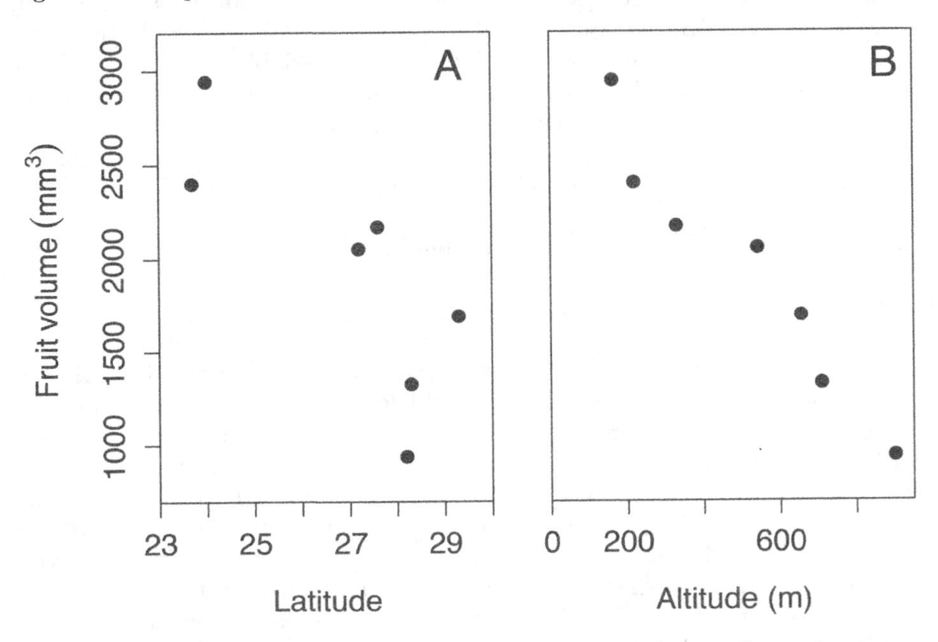

FIGURE 33.1 Relationship between latitude and fruit volume for seven fig fruits collected from Baja, Mexico in 2010, and the relationship between altitude and fruit volume for the same dataset.

As with simple regression (Chapter 32), we can test whether or not the overall model that includes both latitude and altitude as independent variables produces a significantly better fit to the data than just the mean volume. We can also find partial regression coefficients for latitude (b_1) and altitude (b_2), and test whether or not these coefficients are significantly different from 0.

In Chapter 32, we found that a simple regression of latitude against fruit volume had an intercept of $b_0 = 8458.3$ and a regression coefficient of $b_1 = -242.68$,

$$Volume = 8458.3 - 242.68(Latitude).$$

The slope of the regression line (b_1) was significantly different from zero ($P = 0.03562$).

A multiple regression can be used with latitude and altitude as independent variables to predict volume,

$$Volume = b_0 + b_1(Latitude) + b_2(Altitude).$$

We have the values of volume, latitude, and altitude in Table 33.1. We now need to run a multiple regression to find the intercept (b_0) and partial regression coefficients describing the partial effects of latitude (b_1) and altitude (b_2) on volume. In jamovi, running a multiple regression is just a matter of including the additional independent variable (Altitude, in this case). Table 33.2 shows an output table, which gives us estimates of b_0, b_1, and b_2 (column 'Estimate'), along with p-values for the intercept and each partial regression coefficient (column 'Pr(>|t|)').

TABLE 33.2 Output showing the intercept and partial regression coefficients (Estimate), standard errors (Std. Error), t-scores (t value), and p-values (Pr(>|t|)) for a multiple regression model including Latitude and Altitude as independent variables and fig fruit volume as a dependent variable.

	Estimate	Std. Error	t Value	Pr(>|t|)
(Intercept)	2988	1698	1.76	0.1532
Latitude	5.588	71.49	0.07816	0.9415
Altitude	-2.402	0.5688	-4.223	0.01344

There are a few things to point out from Table 33.2. First, note that as with simple linear regression (see Section 32.7), the significance of the intercept and regression coefficients is tested using the t-distribution. This is because we assume that these sample coefficients (b_0, b_1, and b_2) will be normally distributed around the true population parameter values (β_0, β_1, and β_2). In other words, if we were to go back out and repeatedly collect many new datasets (sampling volume, latitude, and altitude *ad infinitum*), then the distribution of b sample coefficients calculated from these datasets would be normally distributed around the true β population coefficients. The t-distribution, which accounts for uncertainty that is attributable to a finite sample size (see Chapter 19), is therefore the correct one to use when testing the significance of coefficients.

Second, the intercept has changed from what it was in the simple linear regression model. In the simple linear regression, it was 8458.30961, but in the multiple regression it is $b_0 = 2988.24539$. The p-value of the intercept has also changed. In the simple linear model, the p-value was significant ($P = 0.0142$). But in the multiple regression model, it is not significant ($P = 0.1532$).

Third, and perhaps most strikingly, the prediction and significance of latitude has changed completely. In the simple linear regression model from Section 32.7.4, fruit volume decreased with latitude ($b_1 = -242.68$), and this decrease was statistically significant ($P = 0.0356$). Now the multiple regression output is telling us that, if anything, fruit volume appears to *increase* with latitude ($b_1 = 5.59$), although this is not statistically significant ($P = 0.9415$). What is going on here? This result illustrates the context dependence of partial regression coefficients in the multiple regression model. In other words, although fruit volume appeared to significantly decrease with increasing latitude in the simple regression model of Chapter 32, this is no longer the case once we account for the altitude from which the fruit was collected. Latitude, *by itself*, does not appear to affect fruit volume after all. It only appeared to affect fruit volume because locations at high latitude also tend to be higher in altitude. And each metre of altitude appears to decrease fruit volume by about 2.4 mm^3 (Table 33.2). This partial effect of altitude on fruit volume is statistically significant ($P < 0.05$). We therefore do not reject the null hypothesis that the intercept (b_0) and partial coefficient of latitude (b_1) is significantly different from 0. But we do reject the null hypothesis that $b_2 = 0$, and we can conclude that altitude has an effect on fig fruit volume.

We can also look at the overall multiple regression model. Figure 33.2 shows what this model output looks like reported by jamovi (The jamovi project, 2024).

Model Fit Measures

			Overall Model Test			
Model	R²	Adjusted R²	F	df1	df2	p
1	0.93035	0.89552	26.71484	2	4	0.00485

FIGURE 33.2 Jamovi output table for a multiple linear regression in which latitude and altitude are independent variables and fig fruit volume is a dependent variable.

As with the simple linear regression output from Section 32.7.4, the overall model test output table includes columns for R^2, F, degrees of freedom, and a p-value for the overall model. There is one key difference between this output table and the overall model output for a simple linear regression, and that is the Adjusted R^2. This is the adjusted coefficient of determination R^2_{adj}, which is necessary to compare regression models with different numbers of independent variables.

33.1 Adjusted coefficient of determination

Recall from Section 32.5 that the coefficient of determination (R^2) tells us how much of the total variation in the dependent variable is explained by the regression model. This was fine for a simple linear regression, but with the addition of new independent variables, the proportion of the variance in y explained by our model is expected to increase even if the new independent variables are not very good predictors. This is because the amount of variation explained by our model can only increase if we add new independent variables. In other words, any new independent variable that we choose to add to the model cannot explain a negative amount of variation; that does not make any sense! The absolute *worst* that an independent variable can do is explain zero variation. And even if the independent variable is just a set of random numbers, it will likely explain *some* of the variation in the dependent variable just by chance. Hence, even if newly added independent variables are bad predictors, they might still improve the goodness of fit of our model by chance. To help account for this spurious improvement of fit, we can use an adjusted R squared (R^2_{adj}). The R^2_{adj} takes into account the R^2, the sample size (N), and the number of independent variables (k),

$$R^2_{\text{adj}} = 1 - (1 - R^2)\left(\frac{N-1}{N-k-1}\right).$$

As k increases, the value of $(N-1)/(N-k-1)$ gets bigger. And as this fraction gets bigger, we are subtracting a bigger value from 1, so R^2_{adj} decreases. Consequently, more independent variables cause a decrease in the adjusted R-squared value. This attempts to account for the inevitable tendency of R^2 to increase with k.

34

Practical. *Using regression*

This chapter focuses on practical exercises to apply the concepts in Chapter 32 and Chapter 33 in jamovi (The jamovi project, 2024). The five exercises in this practical will apply simple linear regression (Exercises 34.1, 34.2, and 34.5) or multiple regression (34.3 and 34.4). The dataset used in this practical is inspired by the work of Dr Carmen Carmona, Dr François-Xavier Joly, and Prof Jens-Arne Subke[1]. Their work focuses on carbon storage in Gabon.

When biomass is burned, a large proportion of its stored carbon is emitted into the atmosphere in the form of carbon dioxide, but some of it remains sequestered in the soil due to incomplete combustion (Santín et al., 2016). This pyrogenic organic carbon can persist in the soil for long periods of time and has positive effects on soil properties (Reisser et al., 2016). In this chapter, we will look at how environmental data might be used to test what factors affect the concentration of pyrogenic carbon in the soil. We will use the fire carbon dataset[2]. This dataset includes variables for soil depth (cm), fire frequency (total number of years in which a fire occurred during the past 20 years), mean yearly temperature (degrees Celsius), mean monthly rainfall (millimetres per squared metre per year, $\mathrm{mm\,m^{-2}\,yr^{-1}}$), total soil organic carbon (SOC, as percentage of soil by weight), pyrogenic carbon (PyC, as percentage of soil organic carbon by weight), and soil pH.

34.1 Predicting pyrogenic carbon from soil depth

In this first activity, we will fit a linear regression to predict pyrogenic carbon (PyC) from soil depth (depth). Before doing this, what is the independent variable, and what is the dependent variable?

Independent variable: ______________________________

Dependent variable: ________________________________

[1] Please note that the data in this practical are for educational purposes only. They are not the data that were actually collected by the researchers.

[2] `https://bradduthie.github.io/stats/data/fire_carbon.csv`

What is the sample size of this dataset?

$N =$ ______________________

Before running any statistical test, it is always a good idea to plot the data. Recall from Section 31.4 how to build a scatterplot in jamovi. Navigate to the 'Exploration' button from the jamovi toolbar, then choose the 'Scatterplot' option from the pull-down menu. Place the independent variable that you identified above on the x-axis, and place the dependent variable on the y-axis. To get the line of best fit, choose 'Linear' under the options below under **Regression line**. Describe the scatterplot that is produced in the jamovi panel to the right.

Recall the four assumptions of linear regression from Section 32.6. We will now check three of these assumptions (we will just have to trust that depth has been measured accurately in the field because there is no way to check). There are two assumptions that we can check using the scatterplot. The first assumption is that the relationship between the independent and dependent variable is linear. Is there any reason to be suspicious of this assumption? In other words, does the scatterplot show any evidence of a curvilinear pattern in the data?

The second assumption that we can check with the scatterplot is the assumption of homoscedasticity. In other words, does the variance change along the range of the independent variable (i.e., x-axis)?

Assuming that these two assumptions are not violated, we can now check the last assumption that the residual values are normally distributed around the regression line. To do this, we need to build the linear regression. From the 'Analyses' tab of jamovi, select the 'Regression' button, then choose 'Linear regression' from the pull-down menu. A new panel called 'Linear regression' will open. The dependent variable 'PyC' should go in the 'Dependent Variable' box to the right. The independent variable 'depth' should go in the 'Covariates' box (Figure 34.1).

We can check the assumption that the residuals are normally distributed in multiple ways. To do this, find the pull-down menu called 'Assumption Checks' in the left panel of jamovi, and check boxes for 'Normality test', 'Q-Q plot of residuals', and 'Residual plots'. Output will appear in the jamovi panel to the right. The first assumption check will be a table providing the results of a Shapiro-Wilk test of normality on the *residuals* (see Section 32.2) of the

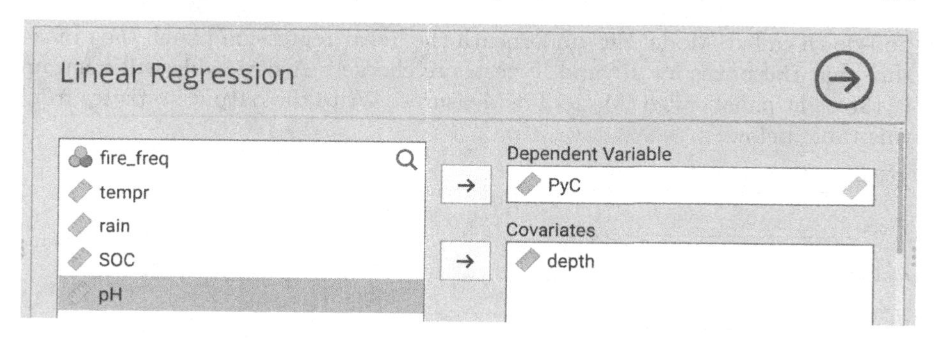

FIGURE 34.1 Jamovi interface for running a linear regression model to predict pyrogenic carbon (PyC) from soil depth (depth).

linear regression model. In your own words, what is this test doing? Drawing a picture might help to explain.

What is the p-value of the Shapiro-Wilk test of normality?

$P =$ ________________________

Based on the above p-value, is it safe to conclude that the residuals are normally distributed?

Conclusion: ____________________________

The assumption checks output also includes a Q-Q plot. Below the Q-Q plot, there is a residual plot that shows 'Fitted' on the x-axis and 'Residuals' on the y-axis. What this tells us is the relationship between the PyC values that are predicted by the regression equation (x-axis, i.e., what our equation predicts PyC will be for a particular depth) and the actual PyC values in the data (y-axis). Visually, this is the equivalent of taking the line of best fit from the first scatterplot that you made and moving it (and the points around it) so that it is horizontal at $y = 0$. It is good to try to take a few moments to understand this because it will help reinforce the concept of residual values, but in practice we can base our conclusion about residual normality on the Shapiro-Wilk test as done above.

Having checked all of the assumptions of a linear regression model, we can finally test whether or not our model is statistically significant. Find the pull-down called 'Model Fit' underneath the linear regression panel, then make sure that the boxes for R^2 and 'F test' are checked. A new table will open up in the right panel called 'Model Fit Measures'. Write the output statistics from this table below:

$R^2 =$ ______________________

$F =$ ______________________

$df1 =$ ______________________

$df2 =$ ______________________

$P =$ ______________________

Based on these statistics, what percentage of the variation in pyrogenic carbon is explained by the linear regression model?

What null hypothesis does the p-value above test? (hint, see Section 32.7.1)

H_0: ______________________

Do we reject or fail to reject H_0?

Lastly, have a look at the output table called 'Model Coefficients - PyC'. This is the same kind of table that was introduced in Section 32.7.4. From this table, what are the coefficient estimates for the intercept and slope (i.e., depth)?

Intercept: ______________________

Slope: ______________________

Find the p-values associated with the intercept and slope. What null hypotheses are we testing when inspecting these p-values? (hint, see Section 32.7.2 and Section 32.7.3)

Intercept H_0: ______________________

Slope H_0: ______________________

Finally, what can we conclude about the relationship between depth and pyrogenic carbon storage?

34.2 Predicting pyrogenic carbon from fire frequency

Now, we can try to predict pyrogenic carbon (PyC) from fire frequency (fire_freq). This exercise will be a bit more self-guided than the previous exercise. To begin, make a scatterplot with fire frequency on the x-axis and PyC on the y-axis. Add a linear regression line, then paste the plot or sketch it below (if sketching, no need for too much detail, just the trend line and 10–15 points is fine).

Next, check the linear regression assumptions of linearity, normality, and homoscedasticity, as we did in the previous exercise. Do all these assumptions appear to be met?

Linearity: ____________________

Normality: ____________________

Homoscedasticity: ____________________

Next, run the linear regression model. To check for the assumption of normality, you should have already specified a regression model with fire frequency as the independent variable and PyC as the dependent variable. Using the same protocol as the previous exercise, what percentage of the variation in PyC is explained by the regression model?

Variation explained: ________________________

Is the overall model statistically significant? How do you know?

Model significance: ____________________________

Are the intercept and slope significantly different from zero?

Intercept: ____________________

Slope: __________________

Write the intercept (b_0) and slope (b_1) of the regression below.

$b_0 =$ ____________________

$b_1 =$ ____________________

Using these values for the intercept and the slope, write the regression equation to predict pyrogenic carbon (PyC) from fire frequency (fire_freq).

Using this equation, what would be the predicted PyC for a location that had experienced 10 fires in the past 20 years (i.e., fire_freq = 10)?

One final note for this exercise. In the Linear Regression panel of jamovi, scroll all the way down to the last pull-down menu called 'Save'. Check the boxes for 'Predicted values' and 'Residuals'. When you return to the 'Data' tab in jamovi, you will see two new columns of data that jamovi has inserted. One column will be the predicted values for the model, i.e., the value that the model predicts for PyC given the fire frequency in the observation (i.e., row). The other column will be the residual value of each observation. Explain what these two columns of data represent in terms of the scatterplot you made at the start of this exercise. In other words, where would the predicted and residual values be located on the scatterplot?

34.3 Multiple regression depth and fire frequency

In this exercise, we will run a multiple regression to predict pyrogenic carbon (PyC) from fire frequency (fire_freq) and depth. Write down what the independent and dependent variable(s) are for this regression.

Independent: ____________________

Dependent: ____________________

To begin the multiple regression, select the 'Regression' button in the Analysis tab of jamovi, then choose 'Linear regression' as you did in the first two exercises. Place the dependent variable in the 'Dependent Variable' box and both independent variables in the 'Covariates' box. As with the previous exercise, check the linear regression assumptions of linearity, normality, and homoscedasticity. Do all these assumptions appear to be met?

Linearity: ____________________

Normality: ____________________

Homoscedasticity: ____________________

Make sure to select R^2, Adjusted R^2, and F test under the Model Fit options. Report these values from the Model Fit Measures output table below.

$R^2 =$ ____________________

Adjusted $R^2 =$ ____________________

$F =$ ____________________

$P =$ ____________________

Explain why the Adjusted R^2 is less than the R^2 value. Which one is most appropriate to use for interpreting the multiple regression?

What is the null hypothesis tested with the F value and the p-value shown in the Model Fit Measures table?

H_0: ____________________

Based on the Overall Model Test output, should you reject or not reject H_0?

Next, have a look at the Model Coefficients - PyC table. What can you conclude about the significance of the Intercept, and the partial regression coefficients for fire frequency and depth?

Using the partial regression coefficient estimates, fill in the equation below,

$$PyC = (\qquad) + (\qquad) fire\_freq + (\qquad) depth.$$

Next, use this to predict the PyC for a fire frequency of 12 and a depth of 60 cm.

PyC = ____________________

Contrast soil depth as a predictor of PyC in this multiple regression model versus the simple linear regression model in the first exercise. Has the significance of soil depth as an independent variable changed? Based on what you know about the difference between simple linear regression and multiple regression, why might this be the case?

34.4 Large multiple regression

Suppose that as scientists, we hypothesise that soil depth, fire frequency, and soil pH will all affect pyrogenic carbon storage. Run a multiple regression model with soil depth, fire frequency, and soil pH all as independent variables and pyrogenic carbon as a dependent variable. Fill in the Model Coefficients output in Table 34.1.

TABLE 34.1 Model Coefficients output table for a multiple regression model predicting pyrogenic carbon from soil depth, fire frequency, and soil pH in Gabon.

	Estimate	Std. Error	t Value	Pr(>|t|)
(Intercept)		0.34591	2.85888	
depth		0.0008	−0.07411	
fire_freq		0.00394	14.42303	
pH		0.05679	−0.27886	

From the Model Fit Measures table, what is the R^2 and Adjusted R^2 of this model?

R^2: ____________________

Adjusted R^2: ____________________

Compare these values to the R^2 and Adjusted R^2 from the multiple regression in the previous exercise (i.e., the one without pH as an independent variable). Is the R^2 value of this model higher or lower than the multiple regression model without pH?

Is the Adjusted R^2 value of this model higher or lower than the multiple regression model without pH?

Based on what you know from Section 33.1, explain why the R^2 and Adjusted R^2 might have changed in different directions with the addition of a new independent variable.

Finally, use the equation of this new model to predict PyC for a soil sample at a depth of 0, fire frequency of 0, and pH of 6.

34.5 Predicting temperature from fire frequency

In this last brief exercise, suppose that we wanted to predict temperature (tempr) from fire frequency (fire_freq). Run some checks of the assumptions underlying linear regression (see Section 32.6). What assumption(s) appear as though they might be violated for this simple regression? Explain how you figured this out.

35

Randomisation

Since introducing statistical hypothesis testing in Chapter 21, this book has steadily introduced statistical tests that can be run for different data types. In all of these statistical tests, the general idea is the same. We calculate some test statistic, then compare this test statistic to a predetermined null distribution to find the probability (i.e., p-value) of getting a test statistic as or more extreme than the one we calculated if the null hypothesis were true. This chapter introduces a different approach. Instead of using a predetermined null distribution, we will build the null distribution using our data. This approach is not one that is often introduced in introductory statistics texts. It is included here for two reasons. First, randomisation presents a different way of thinking statistically without introducing an entirely different philosophical or methodological approach such as likelihood (Edwards, 1972) or Bayesian statistics (Lee, 1997). Second, it helps reinforce the concept of what null hypothesis testing is and what p-values are. Before explaining the randomisation approach, it is useful to summarise the parametric hypothesis tests introduced in earlier chapters.

35.1 Summary of parametric hypothesis testing

For the parametric tests introduced in previous chapters, null distributions included the t-distribution, F-distribution, and χ^2 distribution. For the t-tests in Chapter 22, the test statistic was the t-statistic, which we compared to a t-distribution. The one-sample t-test compared the mean of some variable ($\bar{x}$) to a specific number (μ_0), the independent samples t-test compared two group means, and the paired samples t-test compared the mean difference between two paired groups. All of these tests used the t-distribution and calculated some value of t. Very low or high values of t at the extreme ends of the t-distribution are unlikely if the null hypothesis is true, so, in a two-tailed test, these are associated with low p-values that lead us to reject the null hypothesis.

For the analysis of variance (ANOVA) tests of Chapter 24, the relevant test statistic was the F statistic, with the F-distribution being the null distribution expected if two variances are equal. The one-way ANOVA used the within and among group variances of two or more groups to test the null hypothesis that all group means are equal. The two-way ANOVA of Chapter 27 extended the framework of the one-way ANOVA, allowing for a second variable of groups. This made it possible to simultaneously test whether or not the means of two different group types were the same, and whether or not there was an interaction between group types. All of these ANOVA tests calculated some value of F and compared it to the F-distribution with an appropriate degrees of freedom. Sufficiently high F values were associated with a low p-value and therefore the rejection of the null hypothesis.

The Chi-square tests introduced in Chapter 29 were used to test the frequencies of categorical observations and determine if they matched some expected frequencies (Chi-square goodness of fit test) or were associated in some way with the frequencies of another variable (Chi-square test of association). In these tests, the χ^2 statistic was used and compared to a null χ^2 distribution with appropriate degrees of freedom. High χ^2 values were associated with low p-values and the rejection of the null hypothesis.

For testing the significance of correlation coefficients (see Chapter 30) and linear regression coefficients (see Chapter 32), a t-distribution was used. And an F-distribution was used to test for the overall significance of linear regression models.

For these tests, the approach to hypothesis testing was therefore always to use the t-distribution, F-distribution, or χ^2 distribution in some way. These distributions are more formally defined in mathematical statistics (Miller & Miller, 2004), a field of study that uses mathematics to derive the probability distributions that arise from an outcome of random events (e.g., the coin-flipping of Section 15.1). The reason that we use these distributions in statistical hypothesis testing is that they are often quite good at describing the outcomes that we expect when we collect a sample from a population. But this is not always the case. Recall that sometimes the assumptions of a particular statistical test were not met. In this case, a non-parametric alternative was introduced. The non-parametric test used the ranks of data instead of the actual values (e.g., Wilcoxon, Mann-Whitney U, Kruskal-Wallis H, and Spearman's rank correlation coefficient tests). Randomisation uses a different approach.

35.2 Randomisation approach

Randomisation takes a different approach to null hypothesis testing. Instead of assuming a theoretical null distribution against which we compare our test statistic, we ask, 'if the ordering of the data we collected was actually random, then what is the probability of getting a test statistic as or more extreme than the one that we actually did'. Rather than using a null distribution derived from mathematical statistics, we will build the null distribution by randomising our data in some useful way (Manly, 2007). Conceptually, this is often easier to understand because randomisation approaches make it easier to see why the null distribution exists and what it is doing. Unfortunately, these methods are more challenging to implement in practice because using them usually requires knowing a bit of coding. The best way to get started is with an instructive example.

35.3 Randomisation for hypothesis testing

As in several previous chapters, the dataset used here is inspired by the many species of wasps that lay their eggs in the flowers of the Sonoran Desert rock fig (*Ficus petiolaris*). This tree is distributed throughout the Baja peninsula, and in parts of mainland Mexico. Fig trees and the wasps that develop inside of them have a fascinating ecology, but for now we will just focus on the morphologies of two closely related species as an example. The fig wasps in Figure 35.1 are two unnamed species of the genus *Idarnes*, which we can refer to simply as 'Short-ovipositor 1' (SO1) and 'Short-ovipositor 2' (SO2).

The reason that these two species are called 'SO1' and 'SO2' is that there is actually another species that lays its eggs in *F. petiolaris* flowers, one with an ovipositor that is at least twice as long as those in Figure 35.1 (Duthie et al., 2015).

Suppose that we have some data on the lengths of the ovipositors from each species. We might want to know whether the mean ovipositor length differs between the two species. Figure 35.2 shows histograms of ovipositor lengths collected from 32 fig wasps, 17 SO1s and 15 SO2s.

To test whether or not mean ovipositor length is different between these two fig wasps, our standard approach would be to use an independent samples t-test (see Section 22.2). The null hypothesis would be that the two means are the same, and the alternative (two-sided) hypothesis would be that the two means are not the same. We would need to check the assumption that the data are normally distributed, and that both samples have similar variances.

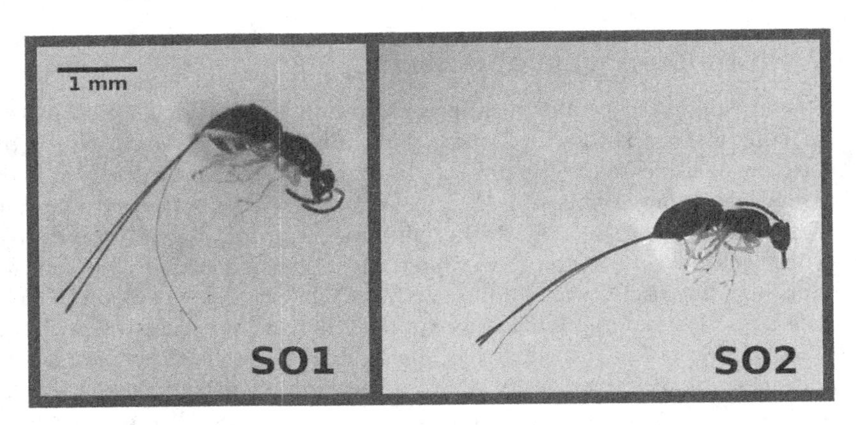

FIGURE 35.1 Two fig wasp species, roughly 3 mm in length, labelled 'SO1' and 'SO2'. Image modified from Duthie, Abbott, and Nason (2015).

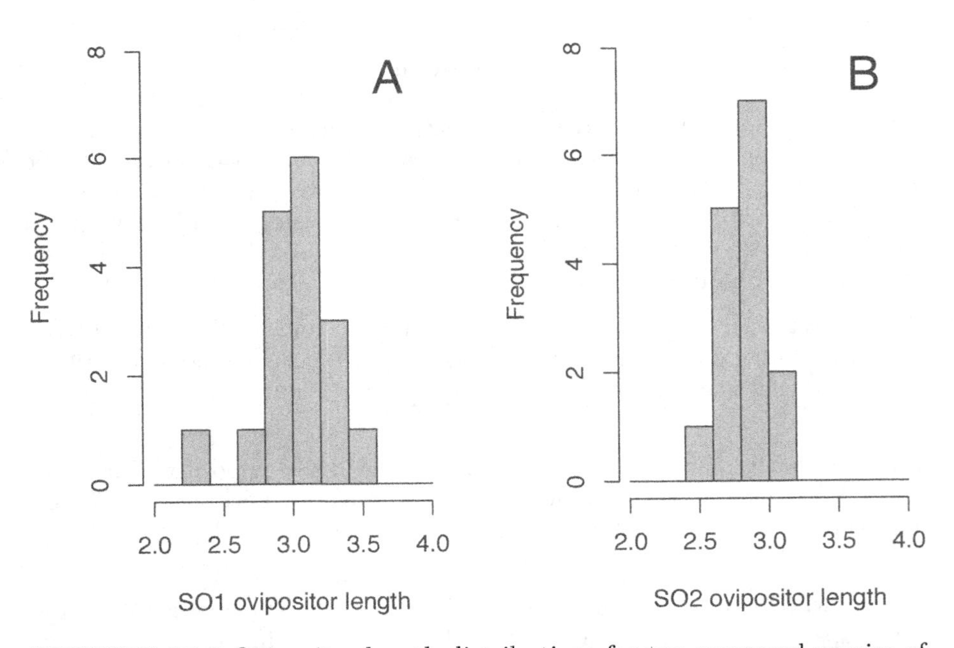

FIGURE 35.2 Ovipositor length distributions for two unnamed species of fig wasps, SO1 (A) and SO2 (B), collected from Baja, Mexico.

Assuming that the assumption of normality is not violated (in which case we would need to consider a Mann-Whitney U test), and that both groups had similar variances (if not, we would use Welch's t-test), we could proceed with calculating our t-statistic for pooled sample variance,

$$t_{\bar{y}_{\mathrm{SO1}}-\bar{y}_{\mathrm{SO2}}} = \frac{\bar{y}_{\mathrm{SO1}} - \bar{y}_{\mathrm{SO2}}}{s_p}.$$

The s_p is just being used as a short-hand to indicate the pooled standard deviation. For the two species of fig wasps, $\bar{y}_{\mathrm{SO1}} = 3.0301176$, $\bar{y}_{\mathrm{SO2}} = 2.8448667$, and $s_p = 0.0765801$. We can therefore calculate $t_{\bar{y}_{\mathrm{SO1}}-\bar{y}_{\mathrm{SO2}}}$,

$$t_{\bar{y}_{\mathrm{SO1}}-\bar{y}_{\mathrm{SO2}}} = \frac{3.03 - 2.845}{0.077}.$$

After we calculate our t-statistic as $t_{\bar{y}_{\mathrm{SO1}}-\bar{y}_{\mathrm{SO2}}} = 2.419$, we would use the t-distribution to find the p-value (or, rather, get jamovi to do this for us). Figure 35.3 shows the t-distribution for 30 degrees of freedom with an arrow pointing at the value of $t_{\bar{y}_{\mathrm{SO1}}-\bar{y}_{\mathrm{SO2}}} = 2.419$.

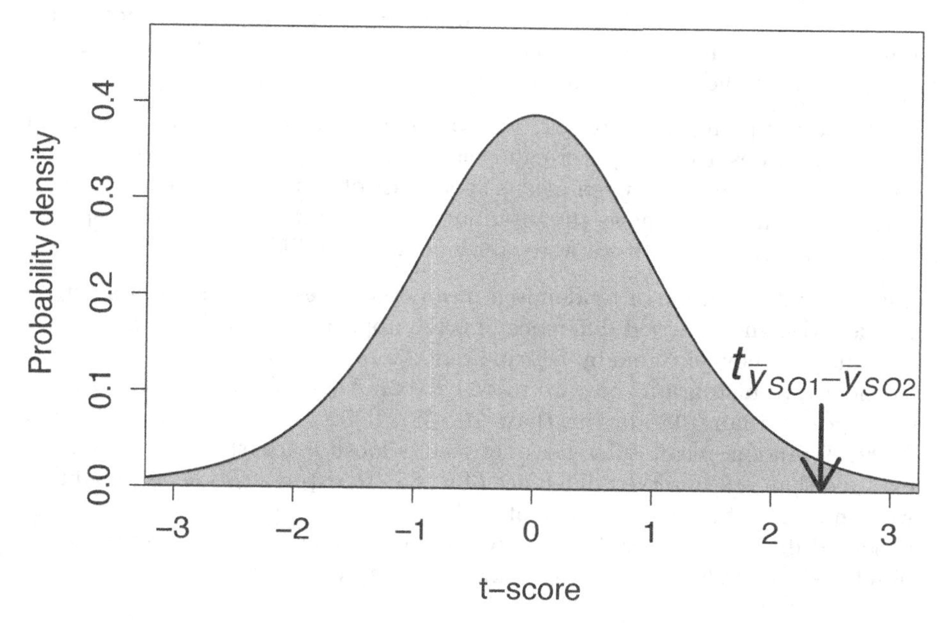

FIGURE 35.3 A t-distribution is shown with a calculated t-statistic of 2.419 indicated with a downward arrow.

If the null hypothesis is true, and $\bar{y}_{\mathrm{SO1}}$ and $\bar{y}_{\mathrm{SO2}}$ are actually sampled from a population with the same mean, then the probability of randomly sampling a value more extreme than 2.419 (i.e., greater than 2.419 or less than −2.419) is

$P = 0.0218352$. We therefore reject the null hypothesis because $P < 0.05$, and we conclude that the mean ovipositor lengths of each species are not the same.

Randomisation takes a different approach to the same problem. Instead of using the t-distribution in Figure 35.3, we will build our own null distribution using the fig wasp ovipositor length data. We do it by randomising group identity in the dataset. The logic is that if there really is no difference between group means, then we should be able to randomly shuffle group identities (species) and get a difference between means that is not far off the one we actually get from the data. In other words, what would the difference between group means be if we just mixed up all of the species (so some SO1s become SO2s, some SO2s become SO1s, some stay the same), then calculated the difference between means of the mixed up groups? If we just do this once, then we cannot learn much. But if we randomly shuffle the groups many, many times (say at least 9999), then we could see what the difference between group means would look like just by chance, i.e., if ovipositor length really was not different between SO1 and SO2. We could then compare our actual difference between mean ovipositor lengths to this null distribution in which the difference between groups means really is random (it has to be, we randomised the groups ourselves!). The idea is easiest to see using an interactive application[1] that builds a null distribution of differences between the mean of SO1 and SO2 and compares it to the observed difference of $\bar{y}_{SO1} - \bar{y}_{SO2} = 0.185$.

With modern computing power, we do not need to do this randomisation manually. A desktop computer can easily reshuffle the species identities and calculate a difference between means thousands of times in less than a second. The histogram below shows the distribution of the difference between species mean ovipositor length if we were to randomly reshuffle groups 99999 times.

Given this distribution of randomised mean differences between species ovipositor lengths, the observed difference of 0.185 appears to be fairly extreme. We can quantify how extreme by figuring out the proportion of mean differences in the above histogram that are more extreme than our observed difference (i.e., greater than 0.185 or less than −0.185). It turns out that only 2085 out of 99999 random mean differences between ovipositor lengths were as or more extreme than our observed difference of 0.185. To express this as a probability, we can simply take the number of differences as or more extreme than our observed difference (including the observed one itself), divided by the total number of differences (again, including the observed one),

$$P = \frac{2085 + 1}{99999 + 1}.$$

When we calculate the above, we get a value of 0.02086. Notice that the calculated value is assigned with a '*P*'. This is because the value **is a p-value**

[1] `https://bradduthie.github.io/stats/app/randomisation/`

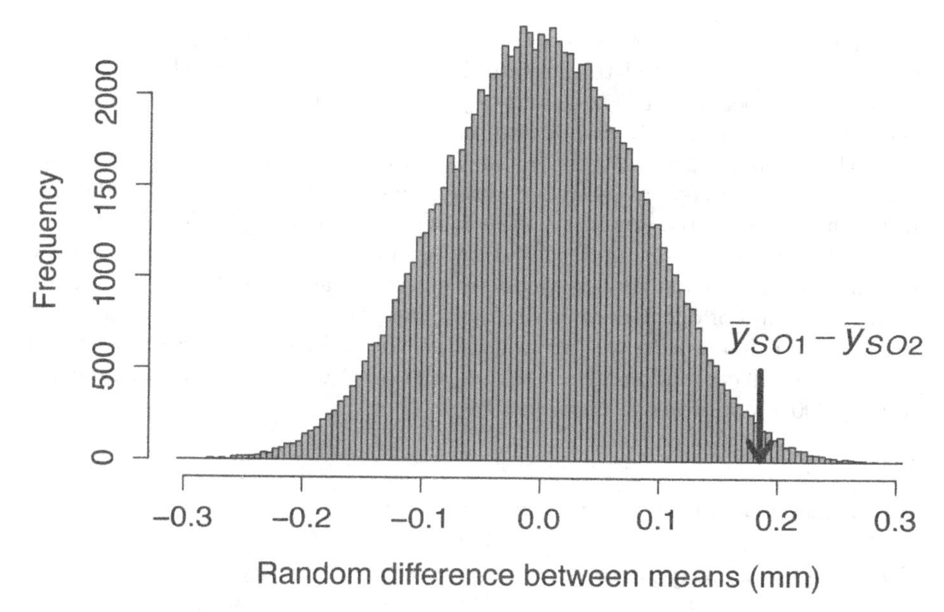

FIGURE 35.4 Distribution of the difference between mean species ovipositor lengths in two different species of fig wasps when species identity is randomly shuffled. The true difference between sampled mean SO1 and SO2 ovipositor lengths is pointed out with the black arrow at 0.185. Ovipositor lengths were collected from wasps in Baja, Mexico in 2010.

(technically, an unbiased estimate of a p-value). Consider how close it is to the value of $P = 0.0218352$ that we got from the traditional t-test. Conceptually, we are doing the same thing in both cases; we are comparing a test statistic to a null distribution to see how extreme the differences between groups really are.

35.4 Randomisation assumptions

Recall from Section 22.4 the assumptions underlying t-tests, which include (1) data are continuous, (2) sample observations are a random sample from the population, and (3) sample means are normally distributed around the true mean. With randomisation, we do not need to assume 2 or 3. The randomisation approach is still valid even if the data are not normally distributed. Samples can have different variances. The observations do not even need to be independent. The validity of the randomisation approach even when standard assumptions do not apply can be quite useful.

The downside of the randomisation approach is that the statistical inferences that we make are limited to our sample, not the broader population (recall the difference between a sample and population from Chapter 4). Because the randomisation method does not assume that the data are a random sample from the population of interest (as is the case for the traditional t-test), we cannot formally make an inference about the difference between populations from which the sample was made. This is not necessarily a problem in practice. It is only relevant in terms of the formal assumptions of the model. Once we run our randomisation test, it might be entirely reasonable to argue verbally that the results of our randomisation test can generalise to our population of interest. In other words, we can argue that the difference between groups in the sample reflects a difference in the populations from which the sample came (Ernst, 2004; Ludbrook & Dudley, 1998).

35.5 Bootstrapping

We can also use randomisation to calculate confidence intervals for a variable of interest. Remember from Chapter 18 the traditional way to calculate lower and upper confidence intervals using a normal distribution,

$$LCI = \bar{x} - (z \times SE),$$

$$UCI = \bar{x} + (z \times SE).$$

Recall from Chapter 19 that a t-score is substituted for a z-score to account for a finite population size.

Suppose we want to calculate 95% confidence intervals for the ovipositor lengths of SO1 (Figure 35.2A). There are 17 ovipositor lengths for SO1.

3.256, 3.133, 3.071, 2.299, 2.995, 2.929, 3.291, 2.658, 3.406, 2.976, 2.817, 3.133, 3.000, 3.027, 3.178, 3.133, 3.210

To get the 95% confidence interval using the method from Chapter 18, we would calculate the mean $\bar{x}_{\mathrm{SO1}} = 3.03$, standard deviation $s_{\mathrm{SO1}} = 0.26$, and the t-score for $df = N - 1$ (where $N = 17$), which is $t = 2.120$. Keeping in mind the formula for standard error $s/\sqrt{N}$, we can calculate,

$$LCI = 3.03 - \left(2.120 \times \frac{0.26}{\sqrt{17}}\right),$$

$$UCI = 3.03 + \left(2.120 \times \frac{0.26}{\sqrt{17}}\right).$$

Calculating the above gives us values of LCI = 2.896 and UCI = 3.164.

Again, randomisation uses a different approach to get an estimate of the same confidence intervals. Instead of calculating the standard error and multiplying it by a z-score or t-score to encompass a particular interval of probability density, we can instead resample the data we have with replacement many times, calculating the mean each time we resample. The general idea is that this process of resampling approximates what would happen if we were to go back and resample new data from our original population many times, thereby giving us the distribution of means from all of these hypothetical resampling events (Manly, 2007). To calculate our 95% confidence intervals, we then only need to rank the calculated means and find the mean closest to the lowest 2.5% and the highest 97.5%.

Remember from Section 15.3 that the phrase 'resampling with replacement' just means that we are going to randomly sample some values, but not remove them from the pool of possible values after they are sampled. If we resample the numbers above with replacement, we might therefore sample some values two or more times by chance, and other values might not be sampled at all. The numbers below resample from the above with replacement.

2.299, 3.027, 2.929, 3.256, 3.071, 2.658, 3.133, 3.027, 2.976, 2.817, 3.406, 3.210, 2.976, 3.133, 3.133, 3.178, 2.299

Notice that some values appear more than once in the data above, while other values that were present in the original dataset are no longer present after resampling with replacement. Consequently, these new resampled data have a different mean than the original. The mean of the original 17 SO1 ovipositor length values was 3.03, and the mean of the values resampled above is 2.972. We can resample another set of numbers to get a new mean.

3.133, 3.027, 2.658, 3.291, 3.133, 2.299, 2.976, 2.658, 2.929, 3.133, 2.929, 2.995, 3.071, 2.299, 3.000, 3.133, 3.133

The mean of the above sample is 2.929. We can continue doing this until we have a high number of random samples and means. Figure 35.5 shows the distribution of means if we repeat this process of resampling with replacement and calculating the mean 10000 times.

The arrows show the locations of the 2.5% and 97.5% ranked values of the bootstrapped means. Note that it does not matter if the above distribution is not normal (it appears a bit skewed). The bootstrap still works. The values using the randomisation approach are LCI = 2.897 and UCI = 3.14. These values are quite similar to those calculated with the traditional approach because we are doing the same thing, conceptually. But instead of finding confidence intervals of the sample means around the true mean using the t-distribution, we are actually simulating the process of resampling from the population and calculating sample means many times.

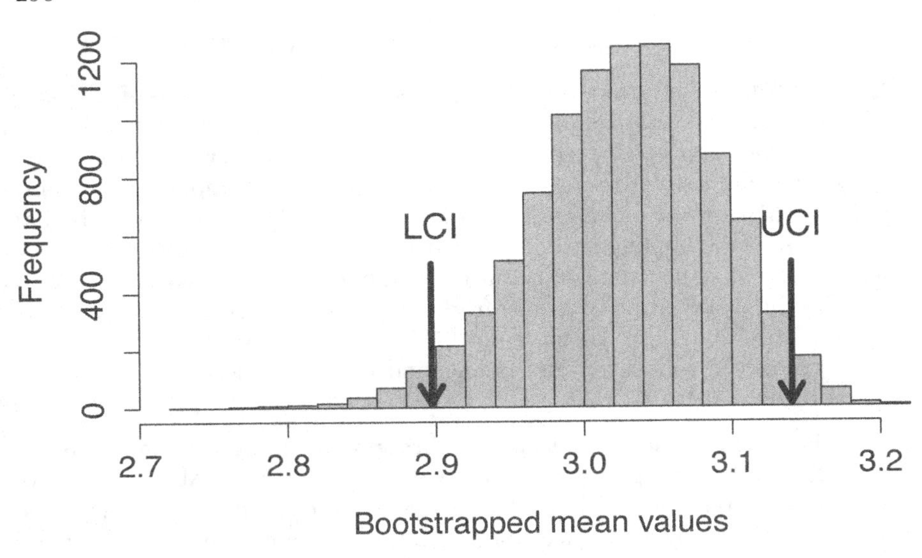

FIGURE 35.5 A distribution of bootstrapped values of ovipositor lengths from 17 measurements of a fig wasp species collected from Baja, Mexico. A total of 10000 bootstrapped means are shown, and arrows indicate the location of the 2.5% (LCI) and 97.5% (UCI) ranked boostrapped mean values.

35.6 Randomisation conclusions

The examples demonstrated in this chapter are just a small set of what is possible using randomisation approaches. Randomisation and bootstrapping are highly flexible tools that can be used in a variety of ways (Manly, 2007). They could be used as substitutes (or complements) to any of the hypothesis tests that we have used in this book. For example, to use a randomisation approach for testing whether or not a correlation coefficient is significantly different from 0, we could randomly shuffle the values of one variable 9999 times. After each shuffle, we would calculate the Pearson product moment correlation coefficient, then use the 9999 randomised correlations as a null distribution to compare against the observed correlation coefficient. This would give us a plot like the one shown in Figure 35.4, but showing the null distribution of the correlation coefficient instead of the difference between group means. Similar approaches could be used for ANOVA or linear regression (Manly, 2007).

Appendix

A

Answers to chapter exercises

Answers to exercises in chapters are provided below. Many questions are open-ended and intended to be thought provoking, and the answers to these questions that are provided below are mostly intended to explain the motivation underlying the question. Numerical answers are also provided. Answers are written in bold. All datasets that are made tidy in the process of doing exercises can be found at `https://osf.io/dxwyv` in the 'exercise_answer_datasets', and in the links below.

A.1 Chapter 3

A.1.1 Exercise 3.1

1. **Seeds are tallied incorrectly.**
2. Tallies are not counted correctly in the lab notebook.
3. **Counts are not correctly input into a spreadsheet.**

You can download the tidy dataset from this exercise here:

`https://bradduthie.github.io/stats/data/Ch3_Exercise_1.csv`

A.1.2 Exercise 3.2

How many columns did you need to create the new dataset? **2 columns.**

Are there any missing data in this dataset? **There are no missing data.**

The tidy dataset should include two columns, one for Species and the other for egg loads. It should have 54 rows (plus the header). You can download the dataset here:

`https://bradduthie.github.io/stats/data/Ch3_Exercise_2.csv`

A.1.3 Exercise 3.3

The tidy dataset should include three columns, one for Species, one for Fruit, and one for Count. It should have 25 rows (plus the header). You can download the dataset here:

`https://bradduthie.github.io/stats/data/Ch3_Exercise_3.csv`

A.1.4 Exercise 3.4

What columns should this new dataset include? **Species, Wasp number, Head length (mm), Head width (mm), Thorax length (mm), Thorax width (mm), Abdomen length (mm), Abdomen width (mm).**

How many rows are needed? **26 rows of data (plus the column header).**

You can download the dataset here:

`https://bradduthie.github.io/stats/data/Ch3_Exercise_4.csv`

What formula will you type into your empty spreadsheet cell to calculate V_{thorax}? **=(4/3) * 3.14 * (D2/2) * ((E2/2)^2)**

What are some reasons that we might want to be cautious about our calculated wasp volumes? **There is error associated with the measurement of fig wasp dimensions (e.g., length, width). There is also error because we are assuming that the head is a sphere and the thorax and abdomen are ellipses.**

A.2 Chapter 8

A.2.1 Exercise 8.1

Mean: **6.52 g C / kg soil**

Minimum: **0.600 g C / kg soil**

Maximum: **16.2 g C / kg soil**

Topsoil Mean: **9.75 g C / kg soil**

Topsoil Minimum: **4.00 g C / kg soil**

Topsoil Maximum: **16.2 g C / kg soil**

Subsoil Mean: **2.43 g C / kg soil**

Subsoil Minimum: **0.600 g C / kg soil**

Subsoil Maximum: **4.60 g C / kg soil**

Based on these samples in the dataset, can we really say for certain that the population mean of topsoil is higher than the population mean of subsoil?

The sample means themselves do not tell us whether the population mean of topsoil will be bigger or smaller than that of subsoil, just that the sample means are different. We can use the data to calculate a standard error associated with the mean, which will give an indication of the level of confidence that is appropriate.

What would make you more (or less) confident that topsoil and subsoil population means are different?

The larger the sample size, the more confidence we have that the sample mean is close to the population mean. Also, the narrower the spread of the data, the more confidence we should have that the sample mean is close to the population mean.

A.2.2 Exercise 8.2

Grand Mean length: **1.59 cm**

Grand Mean height: **1.46 cm**

Grand Mean width: **1.57 cm**

Missing width row: **Row 62**

Missing height row: **Row 22**

Grand Mean length (mm): **15.9**

Grand Mean height (mm): **14.6**

Grand Mean width (mm): **15.7**

Do the differences between means in cm and the means in mm make sense? **Yes. Note that means in millimetres are ten times the value of the means in centimetres, as expected.**

A.2.3 Exercise 8.3

In this case, how might assuming that figs are perfectly spherical affect the accuracy of our estimated fig volume?

As figs are not perfectly spherical, the estimation will be inaccurate. It could be systematically inaccurate, i.e., it would likely consistently over- or under-estimate the true volume of the figs, but it could also be randomly inaccurate as each fig will differ in shape a little so the approximation to a sphere will be differently wrong for each fig. Note that the measuring equipment used (a ruler in this case) will also limit the precision of the estimates.

Mean: **2150 mm^3**

Minimum: **697 mm^3**

Maximum: **4847 mm^3**

Check the option for 'Histogram' and see the new histogram plotted in the window to the right. Draw a rough sketch of the histogram in the area below. **Your drawing should include a histogram with fig volume ranging from under 1000 to about 5000, with most values between 1000 and 2500.**

A.3 Chapter 14

A.3.1 Exercise 14.1

You can download the tidy dataset from this exercise here:

`https://bradduthie.github.io/stats/data/Nymphaea_alba_tidy.csv`

A.3.2 Exercise 14.2

Just looking at the histogram, write down what you think the following summary statistics will be.

While it will never be necessary (or recommended) to try to work out the mean, median, and standard deviation of a distribution directly from a histogram, the point of this is to help you connect the numerical summary statistics with the visualisation of the data in the histogram. You should be able to recognise that the mean and median are likely somewhere between 5 to 6 by recognising this as the centre of the distribution. Working out the standard deviation is a bit more challenging, but the average deviation from the centre looks to be about 2 (i.e., most points are probably about 2 mm from the mean). If your own answer was a bit different, this is not a cause for concern, but if you are able to interpret the histogram successfully, it probably should not be off by more than 3–4 mm.

Based on the histogram, do you think that the mean and median are the same? Why or why not?

The mean and median appear to be quite similar. The distribution is mostly symmetrical, so the mean and median will likely be very close, although it might have a bit of a positive skew to it.

Write a caption for the histogram below.

Figure: Distribution of petiole diameter (mm) from white water lillies (*Nymphaea alba*) collected from 7 Scottish lochs.

Highest: **Lily loch**

Lowest: **Linne**

A.3.3 Exercise 14.3

- N: **140**
- Std. deviation: **1.83**
- Variance: **3.36**
- Minimum: **1.57**
- Maximum: **9.93**
- Range: **8.36**
- IQR: **2.90**
- Mean: **5.52**
- Median: **5.56**
- Mode: **3.2**
- Std. error of mean: **0.155**

Which of the 7 sites in the dataset has the highest mean petiole diameter, and what is its mean?

Site: **Beag**

Mean: **5.94 mm**

Which of the 7 sites has the lowest variance in petiole diameter, and what is its variance?

Site: **Fidhle**

Variance: **2.51 mm^2**

Can you find the first and third quartiles for each site?

Beag: **7.13 mm**

Buic: **6.33 mm**

Choille-Bharr: **6.83 mm**

Creig-Moire: **7.24 mm**

Fidhle: **6.53 mm**

Lily_Loch: **7.24 mm**

Linne: **6.52 mm**

A.3.4 Exercise 14.4

- N: **140**
- Std. deviation: **1.8**
- Variance: **3.4**
- Minimum: **1.6**
- Maximum: **9.9**
- Range: **8.4**
- IQR: **2.9**
- Mean: **5.5**
- Median: **5.6**
- Mode: **3.2**
- Std. error of mean: **0.15**

Were you able to get a similar value from the histogram as calculated in jamovi from the data? What can you learn from the histogram that you cannot from the summary statistics, and what can you learn from the summary statistics that you cannot from the histogram?

Values might or might not be similar (it is not important that you can guess a mean or median to any degree of accuracy just by looking at a histogram). Note, however, that with the histogram, we can see the full shape of the distribution, which is not really possible (or at least, not easy), with the summary statistics alone. The summary statistics, in contrast, can give us specific numbers of central tendency or spread (e.g., mean, median, variance).

A.3.5 Exercise 14.5

Recall back from Chapter 12; what information do these error bars convey about the estimated mean petiole diameter?

Standard errors tell us how far our sample mean is expected to deviate from the true mean. Specifically, the standard error of the mean is the standard deviation of the sample means around the true mean. It is a measure for evaluating the uncertainty of the mean.

What can you say about the mean petiole diameters across the different sites? Do these sites appear to have very different mean petiole diameters?

Different sites appear to have similar means. Given the uncertainty indicated by the standard error, it is unclear if different sites have different mean petiole diameters (note, an ANOVA, which is introduced in Chapter 24, does not reject the null hypothesis that the means are the same).

There were 20 total petiole diameters sampled from each site. If we were to go back out to these 7 sites and sample another 20 petiole diameters, could we really expect to get the exact same site means? Assuming the site means would be at least a bit different for our new sample, is it possible that the sites with the highest or lowest petiole diameters might also be different in our new sample? If so, then what does this say about our ability to make conclusions about the differences in petiole diameter among sites?

If we were to go out and sample another 20 petiole diameters from each site, then we would not expect to get the exact same site means. The site means would be a bit different. The distribution of these sample means (with each repeated resampling of 20 and mean calculation) would be normally distributed around the true mean with a standard deviation approximately equal to the standard error of any given sample (such as the one in the barplots). It is possible that the rank order of mean petiole diameters might change entirely. Given this uncertainty, we cannot really say for sure which site population mean is really the highest or lowest.

A.4 Chapter 17

A.4.1 Exercise 17.1

Now, fill in Table 17.1 with counts, percentage, and the estimated probability of a player selecting a small, medium, or large dam.

TABLE 17.1 Statistics of Power Up! decisions for dam size with answers.

Dam Size	Counts	Percentage	Estimated Probability
Small	**21**	**28.4**	**0.284**
Medium	**13**	**17.6**	**0.176**
Large	**40**	**54.1**	**0.541**

What is the probability that this player chooses a small or a large dam?

To get this probability, calculate the probability that the player chooses a small dam plus the probability that the player chooses a large dam: $P(\text{small or large}) = 0.284 + 0.541 = 0.825$.

Now suppose that 3 new players arrive and decide to play the game. What is the probability that all 3 of these new players choose a large dam?

To get this probability, calculate the probability of choosing a large dam raised to the power of 3: $P(3 \text{ large}) = 0.541 \times 0.541 \times 0.541 = 0.541^3 = 0.158$.

What is the probability that the first player chooses a small dam, the second player chooses a medium dam, and the third player chooses a large dam?

To get this probability, calculate the probability of choosing small dam, times a medium dam, times a large dam: P(Player 1 = small, Player 2 = medium, Player 3 = large) = 0.541 × 0.176 × 0.284 = 0.027.

Imagine that you randomly choose one of the 74 players with equal probability (i.e., every player is equally likely to be chosen). What is the probability that you choose player 20?

To get this probability, we just need to calculate one divided by 74: P(Player 20) = 1/74 = 0.01351.

What is the probability that you choose player 20, *then* choose a different player with a large dam? As a hint, remember that you are now sampling *without replacement*. The second choice cannot be player 20 again, so the probability of choosing a player with a large dam has changed from the estimated probability in Table 17.1.

To get this probability, we need to first recognise that there is a 1/74 probability (0.01351) of choosing Player 20. Player 20 chose a large dam, so the number of remaining players in the dataset is now 73, and now only 39 of them chose large dams. Hence, the probability of choosing a large dam from the remaining players is 39/73 (0.53425). To calculate the probability of both events happening, we need to multiply: P(Player 20, Large) = 0.01351 × 0.53425 = 0.00722.

Now, recreate the table in Figure 17.3 and estimate the probability that an Android user will choose to build a large dam.

To get this probability, divide the number of Android players that choose a large dam (31) by the total number of Android users (56): P(Large | Android) = 31/56 = 0.554.

Is P(Large|Android) much different from the probability that any player chooses a large dam, as calculated in Table 17.1? Do you think that the difference is significant?

This is a small difference, but not that much. It is probably not significant.

A.4.2 Exercise 17.2

Use jamovi to find the mean and the standard deviation of player score (note, we can just say that score is unitless, so no need to include units).

Mean score: 95.9

Standard deviation score: 22.3

What is the probability of a player getting a score between 80 and 120?

$P(80 \leq X \leq 120)$ = **0.6222**

What is the probability of a player getting a score greater than 130?

$P(X \geq 130)$ = **0.0631**

Now try the following probabilities for different scores.

$P(X \geq 120)$ = **0.1399**

$P(X \leq 100)$ = **0.5729**

$P(100 \leq X \leq 120)$ = **0.2872**

What is the probability of a player getting a score lower than 70 or higher than 130?

$P(X \leq 70 \cup X \geq 130) = 1 - 0.8142$ = **0.1858**

There is more than one way to figure this last one out. How did you do it, and what was your reasoning?

We can find the total area under the curve between 70 and 130 using jamovi. Since the entire area under the curve must sum to 1, if we subtract this area (0.8142) from 1, then we are left with the area in the tails of the distribution (i.e., lower than 70 or higher than 130).

A.4.3 Exercise 17.3

How would you describe the shape of the distribution of v1?

The distribution of v1 is approximately uniform.

Sketch what you predict the shape of its distribution will be below.

The 'all_means' values should have the shape of a normal distribution (roughly) as it is the distribution of the sample means. The CLT states that the original distribution (in this case uniform) does not matter; when a set of sample means are individually calculated, the dataset will form a normal distribution.

As best you can, explain why the shapes of the two distributions differ.

A consequence of the central limit theorem is that the distribution of sample means should be normal regardless of the distribution of the sample data. In this case, the sample data were uniformly distributed, but the means of the 40 datasets is normally distributed.

Now try increasing the number of trials to 200. What happens to the histogram? What about when you increase the number of trials to 2000?

The distribution of sample means appears to get closer to a normal distribution.

Try playing around with different source distributions, sample sizes, and numbers of trials. What general conclusion can you make about the distribution of sample means from the different distributions?

Using clt-Demonstrations: Increasing number of trials shows dataset looking more and more normally distributed. General conclusion should be that the distribution of sample means is approximately normal, regardless of the original distribution.

A.5 Chapter 20

A.5.1 Exercise 20.1

Do these data appear to be roughly normal? Why or why not?

The data appear to be normal. The distribution is mostly symmetric with no clear outliers.

Next, calculate the grand mean and standard deviation of tree DBH (i.e., the mean and standard deviation of trees across all sites).

Grand Mean: **36.93**

Grand Standard Deviation: **10.96**

Using the same principles, what is the cumulative 0.4 quantile for the DBH data? **34.11 cm.**

From the Results table on the right, what interval of DBH values will contain 95% of the probability density around the mean? **15.34 – 58.46 cm**

From the Descriptives panel in jamovi (recall that this is under the 'Exploration' button), find the standard error of DBH. Std. error of Mean: **1.000.**

Based on the Results table, what can you infer are the lower and upper 95% confidence intervals (CIs) around the mean?

Lower 95% CI: **34.94**

Upper 95% CI: **38.86**

From this Descriptives table now, write the lower and upper 95% CIs below.

Lower 95% CI: **34.95**

Upper 95% CI: **38.91**

A.5.2 Exercise 20.2

From these quantiles, what is the proper z-score to use in the equations for LCI and UCI above?

z-score: **1.96**

Now, use the values of $\bar{x}$, z, and SE for DBH in the equations above to calculate lower and upper 95% confidence intervals again.

Lower 95% CI: **34.94**

Upper 95% CI: **38.86**

Are these confidence intervals the same as what you calculated in Exercise 19.1?

These confidence intervals are the same as those calculated from the normal distribution, bu they are slightly different from those from the 'Descriptives' menu as those ones are calculated with t-scores (which are more accurate), rather than z-scores.

What are the appropriate df for DBH? **df: $120 - 1 = 119$**

From the Results table, what is the proper t-score to use in the equations for LCI and UCI? **t-score: 1.980.**

Again, use the values of $\bar{x}$, t, and SE for DBH in the equations above to calculate lower and upper 95% confidence intervals.

Lower 95% CI: **34.92**

Upper 95% CI: **38.99**

Reflect on any similarities or differences that you see in all of these different ways of calculating confidence intervals.

The confidence intervals are very similar, but not exactly the same as those calculated with z-scores. This is because the sample size is 120 and degrees of freedom is therefore $120 - 1 = 119$, which is quite large, and as the sample size becomes larger the t-scores and z-scores become more similar. While there is no fixed threshold, usually when the sample size is more than about 30, the difference between the two methods is sufficiently small as to not make an important difference.

A.5.3 Exercise 20.3

From the Descriptives tool in jamovi, write the sample sizes for DBH split by site below.

Site 1182: N = **4**

Site 1223: N = **22**

Site 3008: N = **10**

Site 10922: N = **84**

For which of these sites would you predict CIs calculated from z-scores versus t- scores to differ the most? **Site: 1182 (note that this site has the lowest sample size).**

Now, fill in the table below reporting 95% CIs calculated using each distribution from the 4 sites using any method you prefer.

TABLE 20.1 95% Confidence intervals calculated for tree diameter at breast height (DBH) in centimetres.

Site	N	95% CIs (Normal)	95% CIs (t-distribution)
1182	**4**	**42.73 – 57.57**	**38.09 – 62.21**
1223	**22**	**21.56 – 24.16**	**21.48 – 24.24**
3008	**10**	**51.49 – 61.13**	**50.75 – 61.87**
10922	**84**	**36.09 – 39.25**	**36.07 – 39.27**

Next, do the same, but now calculate 99% CIs instead of 95% CIs.

TABLE 20.2 99% Confidence intervals calculated for tree diameter at breast height (DBH) in centimetres.

Site	N	99% CIs (Normal)	99% CIs (t-distribution)
1182	4	**40.38 – 59.92**	**28.02 – 72.28**
1223	**22**	**21.14 – 24.58**	**20.98 – 24.74**
3008	**10**	**49.96 – 62.66**	**48.32 – 64.30**
10922	**84**	**35.60 – 39.74**	**35.55 – 39.79**

What do you notice about the difference between CIs calculated from the normal distribution versus the t-distribution across the different sites?

The t-distribution gives a slightly wider spread than the normal distribution, and the difference increases as sample size gets smaller.

In your own words, what do these CIs *actually mean?*

Confidence intervals: if you take a sample and calculate the 95% (or 99%) confidence interval, and then go to the same population and resample that population numerous times, then there is a 95% chance the new sample mean will be within the CIs calculated at the 95% level (and 99% chance that it would be within the CIs calculated at the 99% level).

A.5.4 Exercise 20.4

From the Descriptives options, find the number of sites grazed versus not grazed.

Grazed: **4**

Not Grazed: **20**

From these counts above, what is the estimate (p, or more technically $\hat{p}$, with the hat indicating that it is an estimate) of the proportion of sites that are grazed?

p: **4 / (20 + 4) = 0.166667**

We can estimate p using p, and N is the total sample size. Using the above equation, what is the standard error of p?

$$\mathbf{SE(p)} = \sqrt{\frac{\mathbf{0.166667(1 - 0.166667)}}{\mathbf{24}}} = \mathbf{0.0761}$$

Using this standard error, what are the Wald lower and upper 95% confidence intervals around p?

Wald $LCI_{95\%} = 0.166667 - (1.96 \times 0.0761) = \mathbf{0.0175}$

Wald $UCI_{95\%} = 0.166667 + (1.96 \times 0.0761) = \mathbf{0.3158}$

Next, find the lower and upper 99% CIs around p and report them below.

Wald $LCI_{99\%} = 0.166667 - (2.58 \times 0.0761) = \mathbf{-0.0297}$

Wald $UCI_{99\%} = 0.166667 + (2.58 \times 0.0761) = \mathbf{0.3630}$

Do you notice anything unusual about the lower 99% CI? **The lower 99% CI is a negative number, which is not possible for a proportion.**

p: **0.16667**

Clopper-Pearson $LCI_{95\%} = \mathbf{0.04735}$

Clopper-Pearson $UCI_{95\%} = \mathbf{0.37384}$

To calculate 99% CIs, change the number in the Interval box from 95 to 99. Report the 99% CIs below.

Clopper-Pearson $LCI_{99\%} = \mathbf{0.02947}$

Clopper-Pearson $UCI_{99\%} = \mathbf{0.43795}$

What do you notice about the difference between the Wald CIs and the Clopper-Pearson CIs?

The Clopper-Pearson CIs are a bit wider than the Wald CIs, thereby suggesting that a wider range of values is needed to encompass the

means for a given level of confidence. The Clopper-Pearson CIs have higher upper CIs, but also higher lower CIs (and the 99% Clopper-Pearson CI does not overlap zero).

A.5.5 Exercise 20.5

First consider an 80% CI.

$LCI_{80\%}$ = **0.47359**

$UCI_{80\%}$ = **0.75942**

Next, calculate 95% CIs for the proportion of sites classified as Ancient woodland.

$LCI_{95\%}$ = **0.40594**

$UCI_{95\%}$ = **0.81201**

Finally, calculate 99% CIs for the proportion of sites classified as Ancient woodland.

$LCI_{99\%}$ = **0.34698**

$UCI_{99\%}$ = **0.85353**

A.6 Chapter 23

A.6.1 Exercise 23.1

Report these below.

N: **21**

$\bar{x}$: **58.76**

s: **8.687**

What kind(s) of statistical test would be most appropriate to use in this case, and what is the null hypothesis (H_0) of the test?

Test to use: **One sample t-test**

H_0: **The overall student scores were sampled from a population with a mean of 60.1.**

What is the alternative hypothesis (H_A), and should you use a one- or two-tailed test?

H_A: **The overall student scores were sampled from a population with a mean lower than 60.1.**

One- or two-tailed? **One-tailed.**

From the Normality Test table, what is the p-value of the Shapiro-Wilk test? $\boldsymbol{P = 0.112}$**.**

Based on this p-value, should we reject the null hypothesis?

No, we do not reject the null hypothesis that the data are normally distributed.

On the right panel of jamovi, you will see a table with the t-statistic, degrees of freedom, and p-value of the one sample t-test. Write these values down below.

t-statistic: **−0.7067**

degrees of freedom: **20**

p-value: **0.244**

Based on the p-value, should you reject the null hypothesis that your students' mean overall grade is the same as the national average? Why or why not?

We should not reject the null hypothesis because our p-value is greater than our threshold Type I error rate of 0.05 (i.e., $\boldsymbol{P > 0.05}$**). Assuming that the null hypothesis is true, the probability getting a t-statistic as extreme as the one we observed is only about 1 in 4, which is not especially unlikely.**

Based on this test, how would you respond to your colleague who is concerned that your students are performing below the national average?

There is no evidence that students in this class are performing below the national average.

Is there an assumption that might be particularly suspect when comparing the scores of students in a single classroom with a national average? Why or why not?

The students in this classroom are unlikely to be a random sample from the overall population. There may be other factors affecting student test scores that have nothing to do with the quality of the instruction.

A.6.2 Exercise 23.2

Is there any reason to believe that the data are not normally distributed?

No, the Shapiro-Wilk test gives us no reason to reject the null hypothesis that the data are normally distributed ($\boldsymbol{P > 0.05}$**).**

We want to know if student grades have improved. What is the null hypothesis (H_0) and alternative hypothesis (H_A) in this case?

H_0: **The mean change in student grade is 0.**

H_A: **The mean change in student grade is greater than 0.**

Write these values down below.

t-statistic: $\mathbf{-8.18}$

degrees of freedom: **20**

p-value: $\boldsymbol{P < 0.001}$

Based on this p-value, should you reject or fail to reject your null hypothesis? What can you then conclude about student test scores?

Because $\boldsymbol{P < 0.05}$, we reject the null hypothesis. We can conclude that the mean score for Test 1 is less than the mean score for Test 2, so the grades appear to have improved.

A.6.3 Exercise 23.3

We are not interested in whether the scores are higher or lower than 62, just that they are different. Consequently, what should our alternative hypothesis (H_A) be?

H_A: **Test 3 scores were sampled from a population with a mean not equal to 62.**

What is the p-value of the Shaprio-Wilk test this time? $\boldsymbol{P = 0.022}$.

What inference can you make from the Q-Q plot? Do the points fall along the diagonal line?

The points appear to be a bit curved. They do not cleanly fall along the diagonal line.

Based on the Shapiro-Wilk test and Q-Q plot, is it safe to assume that the Test 3 scores are normally distributed?

Based on the Shapiro-Wilk test and the Q-Q plot, we should not assume that the data are normally distributed.

What are the null and alternative hypotheses of this test?

H_0: **Test 3 scores were sampled from a population with a median of 62.**

H_A: **Test 3 scores were sampled from a population with a median not equal to 62.**

What is the test statistic (not the p-value) for the Wilcoxon test? Test statistic: **53.**

Based on what you learned in Section 22.5.1, what does this test statistic actually mean?

It means that if we subtract 62 from the Test 3 values, then rank each by its absolute value, the sum of ranks that came from positive values should equal 53.

Now look at the p-value for the Wilcoxon test. What is the p-value, and what should you conclude from it? $\boldsymbol{P = 0.055}$.

Conclusion: **We do not reject the null hypothesis that the median Test 3 score is 62.**

A.6.4 Exercise 23.4

One- or two-tailed? **Two-tailed.**

Based on the Assumption Checks in jamovi (and Figure 23.5), what can you conclude about the t-test assumptions?

The data appear to be normally distributed, but the groups do not have equal variances.

What is the p-value for Levene's test? $\boldsymbol{P = 0.021}$.

Based on what you learnt in Section 22.2, what is the appropriate test to run in this case? Test: **Welch's independent samples t-test.**

Check the box for the correct test, then report the test statistic and p-value from the table that appears in the right panel.

Test statistic: $\boldsymbol{-0.3279}$

$\boldsymbol{P = 0.745}$

What can you conclude from this t-test?

There is no evidence to reject the null hypothesis that both years were sampled from a population from the same mean score. The overall scores appear to be the same between years.

A.6.5 Exercise 23.5

Below, summarise the hypotheses for this new test.

H_0: **Test 3 scores in 2022 and 2023 were sampled from a population with the same mean.**

H_A: **Test 3 scores in 2022 and 2023 were sampled from a population with different means.**

Is this a one- or two-tailed test?: **Two-tailed test.**

Do the variances appear to be the same for Test 3 scores in 2022 versus 2023? How can you make this conclusion?

The homogeneity of variances test (i.e., Levene's test) show a test statistic of $F = 0.1379$ and a p-value of $P = 0.712$, so there is no evidence to reject the null hypothesis that the two years have different variances.

What is the p-value of the Shapiro-Wilk test? $P < 0.001$

Now, have a look at the Q-Q plot. What can you infer from this plot about the normality of the data, and why?

The data appear to deviate from the diagonal line, suggesting non-normality.

Based on what you found from testing the model assumptions above, and the material in Chapter 22, what test is the most appropriate one to use?

Test: **Mann-Whitney U test.**

Run the above test in jamovi, then report the test statistic and p-value below.

Test statistic: **221.0**

p-value: **0.487**

Based on what you learned in Section 22.5.2, what does this test statistic actually mean?

If we rank the full dataset (all Test 3 scores regardless of year), then sum up the ranks of 2022, the rank sum would be 221.

Finally, what conclusions can you make about Test 3 scores in 2022 versus 2023?

There appears to be no difference in Test 3 scores between 2022 and 2023. We do not reject the null hypothesis that the medians (technically, the distributions) differ between years.

What could you do to test the null hypothesis that the change in scores from Test 1 to Test 2 is the same between years?

Because the paired samples t-test is really just a one-sample t-test, we could first calculate the change from Test 1 to Test 2. That is, create a new column of data that is Change = Test 1 – Test 2. We could then use an independent samples t-test to check if the change differs between 2022 and 2023.

A.7 Chapter 28

A.7.1 Exercise 28.1

What are the null (H_0) and alternative (H_A) hypotheses for the t-test?

H_O: **Mean nitrogen concentration is the same in both sites.**

H_A: **Mean nitrogen concentration is not the same in both sites.**

What can you conclude from these 2 tests?

Normality conclusion: **Do not reject null hypothesis that data are normally distributed.**

Homogeneity of variances conclusion: **Do not reject null hypothesis that groups have the same variances.**

Given the conclusions from the checks of normality and homogeneity of variances above, what kind of test should you use to see if the mean Nitrogen concentration is significantly different in Funda versus Bailundo? Test: **Independent samples Student's t-test.**

Run the test above in jamovi. What is the p-value of the test, and what conclusion do you make about Nitrogen concentration at the two sites? **P = 0.030**

Conclusion: **Mean nitrogen concentration is different in the two sites (reject null hypothesis).**

Write down the test statistic (F), degrees of freedom, and p-values from this table below.

$F =$ **4.98377**

$df1 =$ **1**

$df2 =$ **49**

$P =$ **0.030**

What is the approximate area under the curve (i.e., orange area) where the F value on the x-axis is greater than your calculated F? **About 0.03 for $F =$ 4.98.**

Approximately, what is this threshold F value above which we will reject the null hypothesis? Approximate threshold F: **Somewhere between 4.03 and 4.05.**

What should you conclude regarding the null hypothesis that sites have the same mean? Conclusion: **Reject the null hypothesis that sites have the same mean.**

Look again at the p-value from the one-way ANOVA output and the Student's t-test output. Are the two values the same, or different? Why might this be?

The two p-values are the exact same. The independent Student's t-test and the one-way ANOVA are actually testing the same null hypothesis and making the same assumptions. One test is just using the t-distribution (t-test) to find the probability of rejecting the null hypothesis if it is true (i.e., the p-value). The other test (ANOVA) is using the F-distribution to find the same p-value.

A.7.2 Exercise 28.2

What can you conclude?

Normality conclusion: **Do not reject the null hypothesis of normality.**

Homogeneity of variance conclusion: **Do not reject the null hypothesis that the Profiles have the same variances.**

What are the output statistics in the One-Way ANOVA table?

$F =$ **3.43221**

$df1 =$ **2**

$df2 =$ **48**

$P =$ **0.040**

From these statistics, what do you conclude about the difference in Nitrogen concentration among profiles?

Conclusion: **Profiles do not have the same mean nitrogen concentration (reject null hypothesis).**

Write down the 'Probability' value from the Results table in the panel to the right. Probability: **0.04044.**

From the Results table, what is the critical F value ('Quantile'), above which we would reject the null hypothesis that all groups have the same mean? **Critical F value: 3.19.**

Fill in the table below (Table 28.1) with the information for degrees of freedom, F, and P.

TABLE 28.1 ANOVA output testing the null hypothesis that mean Nitrogen concentration is the same across three different soil profiles in Angola.

	Sum of Squares	df	Mean Square	F	p
Profile	**16888.18606**	**2**	**8444.09303**	**3.43221**	**0.040**
Residuals	**118092.02927**	**48**	**2460.25061**		

A.7.3 Exercise 28.3

Find the p-values associated with Tukey's HSD (P_{Tukey}) for each profile pairing. Report these below.

Tukey's HSD Lower - Middle: $\boldsymbol{P} = \mathbf{0.705}$

Tukey's HSD Lower - Upper: $\boldsymbol{P} = \mathbf{0.193}$

Tukey's HSD Middle - Upper: $\boldsymbol{P} = \mathbf{0.035}$

From this output, what can we conclude about the difference among soil profiles?

Middle and upper profiles appear to have significantly different mean nitrogen concentrations, but other combinations are not significant.

Report the p-values for the Bonferonni correction below.

Bonferonni Lower - Middle: $\boldsymbol{P} = \mathbf{1.000}$

Bonferonni Lower - Upper: $\boldsymbol{P} = \mathbf{0.253}$

Bonferonni Middle - Upper: $\boldsymbol{P} = \mathbf{0.040}$

In general, how are the p-values different between Tukey's HSD and the Bonferroni correction? Are they about the same, higher, or lower?

In general, p-values from the Bonferonni correction are higher.

In general, how are the p-values different between Tukey's HSD and the Bonferroni correction? Are they about the same, higher, or lower?

In general, we have a lower probability of making a Type I error with the Bonferonni test.

A.7.4 Exercise 28.4

How would you describe the distribution? Do the data appear to be normally distributed?

The histogram appears to show a distribution that is very right-skewed. This does not look normally distributed.

From Levene's test, the Shapiro-Wilk test, and the Q-Q plot, what assumptions of ANOVA might be violated?

From the Shapiro-Wilk test ($\boldsymbol{P} < \mathbf{0.001}$) and Levene's test ($\boldsymbol{P} = \mathbf{0.024}$), it appears that the assumptions of normality and equal variances are violated.

Report these values below.

$\chi^2 = \mathbf{0.38250}$

$df = \mathbf{2}$

$P = \mathbf{0.826}$

From the above output, should we reject or not reject our null hypothesis? H_0: **Do not reject null hypothesis.**

Write these null hypotheses down below (the order does not matter).

First H_0: **Mean Nitrogen concentration does not differ among sites.**

Second H_0: **Mean Nitrogen concentration does not differ among profiles.**

Third H_0: **There is no interaction between site or profile in affecting Nitrogen concentration.**

From the assumption checks output tables, is there any reason to be concerned about using the two-way ANOVA?

There is no reason to reject the null hypothesis that data are normally distributed or variances are equal. We can proceed with the two-way ANOVA.

Fill in Table 28.2 with the relevant information from the two-way ANOVA output.

TABLE 28.2 Two-way ANOVA output testing the effects of two sites and three different soil profiles on soil Nitrogen concentration in Angola.

	Sum of Squares	df	Mean Square	F	p
Site	**21522.18384**	1	**21522.18384**	**12.03138**	**0.001**
Profile	**22811.13680**	2	**11405.56840**	**6.37597**	**0.004**
Site * Profile	**16209.13035**	2	**8104.56517**	**4.53063**	**0.016**
Residuals	**80497.68348**	45	**1788.83741**		

From this output table, should you reject or not reject your null hypotheses?

Reject First H_0? **Yes.**

Reject Second H_0? **Yes.**

Reject Third H_0? **Yes.**

In non-technical language, what should you conclude from this two-way ANOVA?

It appears that Nitrogen concentration is different at different sites and across different profiles, and that there is an interaction between site and profile.

Based on what you learned in Chapter 27 about interaction effects, what can you say about the interaction between Site and Profile? Does one Profile, in particular, appear to be causing the interaction to be significant? How can you infer this from the Estimated Marginal Means plot?

We can see the interaction effects in the figure. The middle profile, in particular, appears to be causing the interaction. Both lower and upper profiles are parallel. But the middle is clearly at a different slope than the other two, indicating an interaction effect.

Based on the ANOVA output, what can you conclude?

The two-way ANOVA shows that Site alone has a significant effect on Phosophorus. The Profile and Interaction terms are not significant.

A.8 Chapter 31

A.8.1 Exercise 31.1

What are the null and alternative hypotheses for this Chi-square goodness of fit test?

H_O: **There is no significant difference between expected and observed counts of living and dead bees.**

H_A: **There is a significant difference between expected and observed counts of living and dead bees.**

What is the sample size (N) of the dataset? $\boldsymbol{N}$**: 256.**

Based on this sample size, what are the expected counts for bees that survived and died?

Survived (E_{surv}): **128**

Died (E_{died}): **128**

Write down the observed counts of bees that survived and died.

Survived (O_{surv}): **139**

Died (O_{died}): **117**

What is the χ^2 value? **1.89.**

How many degrees of freedom are there? $\boldsymbol{df} = \mathbf{1}$.

Write these values below, and check to see if the χ^2 and df match the values you calculated above by hand.

$\chi^2 =$ **1.89063**

$df =$ **1**

$P =$ **0.16913**

A.8.2 Exercise 31.2

What are the null and alternative hypotheses in this scenario?

H_O: **There is no significant difference between expected and observed counts of colonies.**

H_A: **There is a significant difference between expected and observed counts of colonies.**

How many colonies are there in this dataset? **8 colonies.**

What is the output from the Goodness of Fit table?

$\chi^2 =$ **3.5**

$df =$ **7**

$P =$ **0.83523**

From this output, what can you conclude about how bees were taken from the colonies?

Bees appear to be taken from colonies in equal frequencies, i.e., with equal probability of sampling among colonies.

A.8.3 Exercise 31.3

What are the null and alternative hypotheses for this test of association?

H_O: **There is no association between bee colony and bee survival.**

H_A: **There is an association between bee colony and bee survival.**

Report the key statistics in the output table below.

Chi-square: **11.31033**

$df =$ **7**

$P =$ **1**

From these statistics, should you reject or not reject the null hypothesis?

H_O: **Do not reject null hypothesis.**

What can you conclude from this test? Explain your conclusion as if you were reporting the results of the test to someone who was unfamiliar with statistical hypothesis testing.

From the statistical analysis, it appears that there is a highly significant association between the level of radiation that bees experience and the frequency with which they survive versus do not survive.

Lastly, did the order in which you placed the two variables matter? What if you switched Rows and Columns? In other words, put 'survived' in the Rows box and 'radiation' in the Columns box. Does this give you the same answer?

The ordering of the two variables does not appear to matter. The Chi-square value and p-values are the same.

A.8.4 Exercise 31.4

Just looking at the scatterplot, does it appear as though bee mass and CO_2 output are correlated? Why or why not?

The scatterplot might indicate a slight negative correlation, but it is difficult to say for sure just based on a visualisation.

What are the null and alternative hypotheses of this test?

H_O: **The correlation coefficient between bee mass and carbon dioxide output is zero.**

H_A: **The correlation coefficient between bee mass and carbon dioxide output is negative.**

Check this box, then find the p-values for the Shapiro-Wilk test of normality in the panel to the right. Write these p-values down below.

Mass $P =$ **0.96248**

CO_2 $P =$ **0.56459**

Based on these p-values, which type of correlation coefficient should we use to test H_0, and why?

We should use the Pearson's product moment correlation coefficient because both variables appear to be normally distributed.

This table reports both the correlation coefficient (here called 'Pearson's r') and the p-value. Write these values below.

$r =$ **-0.18036**

$P =$ **0.002**

Based on this output, what should we conclude about the association between bumblebee mass and carbon dioxide output?

Bumblebee mass and carbon dioxide output are negatively correlated, meaning that as bumblebee mass increases, carbon dioxide output decreases.

A.8.5 Exercise 31.5

What are the null and alternative hypotheses of this test?

H_O: **The correlation coefficient between bee mass and nectar consumption is zero.**

H_A: **The correlation coefficient between bee mass and nectar consumption is not zero.**

Based on the output of these tests, what kind of correlation coefficient should we use for testing the null hypothesis? **Spearman's rank correlation coefficient.**

What is the correlation coefficient and p-value from this test?

$r =$ **0.11954**

$P =$ **0.05611**

Based on these results, should we reject or not reject the null hypothesis?

H_0: **Do not reject null hypothesis.**

Would we have made the same conclusion about the correlation (or lack thereof) between bee mass and nectar consumption? Why or why not?

If we had used a Pearson product moment correlation coefficient instead of the Spearman's rank correlation coefficient, we would have calculated $r = 0.12729$ and a p-value of $P = 0.04186$. Because our p-value would have been less than 0.05, we would have incorrectly rejected the null hypothesis.

A.9 Chapter 34

A.9.1 Exercise 34.1

Before doing this, what is the independent variable, and what is the dependent variable?

Independent variable: **depth**

Dependent variable: **PyC**

What is the sample size of this dataset? $N = 240$.

Describe the scatterplot that is produced in the jamovi panel to the right.

Linear relationship with a lot of scatter. Perhaps a slight downward trend?

In other words, does the scatterplot show any evidence of a curvilinear pattern in the data?

The relationship appears to be linear.

In other words, does the variance change along the range of the independent variable (i.e., x-axis)?

No evidence of heteroscedasticity, so this assumption appears to be valid.

In your own words, what is this test doing? Drawing a picture might be helping to explain.

This test is checking to see if the residual values around the regression line are normally distributed.

What is the p-value of the Shapiro-Wilk test of normality? $\boldsymbol{P} = \mathbf{0.91624}$

Based on the above p-value, is it safe to conclude that the residuals are normally distributed? Conclusion: **Yes, residual values are normally distributed.**

A new table will open up in the right panel called 'Model Fit Measures'. Write the output statistics from this table below:

$R^2 = \mathbf{0.02532}$

$F = \mathbf{6.18319}$

$df1 = \mathbf{1}$

$df2 = \mathbf{238}$

$P = \mathbf{0.01358}$

Based on these statistics, what percentage of the variation in pyrogenic carbon is explained by the linear regression model? **2.532%**

What null hypothesis does the p-value above test?

H_O: **A model that includes depth is not a significantly better predictor of PyC than just the mean PyC.**

Do we reject or fail to reject H_0? **Reject H_0.**

From this table, what are the coefficient estimates for the intercept and the slope (i.e., depth)?

Intercept: **1.61719**

Slope: $-\mathbf{0.00263}$

What null hypotheses are we testing when inspecting these p-values?

Intercept H_0: **$P < 0.0001$, testing null hypothesis that $b_0 = 0$.**

Slope H_0: **$P = 0.01358$, testing null hypothesis that $b_1 = 0$.**

Finally, what can we conclude about the relationship between depth and pyrogenic carbon storage? **Pyrogenic carbon changes with increasing soil depth.**

A.9.2 Exercise 34.2

Do all of these assumptions appear to be met?

Linearity: **No issues.**

Normality: **No issues.**

Homoscedasticity: **No issues.**

Using the same protocol as the previous exercise, what percentage of the variation in PyC is explained by the regression model? **Variation explained: 48.2% (0.482 * 100% = 48.2%).**

Is the overall model statistically significant? How do you know? Model significance: **Yes, because overall model test p-value is $P < 0.05$.**

Are the intercept and slope significantly different from zero?

Intercept: **Yes, significantly different from zero; reject null hypothesis.**

Slope: **Yes, significantly different from zero; reject null hypothesis.**

Write the intercept (b_0) and slope (b_1) of the regression below.

b_0: **0.88911**

b_1: **0.05688**

Using these values for the intercept and the slope, write the regression equation to predict pyrogenic carbon (PyC) from fire frequency (fire_freq).

$Y = 0.88911 + (0.05688 * X)$ OR **PyC = 0.88911 + (0.05688 * fire_freq)**

Using this equation, what would be the predicted PyC for a location that had experienced 10 fires in the past 20 years (i.e., fire_freq = 10)? PyC = **1.45791.**

Explain what these two columns of data represent in terms of the scatterplot you made at the start of this exercise. In other words, where would the predicted and residual values be located on the scatterplot?

The predicted values represent the PyC points that fall along the regression line for a particular fire_freq value. That is, what the model predicts PyC should be at each fire frequency. The residual values are the difference between the actual PyC values and what the predicted PyC values are in the model.

A.9.3 Exercise 34.3

Write down what the independent and dependent variable(s) are for this regression.

Independent: **Fire frequency and depth**

Dependent: **Pyrogenic carbon**

Do all of these assumptions appear to be met?

Linearity: **Yes**

Normality: **Yes**

Homoscedasticity: **Yes**

Report these values from the Model Fit Measures output table below.

$R^2 =$ **0.48202**

Adjusted $R^2 =$ **0.47765**

$F =$ **110.27348**

$P < 0.0001$

Which one is most appropriate to use for interpreting the multiple regression?

The adjusted R-squared takes into account that adding more independent variables will increase the total amount of variation explained in the dependent variable even if the independent variable is not a good predictor. We should therefore use the adjusted R-squared when looking at a multiple regression model.

What is the null hypothesis of this tested with the F value and the P value shown in the Model Fit Measures table?

H_0: **A model that includes the independent variables depth and fire frequency does not explain variation in PyC significantly better than just the mean of PyC.**

Based on the Overall Model Test output, should you reject or not reject H_0? **Reject the null hypothesis.**

What can you conclude about the significance of the Intercept, and the partial regression coefficients for fire frequency and depth?

The intercept and partial regression coefficient of fire frequency are significantly different from 0. But the partial regression coefficient of depth is not significantly different from 0.

Using the partial regression coefficient estimates, fill in the equation below,

PyC = (**0.89456**) + (**0.05680**)fire_freq + (**−0.00008**)depth.

Next, use this to predict the pyrogenic carbon for a fire frequency of 12 and a depth of 60 cm.

PyC = **1.57136**

Has the significance of soil depth as an independent variable changed? Based on what you know about the difference between simple linear regression and multiple regression, why might this be the case?

Yes, the soil depth was significant by itself as a predictor of PyC in the simple linear regression of the first exercise. But in this multiple regression model, the partial regression coefficient is not significant. This might be because when you consider depth in the context of fire frequency, the effect of depth by itself is not significant. Once you account for fire frequency, depth no longer is a meaningful predictor of PyC. If you hold fire frequency constant in a model, then depth by itself does not affect PyC.

A.9.4 Exercise 34.4

TABLE 34.1 Model Coefficients output table for a multiple regression model predicting pyrogenic carbon from soil depth, fire frequency, and soil pH in Gabon.

	Estimate	Std. Error	t Value	Pr(>
(Intercept)	**0.98892**	0.34591	2.85888	**0.00463**
depth	**−0.00006**	0.00080	−0.07411	**0.94098**
fire_freq	**0.05679**	0.00394	14.42303	**< 0.00001**
pH	**−0.01584**	0.05679	−0.27886	**0.78059**

From the Model Fit Measures table, what is the R^2 and Adjusted R^2 of this model?

R^2 = **0.48219**

Adjusted R^2 = **0.47561**

Is the R^2 value of this model higher or lower than the multiple regression model without pH?

The model without pH had an $R^2 = 0.48202$, so it was lower without pH.

Is the Adjusted R^2 value of this model higher or lower than the multiple regression model without pH?

The model without pH had an Adjusted $R^2 = 0.47765$, so it was higher without pH.

Based on what you know from Section 33.1, explain why the R^2 and Adjusted R^2 might have changed in different directions with the addition of a new independent variable.

The R^2 just tells us how much variation in the dependent variable is explained by the model, but there is no penalty for adding more independent variables to the model. Consequently, adding the additional independent variable (pH) can only increase (or at least not decrease) the amount of variation explained. In contrast, the adjusted R^2 penalises the R2 for each additional independent variable in the model, so because we have added the independent variable pH, our adjusted R^2 can go down.

Finally, use the equation of this new model to predict PyC for a soil sample at a depth of 0, fire frequency of 0, and pH of 6.

PyC $= 0.98892 - (0.00006 * 0) + (0.05679 * 0) - (0.01584 * 6)$

PyC $= 0.89388$

A.9.5 Exercise 33.5

What assumption(s) appear as though they might be violated for this simple regression? Explain how you figured this out.

It appears that the assumptions of normality and homoscedasticity are violated. We found this out by running a Shapiro-Wilk test and rejecting the null hypothesis that data are normally distributed, and by using a scatterplot to visualise how the variation in temperature changed along the range of fire frequency.

B

Uncertainty derivation

It is not necessary to be able to derive the equations for propagating error from Chapter 7, but working through the below might be interesting, and provide a better appreciation for why these formulas make sense. Another derivation is available in Box et al. (1978) (page 563), but this derivation is expressed in terms of variances and covariances. What follows is not intended to be rigorous, but to provide an intuitive understanding of why the equation for propagating error when adding or subtracting looks so different from the equation for propagating error when multiplying or dividing.

B.1 Propagation of error for addition and subtraction

For adding and subtracting error, we know that we get our variable Z by adding X and Y. This is just how Z is defined. We also know that Z is going to have some error E_Z, and we know that Z plus or minus its error will equal X plus or minus its error plus Y plus or minus its error,

$$(Z \pm E_Z) = (X \pm E_X) + (Y \pm E_Y).$$

Again, this is just our starting definition, but double-check to make sure it makes sense. We can now note that we know,

$$Z = X + Y.$$

If it is not intuitive as to why, just imagine that there is no error associated with the measurement of X and Y (i.e., $E_X = 0$ and $E_Y = 0$). In this case, there cannot be any error in Z. So, if we substitute $X + Y$ for Z, we have the below,

$$((X + Y) \pm E_Z) = (X \pm E_X) + (Y \pm E_Y).$$

By the associative property[1], we can get rid of the parenthesis for addition and subtraction, giving us the below,

$$X + Y \pm E_Z = X \pm E_X + Y \pm E_Y.$$

Now we can subtract X and Y from both sides and see that we just have the errors of X, Y, and Z,

$$\pm E_Z = \pm E_X \pm E_Y.$$

The plus/minus is a bother. Note, however, that for any real number m, $m^2 = (-m)^2$. For example, if $m = 4$, then $(4)2 = 16$ and $(-4)2 = 16$, so we can square both sides to get positive numbers and make things easier,

$$E_Z^2 = (\pm E_X \pm E_Y)^2.$$

We can expand the above,

$$E_Z^2 = E_X^2 + E_Y^2 \pm 2E_X E_Y.$$

Now here is an assumption that I have not mentioned elsewhere in the book. With the formulas that we have given you, we are assuming that the errors of X and Y are independent. To put it in more statistical terms, the covariance between the errors of X and Y is assumed to be zero. Without going into the details (covariance is introduced in Chapter 30), if we assume that the covariance between these errors is zero, then we can also assume the last term of the above is zero, so we can get rid of it (i.e., $2E_X E_Y = 0$),

$$E_Z^2 = E_X^2 + E_Y^2.$$

If we take the square root of both sides, then we have the equation from Chapter 7,

$$E_Z = \sqrt{E_X^2 + E_Y^2}.$$

B.2 Propagation of error for multiplication and division

Now that we have seen the logic for propagating errors in addition and subtraction, we can do the same for multiplication and division. We can start with the same point that we are getting our new variable Z by multiplying X

[1] https://en.wikipedia.org/wiki/Associative_property

and Y together, $Z = XY$. So, if both X and Y have errors, the errors will be multiplicative as below,

$$Z \pm E_Z = (X \pm E_X)(Y \pm E_Y).$$

Again, all we are doing here is substituting Z, X, and Y, for an expression in parentheses that includes the variable plus or minus its associated error. Now we can expand the right-hand side of the equation,

$$Z \pm E_Z = XY + YE_X + XE_Y + E_X E_Y.$$

As with our propagation of error in addition, here we are also going to assume that the sources of error for X and Y are independent (i.e., their covariance is zero). This allows us to set $E_X E_Y = 0$, which leaves us with the below,

$$Z \pm E_Z = XY + YE_X + XE_Y.$$

Now, because $Z = XY$, we can substitute on the left-hand side of the equation,

$$XY \pm E_Z = XY + YE_X + XE_Y.$$

Now we can subtract the XY from both sides of the equation,

$$\pm E_Z = YE_X + XE_Y.$$

Next, let us divide both sides by XY,

$$\frac{\pm E_Z}{XY} = \frac{YE_X + XE_Y}{XY}.$$

We can expand the right-hand side,

$$\frac{\pm E_Z}{XY} = \frac{YE_X}{XY} + \frac{XE_Y}{XY}.$$

This allows us to cancel out the Y variables in the first term of the right-hand side, and the X variables in second term of the right-hand side,

$$\frac{\pm E_Z}{XY} = \frac{E_X}{X} + \frac{E_Y}{Y}.$$

Again, we have the plus/minus on the left, so let us square both sides,

$$\left(\frac{\pm E_Z}{XY}\right)^2 = \left(\frac{E_X}{X} + \frac{E_Y}{Y}\right)^2.$$

We can expand the right-hand side,

$$\left(\frac{\pm E_Z}{XY}\right)^2 = \left(\frac{E_X}{X}\right)^2 + \left(\frac{E_Y}{Y}\right)^2 + 2\left(\frac{E_X}{X}\right)\left(\frac{E_Y}{Y}\right).$$

Again, because we are assuming that the errors of X and Y are independent, we can set the third term on the right-hand side of the equation to zero. This leaves,

$$\left(\frac{\pm E_Z}{XY}\right)^2 = \left(\frac{E_X}{X}\right)^2 + \left(\frac{E_Y}{Y}\right)^2.$$

Note that $XY = Z$, so we can substitute in the left-hand side,

$$\left(\frac{\pm E_Z}{Z}\right)^2 = \left(\frac{E_X}{X}\right)^2 + \left(\frac{E_Y}{Y}\right)^2.$$

Now we can apply the square on the left-hand side to the top and bottom, which gets rid of the plus/minus,

$$\frac{E_Z^2}{Z^2} = \left(\frac{E_X}{X}\right)^2 + \left(\frac{E_Y}{Y}\right)^2.$$

We can now multiply both sides of the equation by Z^2,

$$E_Z^2 = Z^2\left(\left(\frac{E_X}{X}\right)^2 + \left(\frac{E_Y}{Y}\right)^2\right).$$

We can now take the square root of both sides,

$$E_Z = \sqrt{Z^2\left(\left(\frac{E_X}{X}\right)^2 + \left(\frac{E_Y}{Y}\right)^2\right)}.$$

We can pull the Z^2 out of the square root,

$$E_Z = Z\sqrt{\left(\frac{E_X}{X}\right)^2 + \left(\frac{E_Y}{Y}\right)^2}.$$

That leaves us with the equation that was given in Chapter 7.

Bibliography

Adams, D. C., & Collyer, M. L. (2016). On the comparison of the strength of morphological integration across morphometric datasets. *Evolution*, *70*(11), 2623–2631. https://doi.org/10.1111/evo.13045

Allie, S., Buffler, A., Campbell, B., Lubben, F., Evangelinos, D., Psillos, D., & Valassiades, O. (2003). Teaching measurement in the introductory physics laboratory. *Physics Teacher*, *41*(7), 394–401.

Andersson, P. G. (2023). The Wald confidence interval for a binomial p as an illuminating "bad" example. *American Statistician*, *77*(4), 443–448. https://doi.org/10.1080/00031305.2023.2183257

Askey, R. (1999). Why does a negative x a negative = a positive? *American Educator*, *23*(3), 4–5.

Blanca, M. J., Alarcón, R., Arnau, J., Bono, R., & Bendayan, R. (2018). Effect of variance ratio on ANOVA robustness: Might 1.5 be the limit? *Behavior Research Methods*, *50*(3), 937–962. https://doi.org/10.3758/s13428-017-0918-2

Borowski, E. J., & Borwein, J. M. (2005). *Collins Dictionary of Mathematics* (2nd ed., p. 641). HarperCollins Publishers, London.

Bouma, G. D. (2000). *The Research Process* (4th ed., p. 242). Oxford University Press, Oxford, UK.

Box, G. E. P., Hunter, W. G., & Hunter, S. J. (1978). *Statistics for Experimenters: An Introduction to Design, Data Analysis, and Model Building.* John Wiley & Sons, New York, USA.

Burrows, J. E., Copplestone, D., Raines, K. E., Beresford, N. A., & Tinsley, M. C. (2022). Ecologically relevant radiation exposure triggers elevated metabolic rate and nectar consumption in bumblebees. *Functional Ecology*, *36*(8), 1822–1833. https://doi.org/10.1111/1365-2435.14067

Chang, W., Cheng, J., Allaire, J., Sievert, C., Schloerke, B., Xie, Y., Allen, J., McPherson, J., Dipert, A., & Borges, B. (2024). *Shiny: Web application framework for R.* https://shiny.posit.co/

Cheadle, C., Vawter, M. P., Freed, W. J., & Becker, K. G. (2003). Analysis of microarray data using Z score transformation. *Journal of Molecular Diagnostics*, *5*(2), 73–81. https://doi.org/10.1016/S1525-1578(10)60455-2

Chernoff, E. J., & Zazkis, R. (2022). The simple reason a viral math equation stumped the internet. *The Conversation.* https://theconversation.com/the-simple-reason-a-viral-math-equation-stumped-the-internet-176518

Chiripanhura, B. (2011). Median and mean income analyses. *Economic and Labour Market Review*, *5*, 45–64.

Choi, W., Lee, J. W., Huh, M. H., & Kang, S. H. (2003). An algorithm for computing the exact distribution of the Kruskal-Wallis test. *Communications in Statistics Part B: Simulation and Computation*, *32*(4), 1029–1040. https://doi.org/10.1081/SAC-120023876

Clopper, C. J., & Pearson E. S. (1934). The use of confidence or fiducial limits illustrated in the case of the binomial. *Biometrika*, *26*(4), 404–413.

Courant, R., Robbins, H., & Stewart, I. (1996). *What is Mathematics?* (2nd ed., p. 566). Oxford University Press, Oxford, UK.

de Sousa Teixeira, L. P. (2022). *Geochemical, textural and micromorphological properties of Angolan agroecosystem soils in relation to region, landscape position and land management* [PhD thesis]. University of Stirling, Stirling, UK.

Delacre, M., Lakens, D., & Leys, C. (2017). Why psychologists should by default use Welch's t-test instead of Student's t-test. *International Review of Social Psychology*, *30*(1), 92–101. https://doi.org/10.5334/irsp.82

Doane, D. P., & Seward, L. E. (2011). Measuring skewness: A forgotten statistic? *Journal of Statistics Education*, *19*(2). https://doi.org/10.1080/10691898.2011.11889611

Duthie, A. B., Abbott, K. C., & Nason, J. D. (2015). Trade-offs and coexistence in fluctuating environments: evidence for a key dispersal-fecundity trade-off in five nonpollinating fig wasps. *American Naturalist*, *186*(1), 151–158. https://doi.org/10.1086/681621

Duthie, A. B., & Nason, J. D. (2016). Plant connectivity underlies plant-pollinator-exploiter distributions in *Ficus petiolaris* and associated pollinating and non-pollinating fig wasps. *Oikos*, *125*(11), 1597–1606. https://doi.org/10.1111/oik.02629

Dytham, C. (2011). *Choosing and Using Statistics: A Biologist's Guide* (p. 298). John Wiley & Sons, West Sussex, UK.

Edwards, A. W. F. (1972). *Likelihood: An account of the statistical concept of likelihood and its application to scientific inference* (p. 235). Cambridge University Press, Cambridge, UK.

Elavsky, F., Bennett, C., & Moritz, D. (2022). How accessible is my visualization? Evaluating visualization accessibility with Chartability. *Computer Graphics Forum*, *41*(3), 57–70. https://doi.org/10.1111/cgf.14522

Ellison, A. M. (2004). Bayesian inference in ecology. *Ecology Letters*, *7*(6), 509–520. https://doi.org/10.1111/j.1461-0248.2004.00603.x

Ernst, M. D. (2004). Permutation methods: A basis for exact inference. *Statistical Science*, *19*(4), 676–685. https://doi.org/10.1214/088342304000000396

Fowler, J., Cohen, L., & Jarvis, P. (1998). *Practical Statistics for Field Biology* (2nd ed., p. 259). John Wiley & Sons.

Freedman, D., Pisani, R., & Purves, R. (2011). *Statistics* (4th ed., p. 576). W. W. Norton & Company, New York, USA.

Friedlingstein, P., O'Sullivan, M., Jones, M. W., Andrew, R. M., Gregor, L.,

Hauck, J., Le Quéré, C., Luijkx, I. T., Olsen, A., Peters, G. P., Peters, W., Pongratz, J., Schwingshackl, C., Sitch, S., Canadell, J. G., Ciais, P., Jackson, R. B., Alin, S. R., Alkama, R., … Zheng, B. (2022). Global carbon budget 2022. *Earth System Science Data, 14*(11), 4811–4900. https://doi.org/10.5194/essd-14-4811-2022

Fryer, H. C. (1966). *Concepts and Methods of Experimental Statistics* (p. 602). Allyn & Bacon, Boston, USA.

Fuentes-Montemayor, E., Park, K. J., Cordts, K., & Watts, K. (2022). The long-term development of temperate woodland creation sites: from tree saplings to mature woodlands. *Forestry, 95*, 28–37. https://doi.org/10.1093/forestry/cpab027

Fuentes-Montemayor, E., Watts, K., Sansum, P., Scott, W., & Park, K. J. (2022). Moth community responses to woodland creation: The influence of woodland age, patch characteristics and landscape attributes. *Diversity and Distributions, 28*(9), 1993–2007. https://doi.org/10.1111/ddi.13599

Gelman, A., & Shalizi, C. R. (2013). Philosophy and the practice of Bayesian statistics. *British Journal of Mathematical and Statistical Psychology, 66*(1), 8–38. https://doi.org/10.1111/j.2044-8317.2011.02037.x

Gotelli, N. J. (2001). *A Primer of Ecology* (3rd ed., p. 265). Sinauer Associates, Inc., Sunderland, Massachusetts, USA.

Grafen, A., & Hails, R. (2022). *Modern Statistics for the Life Sciences* (p. 351). Oxford University Press, Oxford, UK.

Greenland, S., Senn, S. J., Rothman, K. J., Carlin, J. B., Poole, C., Goodman, S. N., & Altman, D. G. (2016). Statistical tests, P values, confidence intervals, and power: a guide to misinterpretations. *European Journal of Epidemiology, 31*(4), 337–350. https://doi.org/10.1007/s10654-016-0149-3

Groeneveld, R. A., & Meeden, G. (1984). Measuring skewness and kurtosis. *Journal of the Royal Statistical Society. Series D (The Statistician), 33*(4), 391–399.

Gupta, S. (2020). *Units of measurement: History, fundamentals and redefining the SI base units.* Springer. https://doi.org/10.1007/978-3-030-43969-9

Hardy, G. H. (1908). Mendelian proportions in a mixed population. *Science, 28*(706), 49–50. https://doi.org/10.1126/science.28.706.49

Head, M. L., Holman, L., Lanfear, R., Kahn, A. T., & Jennions, M. D. (2015). The extent and consequences of p-hacking in science. *PLoS Biology, 13*(3), e1002106. https://doi.org/10.1371/journal.pbio.1002106

Hyndman, R. J., & Fan, Y. (1996). Sample quantiles in statistical packages. *American Statistician, 50*(4), 361–365. https://doi.org/10.1080/00031305.1996.10473566

Johnson, D. H. (1995). Statistical sirens: The allure of nonparametrics. *Ecology, 76*(6), 1998–2000.

Kelleher, C., & Wagener, T. (2011). Ten guidelines for effective data visualization in scientific publications. *Environmental Modelling and Software, 26*(6), 822–827. https://doi.org/10.1016/j.envsoft.2010.12.006

Kruskal, W. H. (1952). A nonparametric test for the several sample problem. *The Annals of Mathematical Statistics*, *23*(4), 525–540.

Kruskal, W. H., & Wallis, W. A. (1952). Use of ranks in one-criterion variance analysis. *Journal of the American Statistical Association*, *47*(260), 583–621. https://doi.org/10.1080/01621459.1952.10483441

Lande, R. (1977). On comparing coefficients of variation. *Systematic Zoology*, *26*(2), 214–217.

Law, A., Bunnefeld, N., & Willby, N. J. (2014). Beavers and lilies: Selective herbivory and adaptive foraging behaviour. *Freshwater Biology*, *59*(2), 224–232. https://doi.org/10.1111/fwb.12259

Lee, P. (1997). *Bayesian statistics: An introduction, 344 pp.* Edward Arnold, London.

Ludbrook, J., & Dudley, H. (1998). Why permutation tests are superior to t and F tests in biomedical research. *American Statistician*, *52*(2), 127–132.

Lumley, T., Diehr, P., Emerson, S., & Chen, L. (2002). The importance of the normality assumption in large public health data sets. *Annual Review of Public Health*, *23*, 151–169. https://doi.org/10.1146/annurev.publheath.23.100901.140546

Manly, B. F. J. (2007). *Randomization, Bootstrap and Monte Carlo Methods in Biology* (3rd ed.). Chapman & Hall/CRC, Boca Raton, USA.

Mayo, D. G. (1996). *Error and the Growth of Experimental Knowledge* (p. 493). University of Chicago Press, Chicago, USA.

Mayo, D. G. (2019). P-value thresholds: Forfeit at your peril. *European Journal of Clinical Investigation*, *49*(10), 1–4. https://doi.org/10.1111/eci.13170

Mayo, D. G. (2021). Significance tests: Vitiated or vindicated by the replication crisis in psychology? *Review of Philosophy and Psychology*, *12*(1), 101–120. https://doi.org/10.1007/s13164-020-00501-w

McDonald, J. B., Sorensen, J., & Turley, P. A. (2013). Skewness and kurtosis properties of income distribution models. *Review of Income and Wealth*, *59*(2), 360–374. https://doi.org/10.1111/j.1475-4991.2011.00478.x

Mclean, R. A., Sanders, W. L., & Stroup, W. W. (1991). A unified approach to mixed linear models. *American Statistician*, *45*(1), 54–64.

McShane, B. B., Gal, D., Gelman, A., Robert, C., & Tackett, J. L. (2019). Abandon statistical significance. *American Statistician*, *73*, 235–245. https://doi.org/10.1080/00031305.2018.1527253

Miller, I., & Miller, M. (2004). *John E. Freund's mathematical statistics* (7th ed., p. 614). Pearson Prentice Hall, Upper Saddle River, New Jersey, USA.

Morrissey, M. B., & Ruxton, G. D. (2018). Multiple regression is not multiple regressions: The meaning of multiple regression and the non-problem of collinearity. *Philosophy, Theory, and Practice in Biology*, *10*(20180709). https://doi.org/10.3998/ptpbio.16039257.0010.003

Narum, S. R. (2006). Beyond Bonferroni: Less conservative analyses for conservation genetics. *Conservation Genetics*, *7*(5), 783–787. https://doi.org/10.1007/s10592-005-9056-y

Navarro, D. J., & Foxcroft, D. R. (2022). *Learning Statistics with Jamovi* (pp. 1–583). (Version 0.75). https://doi.org/10.24384/hgc3-7p15

Navidi, W. C. (2006). *Statistics for engineers and scientists* (Vol. 2). McGraw-Hill, New York, USA.

Pandey, S., & Bright, C. L. (2008). What are degrees of freedom? *Social Work Research*, *32*(2), 119–128. https://doi.org/10.1080/00031305.1974.10479077

Pastor, J. (2008). *Mathematical Ecology of Populations and Ecosystems* (pp. 1–50). John Wiley & Sons, Inc., West Sussex, England.

Pélabon, C., Hilde, C. H., Einum, S., & Gamelon, M. (2020). On the use of the coefficient of variation to quantify and compare trait variation. *Evolution Letters*, *4*(3), 180–188. https://doi.org/10.1002/evl3.171

Preston, C. M., & Schmidt, M. W. I. (2006). Black (pyrogenic) carbon: A synthesis of current knowledge and uncertainties with special consideration of boreal regions. *Biogeosciences*, *3*(4), 397–420. https://doi.org/10.5194/bg-3-397-2006

Quinn, T. J. (1995). Base units of the Système International d'Unités, their accuracy, dissemination and international traceability. *Metrologia*, *31*(6), 515–527. https://doi.org/10.1088/0026-1394/31/6/011

R Core Team. (2022). *R: A language and environment for statistical computing.* R Foundation for Statistical Computing. https://www.R-project.org/

Rabinovich, S. G. (2013). *Evaluating Measurement Accuracy: A Practical Approach.* Springer Science & Business Media. https://doi.org/10.1007/978-3-319-60125-0

Rahman, N. A. (1968). *A Course in Theoretical Statistics* (p. 542). Charles Griffin & Company, London.

Reed, J. F. (2007). Better binomial confidence intervals. *Journal of Modern Applied Statistical Methods*, *6*(1), 153–161. https://doi.org/10.22237/jmasm/1177992840

Reichmann, W. J. (1970). *Use and Abuse of Statistics* (3rd ed., p. 345). Penguin Books, London.

Reisser, M., Purves, R. S., Schmidt, M. W. I., & Abiven, S. (2016). Pyrogenic carbon in soils: A literature-based inventory and a global estimation of its content in soil organic carbon and stocks. *Frontiers in Earth Science*, *4*, 1–14. https://doi.org/10.3389/feart.2016.00080

Rencher, A. C. (2000). *Linear Models in Statistics* (p. 578). John Wiley & Sons, Inc., Provo, Utah, USA.

Rihs, M., & Mayer, B. (2018). *distrACTION-calculating and plotting distributions.* jamovi.org.

Rodgers, J. L., & Nicewander, W. (1988). Thirteen ways to look at the correlation coefficient. *American Statistician*, *42*(1), 59–66.

Rowntree, D. (2018). *Statistics Without Tears* (p. 199). Penguin, Milton Keynes, UK.

RStudio Team. (2020). *RStudio: Integrated development environment for r.* RStudio, PBC. http://www.rstudio.com/

Ruxton, G. D. (2006). The unequal variance t-test is an underused alternative to Student's t-test and the Mann-Whitney U test. *Behavioral Ecology, 17*(4), 688–690. https://doi.org/10.1093/beheco/ark016

Santín, C., Doerr, S. H., Kane, E. S., Masiello, C. A., Ohlson, M., Rosa, J. M. de la, Preston, C. M., & Dittmar, T. (2016). Towards a global assessment of pyrogenic carbon from vegetation fires. *Global Change Biology, 22*(1), 76–91. https://doi.org/10.1111/gcb.12985

Schilling, M. F., & Doi, J. A. (2014). A coverage probability approach to finding an optimal binomial confidence procedure. *American Statistician, 68*(3), 133–145. https://doi.org/10.1080/00031305.2014.899274

Schloerke, B., & Chang, W. (2023). *Shinylive: Run "shiny" applications in the browser.* https://CRAN.R-project.org/package=shinylive

Schmider, E., Ziegler, M., Danay, E., Beyer, L., & Bühner, M. (2010). Is it really robust?: Reinvestigating the robustness of ANOVA against violations of the normal distribution assumption. *Methodology, 6*(4), 147–151. https://doi.org/10.1027/1614-2241/a000016

Slakter, M. J. (1968). Accuracy of an approximation to the power of the chi-square goodness of fit test with small but equal expected frequencies. *Journal of the American Statistical Association, 63*(323), 912–918. https://doi.org/10.1080/01621459.1968.11009319

Sokal, R. R., & Rohlf, F. J. (1995). *Biometry* (3rd ed., p. 887). W. H. Freeman & Company, New York, USA.

Spiegelhalter, D. (2019). *The Art of Statistics Learning from Data* (p. 426). Penguin, Milton Keynes, UK.

Stanton-Geddes, J., De Freitas, C. G., & De Sales Dambros, C. (2014). In defense of P values: Comment on the statistical methods actually used by ecologists. *Ecology, 95*(3), 637–642. https://doi.org/10.1890/13-1156.1

Stewart, I. (2008). *Taming the Infinite* (p. 384). Quercus, London, UK.

Stock, M., Davis, R., De Mirandés, E., & Milton, M. J. T. (2019). Corrigendum: The revision of the SI – the result of three decades of progress in metrology. *Metrologia, 56*(4). https://doi.org/10.1088/1681-7575/ab28a8

Suárez, M. (2020). *Philosophy of Probability and Statistical Modelling* (R. Northcott & J. Stegenga, Eds.; p. 72). Cambridge University Press, Cambridge, UK. https://doi.org/10.1017/9781108985826

Tate, M. W., & Hyer, L. A. (1973). Inaccuracy of the X^2 test of goodness of fit when expected frequencies are small. *Journal of the American Statistical Association, 68*(344), 836–841. https://doi.org/10.1080/01621459.1973.10481433

The jamovi project. (2024). *Jamovi (version 2.5).* https://www.jamovi.org

Thulin, M. (2014). The cost of using exact confidence intervals for a binomial proportion. *Electronic Journal of Statistics, 8*(1), 817–840. https://doi.org/10.1214/14-EJS909

Tukey, J. W. (1949). Comparing individual means in the analysis of variance. *Biometrics, 5*(2), 99–114.

Upton, G., & Cook, I. (2014). *Dictionary of Statistics* (3rd ed., p. 488). Oxford University Press, Oxford, UK.

Wardlaw, A. C. (1985). *Practical Statistics for Experimental Biologists* (p. 290). John Wiley & Sons, Chichester, UK.

Wasserstein, R. L., & Lazar, N. A. (2016). The ASA's statement on p-values: Context, process, and purpose. *American Statistician, 70*(2), 129–133. https://doi.org/10.1080/00031305.2016.1154108

Weiblen, G. D. (2002). How to be a fig wasp. *Annual Review of Entomology, 47*, 299–330.

Welch, B. L. (1938). The significance of the difference between two means when the population variances are unequal. *Biometrika, 29*(3/4), 350. https://doi.org/10.2307/2332010

Wickham, H. (2014). Tidy data. *Journal of Statistical Software, 59*(10), 1–23. https://doi.org/10.18637/jss.v059.i10

Wickham, H. (2021). *Mastering shiny.* O'Reilly Media. https://mastering-shiny.org/

Xie, Y. (2015). *Dynamic documents with R and knitr* (2nd ed.). Chapman & Hall/CRC. http://yihui.name/knitr/

Xie, Y. (2016). *Bookdown: Authoring books and technical documents with R markdown.* Chapman & Hall/CRC, Boca Raton, USA. https://bookdown.org/yihui/bookdown

Xie, Y. (2023). *Bookdown: Authoring books and technical documents with R markdown.* https://github.com/rstudio/bookdown

Yee, A. J. (2019). *Google Cloud Topples the Pi Record.* http://www.numberworld.org/blogs/2019_3_14_pi_record/

Index

Printed in the United States
by Baker & Taylor Publisher Services